Vehicle Noise and Vibration

Conference Organizing Committee

L Bernard
Centro Ricerche Fiat

B Challen (Chairman)
Shoreham Services Engineering Consultancy

S Dadgostar
Ford Motor Company Limited

J Dixon
University of Southampton

R Freymann
BMW AG

A Middleton
Anthony Best Dynamics Limited

M Russell
Lucas Diesel Systems

A Walker
MIRA

J Wolschendorf
FEV Motorentechnik GmbH and Co KG

IMechE Conference Transactions

European Conference on

Vehicle Noise and Vibration

12–13 May 1998
IMechE Headquarters, London, UK

Organized by the Automobile Division of the
Institution of Mechanical Engineers (IMechE)

Co-sponsored by

Engineering Integrity Society
Institute of Acoustics
FISITA

IMechE Conference Transactions 1998–5

Professional Engineering Publishing

Published by Professional Engineering Publishing Limited for
The Institution of Mechanical Engineers, Bury St Edmunds and London, UK.

First Published 1998

ISSN 1356–1448
ISBN 1 86058 145 5

A CIP catalogue record for this book is available from the British Library.

Printed by The Ipswich Book Company, Suffolk, UK.

Contents

Powertrain

Systems

Body and Intake/Exhaust

Test Techniques

Related Titles of Interest

Title	Editor/Author	ISBN
Handbook of Vehicle Design Analysis	John Fenton	0 85298 963 6
Vehicle Handling Dynamics	John R Ellis	0 85298 885 0
Vehicle NVH and Refinement	IMechE Conference 1994–3	0 85298 905 9
Automotive Modelling and NVH	IMechE Seminar 1997–15	1 86058 124 2
Bus and Truck Technology (Autotech '97)	IMechE Seminar 1997–11	1 86058 118 8
Automotive Electronics (Autotech '97)	IMechE Seminar 1997–10	1 86058 115 3
Automotive Manufacturing (Autotech '97)	IMechE Seminar 1997–9	1 86058 113 7
Automotive Environmental Impact and Safety (Autotech '97)	IMechE Seminar 1997–8	1 86058 116 1
Automotive Vehicle Technologies (Autotech '97)	IMechE Seminar 1997–7	1 86058 112 9
Automotive Engines and Powertrains (Autotech '97)	IMechE Seminar 1997–6	1 86058 114 5
Automotive Materials Technology (Autotech '97)	IMechE Seminar 1997–5	1 86058 117 X

For the full range of titles published by Professional Engineering Publishing contact:

Sales Department
Professional Engineering Publishing Limited
Northgate Avenue
Bury St Edmunds
Suffolk
IP32 6BW
UK

Tel: 01284 763277
Fax: 01284 704006

Powertrain

C521/001/98

Gear rattling noise analysis for a diesel engine

Y MIURA BMEng, MJSME, MJSAE and **S NAKAMURA** BMEng, MJSAE
Hino Motors Limited, Tokyo, Japan

Synopsis

The reduction of the gear rattling noise of diesel engines at idling condition is required to keep quiet environment and comfort of drivers and passengers in trucks and buses. Decreased gear backlash is generally adopted in reducing gear rattling noise. In this paper, the newly devised measurements of momentary gear teeth velocity and gear meshing error have clarified the phenomena of rattling between the crankshaft gear and the camshaft gear of a diesel engine. Also, it was found that gear rattling noise is reduced by changing meshing position between the two gears.

1. INTRODUCTION

A social demand for traffic noise reduction has become more and more severe in recent years. The reduction of traffic noise has been one of the main concerns in the area of environmental noise pollution. The noise regulation for some vehicle types in Japan has been reinforced in 1998. Also for other vehicle types, it has been decided to be reinforced by 2002. The reduction of vehicle noise at idling condition is required to keep pleasant environment in the town area and securing comfortable ride for drivers and passengers in trucks and buses. As the gear rattling noise is unpleasant, it becomes a problem concerning user's complaints. Under the idling condition in diesel engines, the gear rattling noise can easily be a problem because of large fluctuation in engine revolution at idling, and thus the gear rattling noise reduction is important. The gear rattling noise is often caused by large gear backlash. Typical countermeasures are to decrease tolerance of the gear backlash and/or adopt gear backlash removal mechanism (scissors gear (1) and friction gear (2), etc.). However, this mechanism might not be able to be adopted due to the restriction in the layout and the cost. Therefore, in order to find another noise reduction measures, it is important to clarify the generation mechanism of the gear rattling noise. The authors devised to measure the momentary gear teeth velocity of each gear at rattling and clarified the mechanism of the rattling noise generation. Moreover, it was found that the gear rattling noise at idling is remarkably influenced by the gear meshing error, and the design concept that the gear rattling noise can be reduced by decreasing the phase difference velocity of meshing gears was obtained.

2. ANALYSES OF GEAR RATTLING NOISE

2.1 Generation phenomena of gear rattling noise

As for a V8 diesel engine, unpleasant noise in front of a vehicle radiated from timing gear cover of the engine in idling condition became a problem. The noise was conjectured to be the rattling noise of engine timing gears from its sound quality. The engine analyzed in this paper was a 20 litter V8 type direct injection diesel engine. The timing gear train of the test engine is shown in Figure 1. A crankshaft gear located at the front end of the crankshaft drives a camshaft gear installed on the camshaft. The gear ratio between the two gears is 2.0. They are helical gears with a tool module of three whose tooth profiles are involute. The numbers of teeth are shown in parentheses in Figure 1. The injection pump drive gear of the engine have adopted the scissors gear structure to remove gear backlash with the cam idler gear. Hereafter, the name of each gear is defined as shown in Figure 1.

Figure 2 shows the time histories of sound pressure near the surface of the timing gear train cover, driving torque of the injection pump, and engine revolution fluctuation when the gear rattling noise occurred. The engine revolution fluctuation is about eight percent peak to peak, which is normal. The sound pressure synchronizes with the combustion of each cylinder especially at the combustions of the fourth and the sixth cylinder. The sound pressure also correlates with the peak value of the injection pump driving torque. However, the reason of the increased sound pressure at the combustions of the fourth and the sixth cylinder cannot be explained by the peak values of the injection pump driving torque. This implies the existence of some unrevealed mechanisms.

2.2 Generation mechanism of gear rattling noise

The authors newly devised the measurement method of momentary gear teeth velocity and attempted to clarify the mechanism of the gear rattling noise generation by analyzing the speed difference of each gear at rattling. Currently, as a typical method to momentarily analyze gear teeth velocity, signals from electromagnetic pick-up installed on the timing gear cover opposite to the gear teeth are inputted to a frequency-to-voltage (F/V) converter. The gear teeth velocity measured by the F/V converter has one tooth delay from the actual gear teeth velocity because the F/V converter outputs the voltage in proportion to the gear teeth velocity after measuring the pitch time between two teeth. In the case of the engine, as the numbers of teeth in each gear measured are different, the relative momentary gear teeth velocity of each gear cannot be compared due to the one tooth delay from the actual gear teeth velocity. The schematic diagram of the momentary gear teeth velocity measurement apparatus is shown in Figure 3 (a). Electromagnetic pick-ups were installed in places against the camshaft gear (76 teeth) and the injection pump drive gear (52 teeth), and an additional gear (137 teeth) at the front end of the crankshaft instead of the crankshaft gear in order to improve measurement accuracy because the number of teeth on the crankshaft gear is small (38 teeth). The signals of electromagnetic pick-ups were inputted to an analog-to-digital (A/D) converter and then the digitized gear pulse signal was analyzed by a computer. The momentary gear teeth velocity can be calculated from the cycle of the gear pulse signal because it can be regarded as a sinusoidal curve. By calculating momentary gear teeth velocities at the ascending and descending cycles of the sinusoidal curve as shown in Figure 3 (b), the accuracy of the gear teeth velocity measurement can be doubled thus the accuracy of relative gear teeth velocity between two gears can be improved. Namely, the gear teeth velocity as a peripheral velocity on the reference pitch circle at the time A is as follows:

$$\pi * D/Z/\Delta ta \cdots\cdots\cdots mm/s$$

where, D is the reference pitch diameter and Z is the number of the gear teeth.
Also, the gear teeth velocity at the time B is as follows:

$$\pi * D/Z/\Delta tb \cdots\cdots\cdots mm/s$$

Then, the accuracy of the momentary gear teeth velocity of each gear can be improved: the number of data from the crankshaft gear is 274, from the camshaft gear is 76, and from the injection pump drive gear is 52 points per revolution of the crankshaft gear with no delay for the crank angle.

The measurement results of the sound pressure near the surface of the timing gear train cover, and the momentary gear teeth velocities of the crankshaft gear, the camshaft gear and the injection pump drive gear are shown in Figures 4 (a) and (b). Figure 4 (a) shows time histories before and after the fourth cylinder combustion with the remarkable gear rattling noise as that indicated in Figure 2. Figure 4 (b) is time histories before and after the first cylinder combustion with the small gear rattling noise. The cyclic changes of angular velocity of gear in Figures 4 (a) and (b) are related to the engine revolution fluctuation of the fourth order. As the camshaft gear and the injection pump drive gear have 76 and 52 teeth, respectively, the scale of the gear teeth velocity of the injection pump drive gear is multiplied by 76/52 in Figures 4 (a) and (b) in order to compare the relative velocities. The reason why the gear teeth velocities of the camshaft and the injection pump drive gears hardly show differences is that the scissors gear structure which absorbs velocity differences is adopted to the injection pump drive gear. Since the gear teeth velocity of the camshaft gear and the injection pump drive gear exceeds that of the crankshaft gear from the crank angle of $3\pi/4$, the meshing separation between the crankshaft gear and the camshaft gear is conjectured. Then the gear teeth velocity difference of the two gears are reversed before the crank angle of π which leads to the collision of the teeth. Thus it is considered that the gear rattling occurs when the velocity difference of both gears is large.

It is important to confirm the actual moment of collision between two gears. The gear teeth velocity were integrated for further analyses. Figure 5 shows the phase differences relative to uniform velocity of each gear before and after the fourth cylinder combustion. The phase differences of the camshaft and the injection pump drive gears hardly show differences in Figure 5 because of the scissors gear structure of the injection pump drive gear. It is clarified from Figures 4 (a) and 5 that the collision between the crankshaft gear and the camshaft gear occurs immediately before the top dead center of the fourth cylinder where the difference of the gear teeth velocity of two gears is large.

The gear rattling noise of the engine is remarkable at the combustion of the fourth and the sixth cylinders as shown in Figure 2. The separation of meshing between the crankshaft and the camshaft gears occurs at the crank angle of $\pi/4$ before the top dead center on both cylinders. Since gear rattling noise has been considered to be due to meshing error between the crankshaft gear and the camshaft gear, the meshing error should be measured. The schematic diagram of phase difference measurement between the crankshaft gear and the camshaft gear is shown in Figure 6. The engine, installed with only the crankshaft and the camshaft with respective gears, was motored by the direct current dynamometer. Rotary encoders of 300 and 600 pulses per revolution were connected to the front ends of the crankshaft and the camshaft, respectively. Then the outputs from the encoders were inputted to the phasemeter. Figure 7 (a) shows the phase difference between the crankshaft gear and the camshaft gear, and the differential of the phase difference, that is, the phase difference velocity concerning the original meshing as illustrated in the figure. The phase difference velocity between the crankshaft gear and the camshaft gear has two peaks at the crank angle of $\pi/4$ before the top dead center of the fourth and the sixth cylinders. These peaks coincide with the ridge of the engine revolution fluctuation as shown in Figure 4 (a). The separation between the crankshaft gear and the camshaft gear concerning the gear rattling noise generation is influenced by the phase difference velocity rather than the phase difference.

The generation mechanism of the gear rattling noise includes two phenomena: separation and collision. Namely, the crankshaft and the camshaft gear teeth separate when the peak of the phase difference velocity caused by meshing error between both gears coincides with the ridge of the engine revolution fluctuation and then collide when the relative velocity of two gears is large. The impact of collision is increased further if the degree of separation is large and the collision occurs at the peak of the injection pump drive torque. Thus, the gear rattling noise is primarily influenced by the relative velocity of two gears and secondarily by the injection pump drive torque.

3. EFFECT OF CHANGING GEAR MESHING POSITION

The meshing error between the crankshaft gear and the camshaft gear contributes largely to the gear rattling noise generation as mentioned above. Therefore, in order to verify that the peak of the phase difference velocity between the crankshaft gear and the camshaft gear causes the separation of meshing of both gears, the influence of meshing on the rattling noise was tested by changing the meshing position between the two gears. The phase difference velocity of the two gears was decreased when meshing position of both gears was forwarded by 19 teeth among various meshing condition. In the following, the rattling noise and the gear teeth velocity of each gear were analyzed in this modified meshing condition.

Figure 7 (b) shows the phase difference, and the phase difference velocity at the modified meshing as illustrated in the figure. The peak to peak value of the phase difference in a cycle (the crank angle of 4π) in Figure 7 (b) has little difference from the original meshing as shown in Figure 7 (a). On the other hand, the phase difference velocity at the crank angle of $\pi/4$ before the top dead center of the fourth cylinder is about half of the value of the original meshing and divided into two peaks. Figure 8 (a) shows time histories of the modified meshing before and after the fourth cylinder combustion. The separation between the crankshaft gear and the camshaft gear occurs before the crank angle of $3\pi/4$, which was earlier than that of the original meshing. The difference of the gear teeth velocity between the two gears at rattling and the sound pressure of the gear rattling noise are smaller than those of the original meshing as shown in Figure 4 (a). The noise level of the modified meshing near the surface of the timing gear train cover is reduced by 1.7dB(A) as shown in Figure 8 (b).

It was verified that it is effective to reduce the phase difference velocity in order to reduce gear rattling noise.

4. CONCLUSIONS

The conclusions of this study can be made as follows:

(a) The generation mechanism of the gear rattling noise includes two phenomena, separation and collision: the crankshaft and the camshaft gear teeth separate when the peak of the phase difference speed caused by meshing error between both gears coincides with the ridge of the engine revolution fluctuation and then collide when the relative velocity of two gears is large.

(b) The newly devised measurements of the momentary gear teeth velocity and the gear meshing error are useful for the gear rattling noise analysis.

(c) For the gear rattling noise reduction, the phase difference velocity of meshing gears should be reduced.

The rattling noise generation mechanism concluded experimentally in this paper should further be verified by the simulation calculation in the future.

5. ACKNOWLEDGEMENT

The authors wish to thank their colleagues who assisted this work. Also, the authors are thankful to the directors of Hino Motors, Ltd. for their permission to publish this paper.

REFERENCES

1. Michael Wilhelm et al. : Structure Vibration Excitation by Timing Gear Impacts, SAE Paper 900011, 1990
2. Hiroshi Kodama et al. : New 2.5L In Line Turbo Diesel Engine, in Japanese. JSAE Paper 9539329, 1995

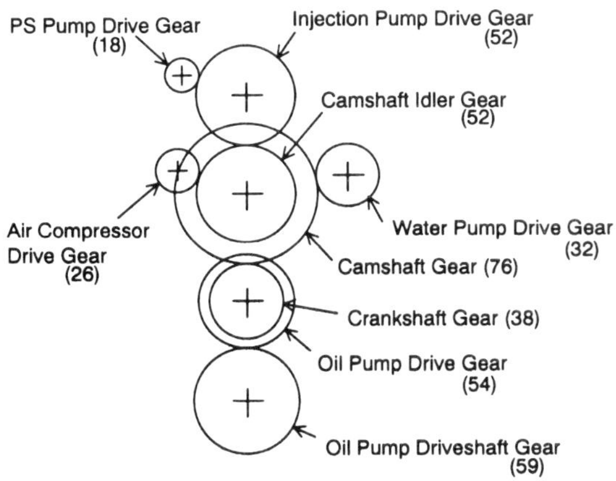

Fig.1 Timing gear train of the test engine

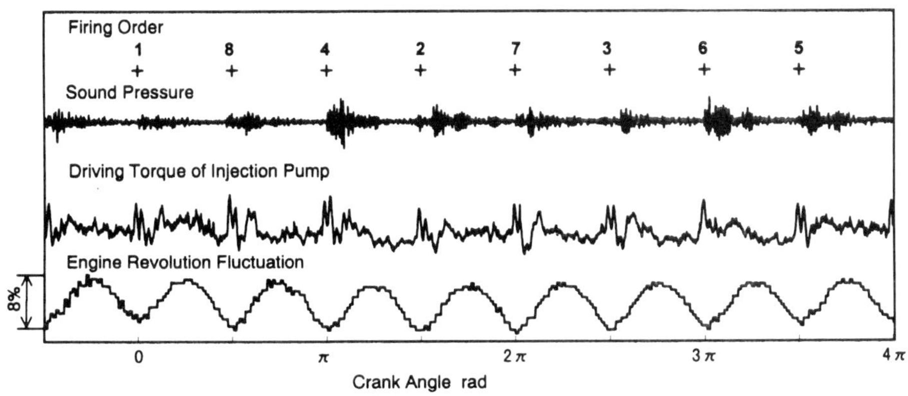

Fig.2 Time histories of sound pressure, driving torque of injection pump and engine revolution fluctuation

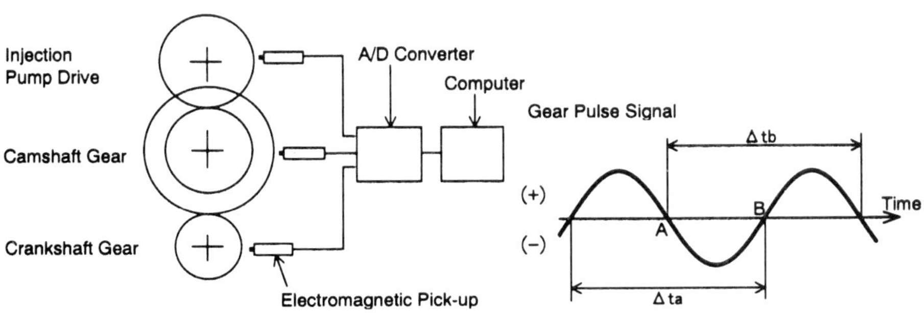

(a) Schematic diagram of measurement apparatus **(b) Calculation concept**

Fig.3 Momentary gear teeth velocity measurement

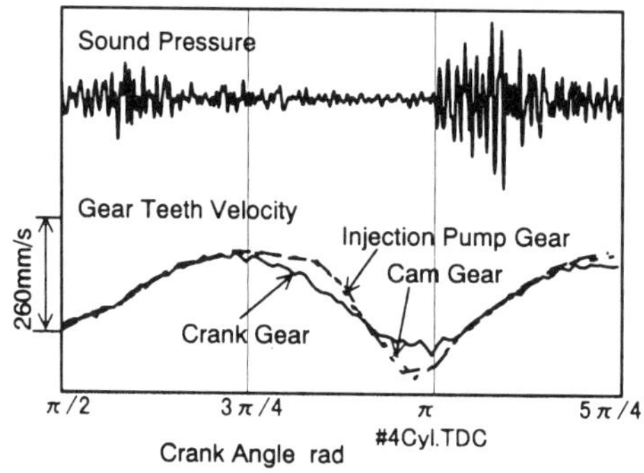

(a) Remarkable gear rattling

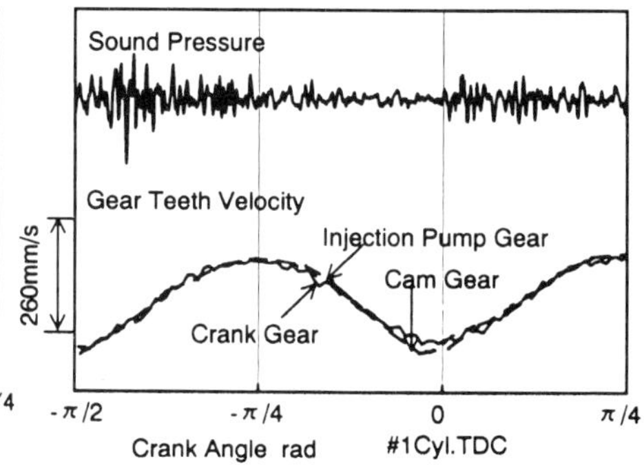

(b) Small gear rattling

Fig.4 Time histories of sound pressure and momentary gear revolution speed of the original meshing

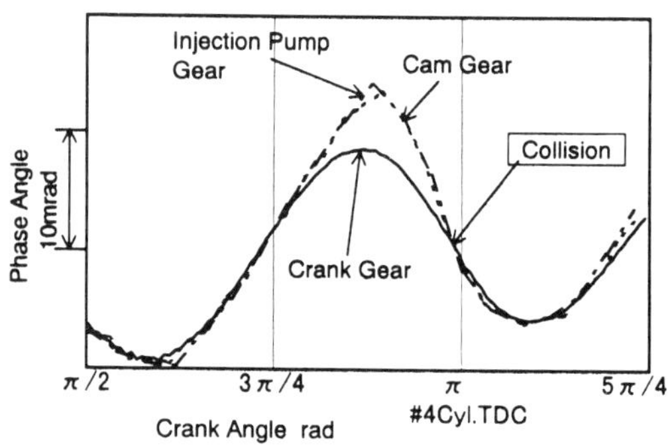

Fig.5 Phase angle of the gears of the original meshing

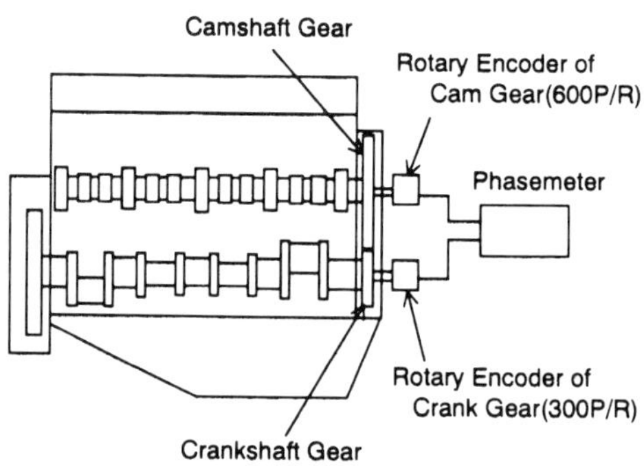

Fig.6 Schematic diagram of phase difference measurement between the crankshaft gear and the camshaft gear

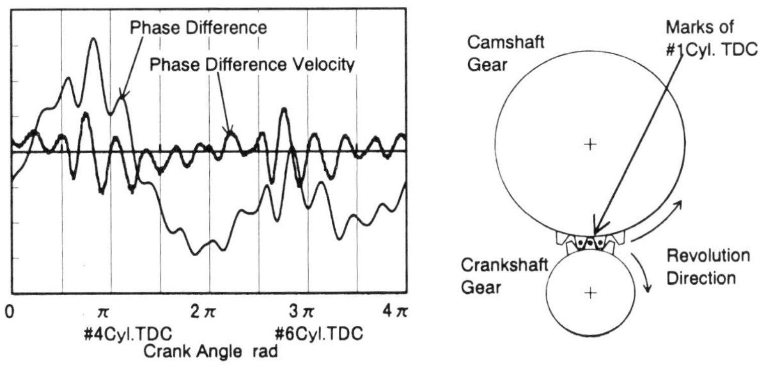

(a) Original meshing

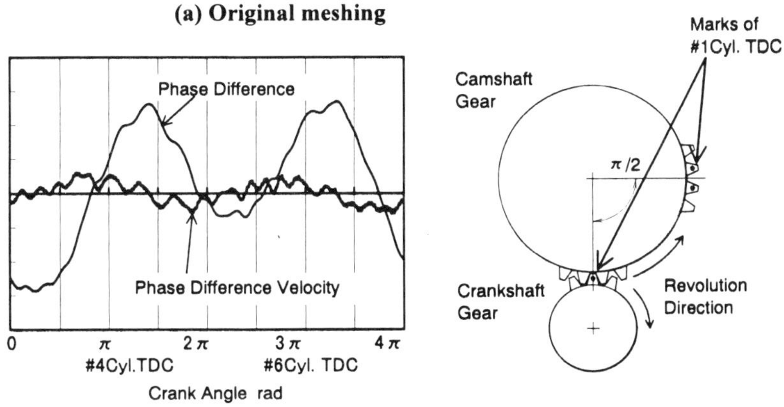

(b) Modified meshing (forwarded by 19 teeth)

Fig.7 Phase difference between the crankshaft gear and the camshaft gear

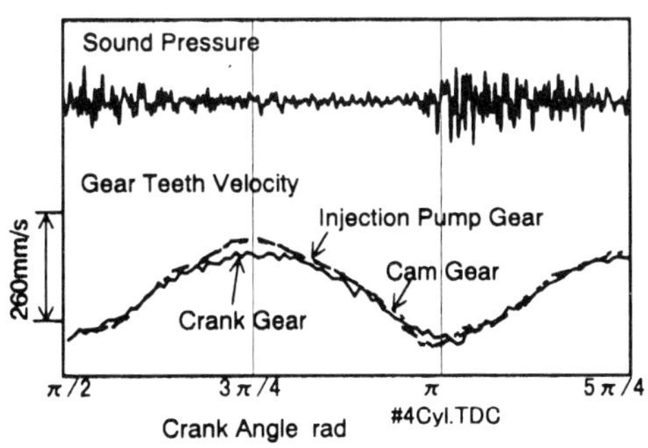

(a) Time histories

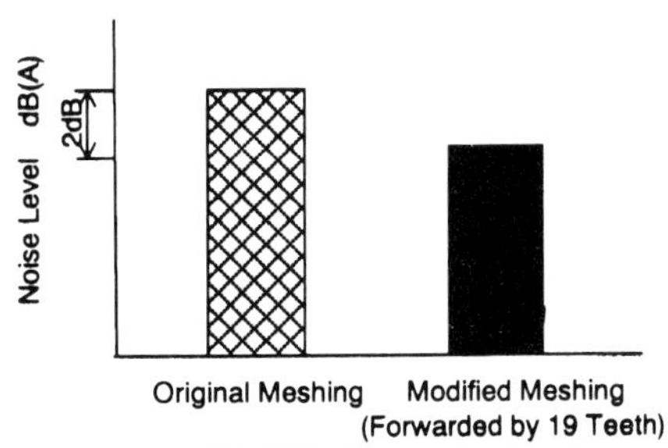

(b) Effect on noise level

Fig.8 Time histories of sound pressure and momentary gear revolution speed, and effect on noise level concerning the modified meshing

C521/004/98

Transmission modelling for gear rattle analysis

Y M WANG PhD, MASME, MSAE
Department of Mechanical Engineering, University of Maryland, USA

Synopsis

This paper describes a model for torsional vibration of automotive manual transmissions. The model is developed for the purpose of analysis and prediction of gear rattles. Elements of the model include engine, clutch, gear mesh, gear backlash, and shaft flexibility. A numerical method – *finite element method in the time domain (FET)* – is employed for the dynamic analysis. Results are presented which show the effectiveness of the modeling and analysis scheme.

1 INTRODUCTION

In recent years the automotive design evolution has converged to a level that is lessening obvious differences among automobile models within market segments. It is no longer enough to build just sturdy, reliable, attractively priced, or even luxurious cars anymore – apparently they have to sound right too (e.g., the door slam). Automotive manufactures over the world now have strong incentives to attain high automotive sound quality in order to differentiate what might otherwise become common products. Advertisements such as those shown on television for Dodge Caravan and Toyota Corolla are often touting the superior qualities of the acoustic environments of passenger compartments.

In the latest automotive vehicle refinement programs of the leading automobile manufactures in the world, major efforts are continuously made to minimize undesirable sound qualities for passenger cars and light trucks equipped with manual transmissions. Unlike automatic transmissions, manual transmissions do not have the high viscous damping inherent to a hydrodynamic torque converter to suppress the impacting of gear teeth oscillating through their mesh backlash. Therefore, a significant level of noise can be produced by the gear rattle and transmitted both inside the passenger compartment and outside the vehicle. Gear rattle, idle shake, and other noise generated by low frequency vibration phenomena in the automobile driveline have become an important concern to automobile manufactures in their pursuit of an increased level of perception of high sound quality.

The torsional vibration of driveline is a major source of gear rattle noise. The manual transmission produces gear rattle by the impacting of gears oscillating through their *mesh backlash*. The impact collisions are transmitted to the transmission housing via shafts and bearings. The vibrations are then converted into an audible rattle. The noise is most noticeable outside the vehicle on warm days. It tends to increase in intensity at higher transmission temperatures and thinner engine oil viscosities. Not only is the noise audibly objectionable, but it may also be misconstrued as an impending transmission failure leading to warranty returns. Gear rattle may occur in neutral, drive, and coast conditions.

Much of the past research in the area has concluded that an essential solution to the problem is to use a comprehensive computer modeling and simulation tool to aid the transmission design and experiments. There have been two major obstacles to such an approach: (1) Progress in understanding of the basic gear rattle phenomenon has been limited and slow. This is because the engine-clutch-transmission system involves some strong *non-linearities* including gear backlash, multi-valued springs, dry friction, hysteresis, and the like. The *non-analytic* nature of these nonlinearities presents particular difficulties at the mathematical modeling and the simulation stages, including ill-conditioning and numerical "stiffness." (2) The gear rattle is a system problem and not only a problem of gear teeth. Even though the research and industrial community has discussed the difficulties in various stages of the problem, yet no thorough framework covering the entire investigation process of such problems currently exists. This is largely due to the complexity of the powertrain system, which may make a computer analysis tool inefficient, particularly when many different elements and clearances are encountered (e.g., gears, bearings, splines, synchronizers, and the clutch).

The major goal of this article is to present a modeling scheme for a comprehensive understanding of the gear rattle dynamics including engine excitation, parametric excitation of gear meshing, and the strong nonlinearities of the driveline elements (including gear backlash, multi-stage clutch stiffness, and frictional hysteresis). The aim of the modeling effort is to develop an effective computer simulation system capable of discovering the basic rattle characteristics and finding optimal design solutions.

2 THE POWERTRAIN MODEL

2.1 The Powertrain Components

For gear rattle problems, the principal considerations in dynamics modeling are the torsional degrees of freedom of the system. With the 5-speed Neon T350 transmission (Fig. 1) as an example, the essential elements in the powertrain dynamics include the following:

Engine Excitation: The engine is the principal source of torsional excitation in the driveline. Rotational fluctuations of the engine crankshaft are mainly caused by the firing pulses, combustion nonuniformality, and unbalanced inertia forces [1]. The dominant component of the engine torsional signature is the ignition firings, which show up twice per engine revolution in a standard four-

T350 Transmission

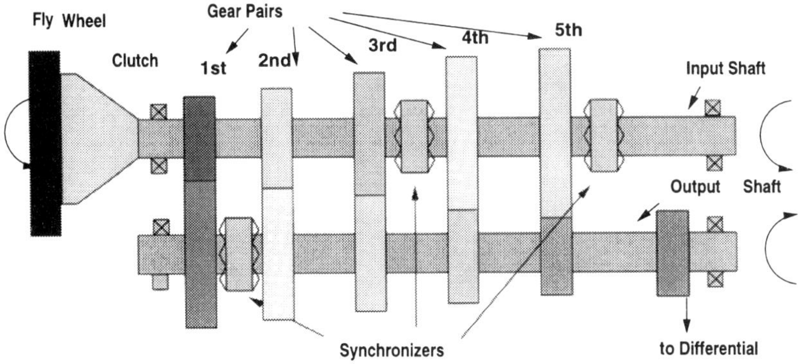

Figure 1: Dodge/Plymouth Neon transmission T350 model.

cylinder engine (the second-order). At wide open throttle, the engine torque fluctuation is nearly constant regardless of engine speed, of a typical amplitude of about 10% (20% for diesel engines) of the mean engine torque [1], as illustrated in Fig. 2.

Clutch: The clutch plays an important role in transmission rattle behavior and also in driveability. It typically consists of several strong nonlinearities: multi-stage torsional spring rate, dry friction damping, and preloads on the torsional springs (Fig. 4). In the current practice, critical clutch characteristics are often evaluated by a tuning process using audible subjective ratings by experienced engineers, for the best combination of the nonlinearities to "dampen" particular rattle/vibration problems. It is a practical but limited approach.

Transmission: The transmission includes three major types of components: input and output shafts, speed gears, and synchronizers. In modern transmissions, all forward speed gears are helical gears with spline pinions. At any speed shift, e.g., the 1*st* shift, all the pinions are in rotation and are always meshing with their respective gears; but only the particular pair of gears of the shift (e.g., 1*st* shift) transmit power, by engaging the respective synchronizer. Three *synchronizers* are used in the T350 transmission for achievement of high gear shift quality (Fig. 1).

Gears: The power-transmitting gears are **laden gears** and, together with the shaft-synchronizer assemblies, they provide the baseline torsional vibration characteristics. The laden gears are considered to be always in contact with a linear meshing stiffness. The meshing stiffness however is time-varying as shown in Fig. 3, due to the conjugate action of the involute helical gear teeth [2].

In the *in-drive* mode, there are 4 pairs of **unladen gears** and they become a problem. Without any loading but in meshing, the gear and the spline teeth may be driven across the backlash, causing impacts and rattle noise. The unloaded gear pairs are all potential rattle sources. Rattle may also occur *at idle* when the vehicle is at rest, the transmission is in neutral, and the clutch is engaged. In the T350 transmission, gears for the 1*st* and 2*nd* speeds are driven by their spline pinions on the input-shaft at idle, becoming potential neutral rattle sources.

2.2 Models of Torsional Vibration and Gear Rattle

Gear rattle phenomena of a manual powertrain are typified by complex interactions between the torsional vibration characteristics of the driveline and the coupled vibro-acoustic responses of unladen

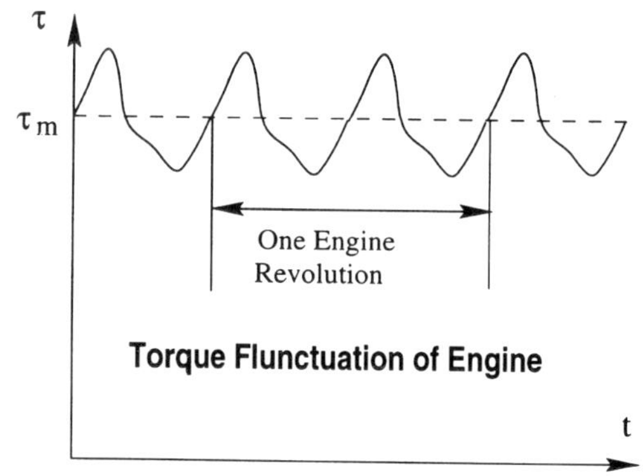

Figure 2: Engine mean and fluctuating torque.

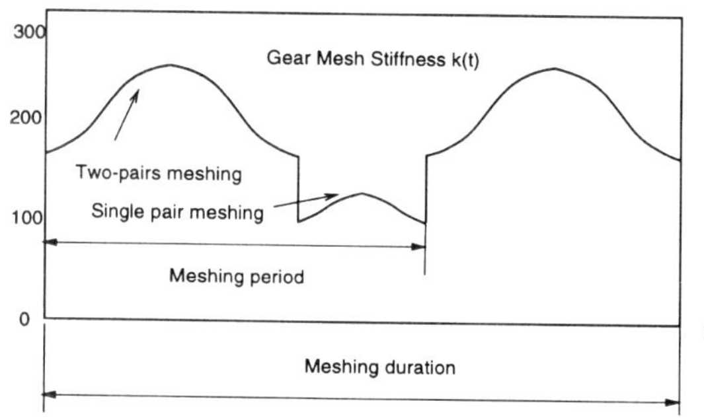

Figure 3: Time-varying gear meshing stiffness.

gears. The first step of our research is to determine an appropriate model for efficient simulation while retaining all of the essential dynamic characteristics. The primary objective is to examine the dynamic interactions between the rattling gears and the key driveline components (e.g., the clutch).

An insight into this process can be gained by examining the path of torque transmission in the system. Use the *1st* gear shift as an example as shown Figure 5. The engine power is transmitted to the input shaft through the flywheel and the fully engaged clutch. Then after velocity reduction through the first-speed gears, the power is transmitted to the output shaft and subsequently to the differential. The engine torque fluctuation generates torsional vibrations in the components on the power transmission path. The speed-reduction gears are continuously loaded and their teeth maintain continuous meshing in spite of the presence of backlash. Therefore, the *laden components* of the power path (from the flywheel to the end of the output shaft) are largely *linear*, except for a small frictional damping of the clutch. On the other hand, with the employment of all spline pinions and synchronizers, other 4 gear pairs (*2nd*, *3rd*, *4th*, and *5th*) are also meshing *but* without loading. Through their backlash, the *unladen gears* will yield vibro-impact motions and rattle noise.

The relationship between the torsional vibration and gear rattle is further illustrated in the cause-and-effect diagram of Fig. 6. The laden components of the transmission form the **baseline system** (linear except for small dry friction in the clutch), while the unladen gears become **appendages** to the baseline system. The engine excites the laden baseline system, which in turn vibrates the

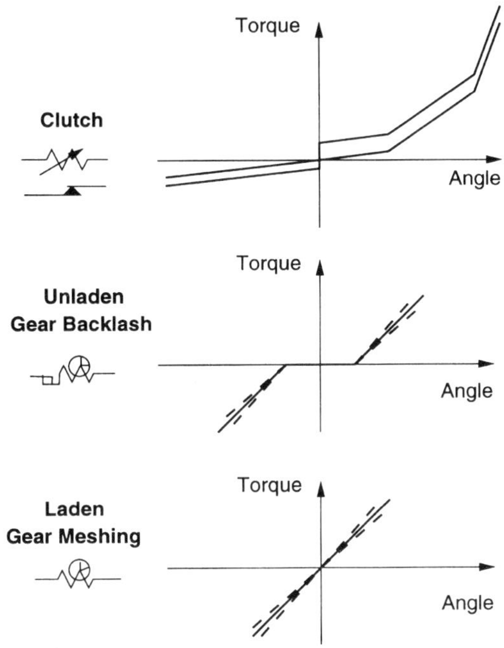

Figure 4: Various stiffness characteristics of the powertrain.

unladen gears. Thus, gear rattle phenomena are the results of the interactions between the baseline torsional vibrations and the vibro-impact characteristics of the unladen gears.

Therefore, a physical model of the transmission system is obtained according to gear configurations of each shift. Fig. 7 shows the model for the $1st$ shift. The baseline laden system consists of 7 degrees of freedom, with linear stiffness in drive shafts, linear time-varying stiffness in loaded gears, and piecewise linear stiffness in the clutch (Fig. 4). The 4 unladen appendages are strongly nonlinear with backlash and time-varying stiffness in their gear meshing. In the model, the inertial properties of the differential, CV joints, and drive-shafts are lumped to the end of the output shaft.

An engine torque applied at the flywheel is $\tau(t) = \tau_m + \tau_f(t)$ with a fluctuation period T in τ_f. An opposite mean torque τ_m is applied at the end of output shaft. The inclusion of the mean torque is essential, because the clutch stiffness value depends on the mean torque and the internal parametric excitation of time-varying meshing stiffness interacts with the mean load (to be discussed in the next section). Drag torques ($\tau_2, \cdots, \tau_5$) are also specified due to churning and/or bearing drag.

3 VIBRATION AND RATTLE ANALYSIS

The physical models of the manual transmission described above consist of 5 separate models for the 5-speed *in-drive* modes and a neutral model for the *at-idle* mode. The next step is the numerical modeling and analysis of the dynamics of each model, including describing the characteristics of the baseline vibration of the laden system, the rattle vibrations of unladen gears, and the interactions between the baseline system and the rattling gears.

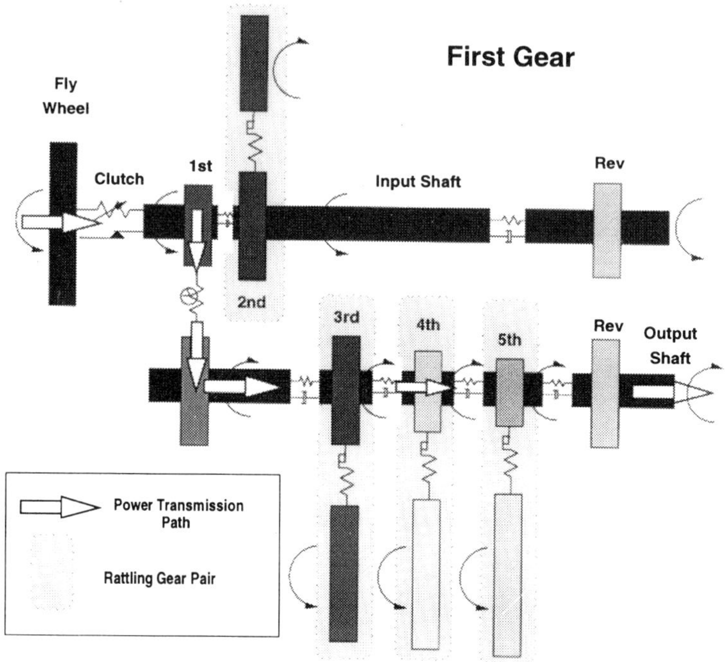

Figure 5: The powertrain model for the first speed.

It has been widely recognized in the literature that the unladen gears undertake the main role of rattle impacts [3, 4]; Their vibrations have little effects on the motion of the baseline system. This fact can be utilized to study the overall system behavior more effectively. By an appropriate lumping of unladen gear inertia, vibration of the laden system can be characterized by its response to the engine excitation. The resulting baseline torsional response then becomes the excitation to the unladen gears. By focusing on the dominant effects, a suitable model for effective numerical analyses can be obtained. This would also greatly facilitate parametric studies and design optimization.

3.1 Baseline Torsional Vibration

The first goal of numerical analysis is to characterize the baseline vibration, i.e., the response of the laden system to engine excitation. This is an important understanding for clutch design, in which torsional springs are employed to attenuate the baseline vibrational response. As shown in Fig. 7 for the $1st$ shift, when the clutch friction is not included, the laden system is essentially linear and is described by

$$\boldsymbol{M}\ddot{\boldsymbol{X}} + \boldsymbol{C}\dot{\boldsymbol{X}} + \boldsymbol{K}(t)\boldsymbol{X} = \boldsymbol{F}_m + \boldsymbol{F}_f(t) \tag{1}$$

where $\boldsymbol{X}$ is a linear displacement vector representing the 7 degrees of freedom. For example, $X_1 = R_1\theta_1$ and $X_2 = R_2\theta_2$ for the laden gear pair of base radii R_1 and R_2. $\boldsymbol{M}$, $\boldsymbol{C}$, and $\boldsymbol{K}$ denote the mass, viscous damping and linear stiffness matrices, respectively. Note that meshing stiffness of the laden gears is not constant but rather time-varying (Fig. 3). Meshing stiffness functions for helical gears are well known [2, 5]. Thus, the baseline system is a *time-varying linear* system, which can

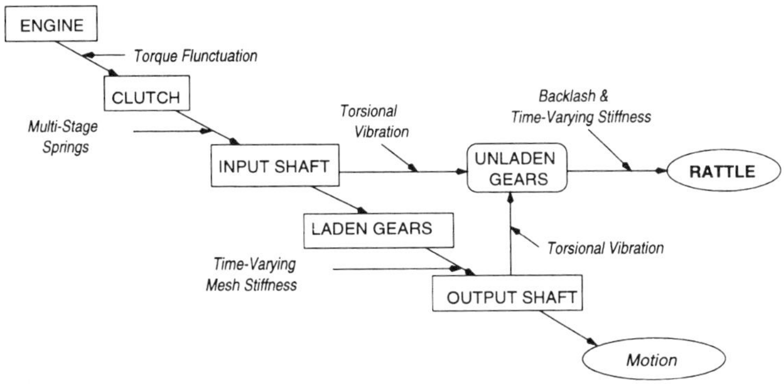

Figure 6: The cause and effect diagram of gear rattle system for the first-speed).

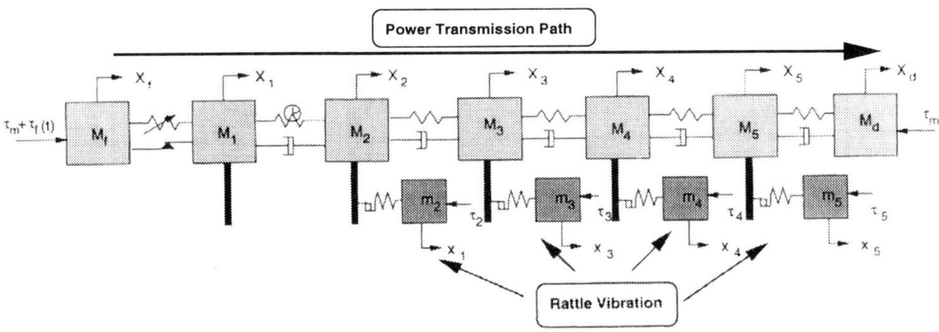

Figure 7: The vibrational model for rattle analysis for the first-speed.

be best analyzed in the time domain to obtain its fundamental vibration properties [2].

3.2 Rattle Gear Vibration

The unladen gears are the sources of rattle noise and their vibro-impact behavior is the focus of the dynamics study. Any pair of the unsynchronized gears are modeled as in Fig. 8. The gear backlash is defined by a piecewise meshing load between meshing gear teeth. The meshing stiffness is again time-varying (Fig. 3) [2]. Therefore, the rattling gear pair represent a *time-varying piecewise-linear* system.

In the model, the pinion (θ_p) is a spline gear on either the input or the output shaft, driving the unloaded gear (θ_g). Let q denote the relative displacement along the line of conjugate action, i.e., $q = x(t) - X(t) = R_g\theta_g - R_p\theta_p$. Then, the unladen gear pair is reduced to a single degree of freedom system described by

$$m\ddot{q} + f(t, q) + \tau = -m\ddot{X}(t) \qquad (2)$$

where $m = I_q/R_q^2$ and τ is the drag torque. The meshing/backlash load f and its derivative $\partial f/\partial q$

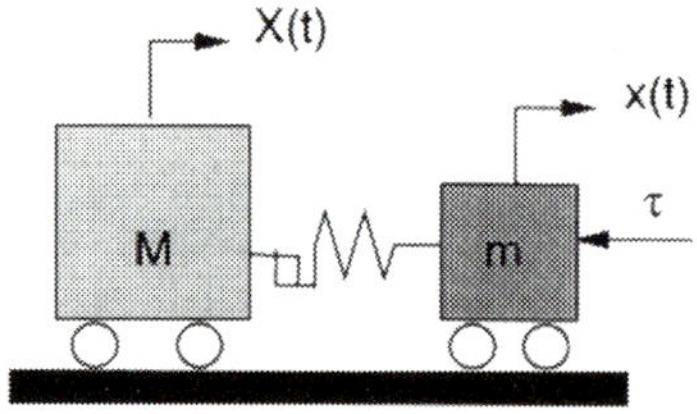

(a) Rattle gear pair model

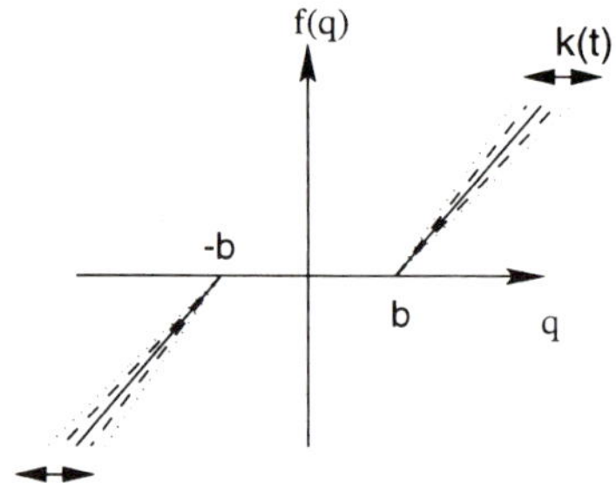

(b) Time-Varying-Piecewise-Linear

Figure 8: Physical model of a pair of unladen gears with backlash.

are defined as

$$f(t,q) = k(t)\begin{cases} q-b & \text{for } q \geq b \\ 0 & \text{for } |q| < b \\ q+b & \text{for } q \leq -b \end{cases} \quad \text{and} \quad \frac{\partial f(t,q)}{\partial q} = k(t)\begin{cases} 1 & \text{for } q \geq b \\ 0 & \text{for } |q| < b \\ 1 & \text{for } q \leq -b \end{cases} \tag{3}$$

with a time-varying meshing stiffness $k(t)$ and backlash b. The motion $X(t)$ of the pinion is that of the input (or output) shaft and is previously obtained in the baseline analysis. The motion is the excitation input to the gear rattle.

3.3 Finite Element in Time Method

Gear rattle problems are best analyzed in the time domain even though other domains may be used to obtain some fundamental vibration properties. In the past, a wide variety of direct time integration and harmonic balance techniques have been used [6], but computational difficulties are often encountered due to the "stubborn" nature of the backlash nonlinearity [7]. In the present study, an alternative method, the finite element method in the time domain (FET) is used. Our extensive experience with the new method have indicated that it offers several potential advantages and is well suited for handling the backlash and friction nonlinearities. The FET method is described in detail in [8].

The FET method is based on a weak form of Hamilton's law of varying action [9, 10], commonly known as Hamilton's weak principle. The FET modeling yields a set of *nonlinear* algebraic equations for the rattle gears as

$$\boldsymbol{A}_r \boldsymbol{Z}_r + \boldsymbol{G}_r(\boldsymbol{Z}_r) + \boldsymbol{P}_r = \boldsymbol{B}_r \text{ (where } \boldsymbol{G}_r = \sum_i \int \boldsymbol{N}_i^T f(t,q)dt) \tag{4}$$

with a set of nodal displacements $\boldsymbol{Z}$ for an approximate solution to q. These equations will be solved with the Newton-Raphson method with linearized incremental equations $\boldsymbol{Z}_r^{(n+1)} = \boldsymbol{Z}_r^{(n)} + \Delta \boldsymbol{Z}_r^{(n)}$ satisfying

$$\boldsymbol{A}_r^{(n)} \boldsymbol{Z}_r^{(n)} + \boldsymbol{G}_r^{(n)} + \boldsymbol{P}_r - \boldsymbol{B}_r^{(n)} + \boldsymbol{J}^{(n)} \Delta \boldsymbol{Z}_r^{(n)} = \boldsymbol{0} \tag{5}$$

$$\boldsymbol{J}^{(n)} = \boldsymbol{A}_r^{(n)} + (\partial \boldsymbol{G}_r / \partial \boldsymbol{Z}_r)^{(n)} \text{ (where } \partial \boldsymbol{G}_r/\partial \boldsymbol{Z}_r = \textstyle\sum_i \int (\boldsymbol{N}_i^T \partial f / \partial q) dt) \tag{6}$$

The Jacobian matrix $\boldsymbol{J}$ is analytically described and is highly structured and very sparse. Hence, the linearized equations can be solved using techniques which exploit the structure and the sparsity [8].

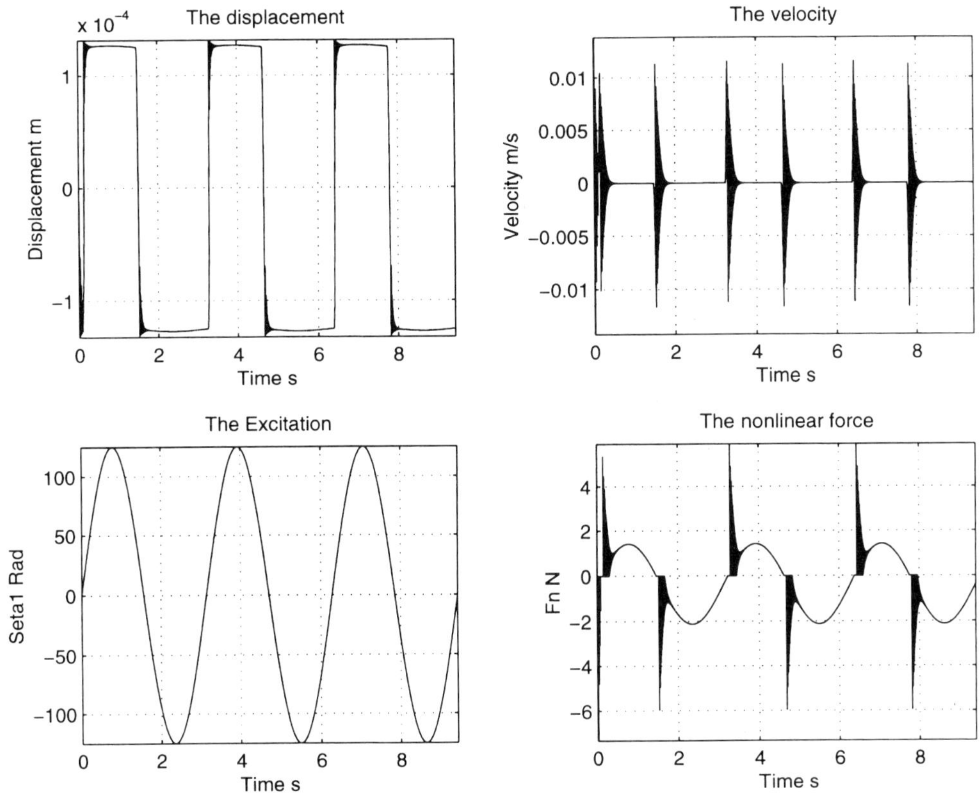

Figure 9: Relative motion of impacting gears with backlash.

4 EXAMPLE

A vibro-impact system of rattle gear pair is considered to illustrate the numerical approach. Parameters of the unladen gear include: inertia = 0.0002 kg·m^2, pitch circle diameter = 60 mm, support bearing diameter = 30 mm, gear width = 20 mm, backlash = 0.125 mm, gear ratio = 2, and damping ratio = 0.01. The driving gear is assumed in a pure second order fluctuation with a peak angular velocity of 250 rad/s and a peak acceleration of 500 rad/s^2. A constant meshing stiffness of 10^6 N/m is used. This is a case studied in [11], where the coefficient of restitution model was used for gear tooth impact.

The numerical results obtained by the finite element method are shown in Figure 9 for three revolutions of the driving gear. The figure shows clearly that the unladen gear moves from side to side of backlash alternately impacting with the coast and drive sides of the mesh periodically. At each alternation, a number of closely grouped impacts occur. Then, the unladen gear is in constant contact with the driving gear mesh side, until an alternate set of impacts occur. This is a significant difference from the motions predicted with a model based on a coefficient of restitution [11].

5 CONCLUSIONS

A manual transmission model has been developed which accounts for the major intricate properties for studying gear rattles. The model contains an approach to numerical solutions of the complex and highly nonlinear system. The computer modeling program allows engineers to input different levels of design parameters to determine their effect on gear rattle. It will be used for simulation of the dynamics of the powertrain and design optimization to reduce gear rattle.

ACKNOWLEDGMENT

The project of this work is supported in part by the National Science Foundation under grants MSS-9308663 and CMS-9713914 and by the Chrysler College Challenge Fund, which are gratefully acknowledged. The author would like to thank D. C. Wieczorek, J. Ward, and F. R. Shaver of the Chrysler Corporation and B. Joseph of New Venture Gear Inc. for their contributions.

References

[1] R. Shaver. *Manual Transmission Clutch Systems*, volume AE-17. SAE International, 1997.

[2] Y. Cai. Simulation on the rotational vibration of helical gears in consideration of the tooth separation phenomenon. *Journal of Mechanical Design*, 117(3):460 – 469, 1995.

[3] S. Meisner and B. Campbell. Development of gear rattle analytical simulation methdology. SAE Paper 951317, SAE International, 1995.

[4] S. A. Amphlett and J. P. March. Study of low frequency vehicle phenomena using advanced modeling techniques. SAE Paper 951270, SAE International, 1995.

[5] M. Maatar and P. Velex. An analytical expression for the time-varying contact length in perfect cylindrical gears. *Journal of Mechanical Design*, 118(4):586 – 589, 1996.

[6] C. Padmanabhan et al. Computation issues associated with gear rattle analysis. *Journal of Mechanical Design*, 117(1):185 – 192, 1995.

[7] R. C. Aiken. Classifications of other stubborn problem types. In R. C. Aiken, editor, *Stiff Computation*, pages 16 – 21. Oxford University Press, 1985.

[8] Y. Wang and Z. Wang. Modeling and analysis of automotive transmission rattle. In *Proceedings of 1997 SAE Noise and Vibration Conference*, volume 3, pages 1457 – 1462, Traverse City, MI, May 1997. SAE.

[9] M. Borri et al. Dynamic response of mechanical systems by a weak hamiltonian formulation. *Computers and Structures*, 20(1-3):495 – 508, 1985.

[10] Y. Wang. Dynamics of unsymmetric piecewise-linear/nonlinear systems using finite elements in time. *Journal of Sound and Vibration*, 185(1):155 – 170, 1995.

[11] N. N. Powell, S. A. Amphlett, and J. P. March. The modeling of automotive transmission rattle. IMechE Paper C492/010, Institution of Mechanical Engineers, 1995.

C521/023/98

Present and future perspectives of powertrain refinement

J P MARCH MA, CEng, MIMechE, MSAE and **M D CROKER** MEng, CEng, MIMechE
Ricardo Consulting Engineers Limited, Shoreham-by-Sea, UK

SYNOPSIS: The consideration given to incorporating specific vehicle noise and vibration reducing features into the base powertrain design has increased significantly in recent years, to the extent that many new engines are both structurally optimized and contain a wide range of additional or enhanced components, all aimed at noise and vibration reduction. This paper reviews the status of recent powertrain noise, vibration and harshness (NVH) reduction technology and looks forward to the threats and opportunities presented by forthcoming developments in the automotive industry. Particular emphasis will be placed upon the impact of future gaseous emissions requirements on engine NVH and also the opportunities presented by emerging gasoline and diesel fuel combustion technologies. The paper concludes with the author's view of the anticipated critical target areas for future powertrain noise and vibration development programmes.

1. CONTEMPORARY NOISE AND VIBRATION CHARACTERISTICS

In reviewing contemporary passenger car powertrain noise and vibration characteristics, it is tempting to look back at how noise and vibration levels have changed with time (1). The interpretation of the results from such an analysis is often confused by the considerable change in the fundamental design requirements of new engines in recent years. Although in certain cases noise and vibration still remains a secondary consideration compared to achieving the fundamental emissions, performance, packaging, cost and weight requirements of a new engine design, there are an increasing number of new engines where it has been recognised that a good subjective perception of low noise and vibration is a key attribute in product marketing, especially at the actual point of sale. As a result considerable progress has been made in both understanding the powertrain noise and vibration generating mechanisms and in minimising their effects, to achieve an installed powertrain which provides vastly improved levels of vehicle refinement compared to its predecessors.

The most common benchmark measurement of refinement is the power averaged sound pressure level from a 4-microphone array at 1m around the powertrain, such as shown in Figure 1 comparing a range of contemporary passenger car powertrains (of less than 3.0ℓ swept volume).

With the dramatic reductions in intake and exhaust noise achieved by advanced computer based modelling methods (2), surface radiated noise is now frequently the most significant contribution from the powertrain to vehicle pass-by noise levels. Base powertrain noise reduction has been one of the key targets for achieving current legislated pass-by noise limits, and the contribution of the powertrain relative to other sources (e.g. tyres) has decreased considerably as a result. However, the powertrain generally remains a major source (Figure 2) which cannot be ignored if further reductions in pass-by noise levels are to be achieved.

The contribution of the powertrain to vehicle interior refinement is most complex and requires consideration of both the structureborne and airborne noise and vibration transfer paths in all their manifestations (Figure 3). Despite significant improvements in vehicle sealing and sound packs, airborne noise remains an issue at high frequencies, particularly for sound quality problems relating to tonal noise (e.g. from gear , alternator and chain whine) and impulsive noise (e.g. from diesel combustion and gear rattle).

The reduction achieved in airborne noise has resulted in low frequency structureborne noise and vibration now being the most frequently encountered mechanism behind powertrain related vehicle interior refinement concerns (Figure 4).

2. CONTEMPORARY POWERTRAIN DESIGN FOR REFINEMENT

Close examination of most contemporary powertrain designs reveals features, either specific components or localised developments of the base powertrain design, which imply that considerable effort has been applied to noise and vibration reduction, within the primary constraints of cost, weight and packaging. That is, with a few noticeable exceptions (3), the current strategic approach to improving new powertrain refinement remains dominated by 'bottom-up' development as opposed to 'top-down' design.

Some of the more common and effective refinement related design features of contemporary powertrains are described and discussed below.

2.1 Mounting configuration

Due to the consequent reduction in vertical forcing of the vehicle body (Figure 5), the adoption of 3 and 4 point mounting arrangements for transverse front wheel drive 4-cylinder engine installations with the two static load bearing mounts aligned as close as possible along the torque roll axis (TRA) has led to substantial improvements in idle refinement for this class of vehicle, both in terms of idle noise and contact point vibration (4, 5). Optimisation of the complete dynamic behaviour of such systems, including response under severe tip-in and tip-out manoeuvres, is made much simpler by the use of dynamic simulation (6).

2.2 Powertrain structure

Significant progress has been made in the use of forced finite element models to optimise the powertrain to minimise the influence of global and local structural resonances (7), particularly in the following areas:

- Improved geometric matching and stiffness of the engine transmission interface (8) to place the frequency of global powertrain bending modes outside of the range of excitation by the firing order of the engine (Table 1)
- Use of short, stiff powertrain mounting brackets, connected to rigid areas of the powertrain structure to achieve frequencies of local mounting bracket modes outside of the range of excitation by the first harmonic of the firing order of the engine (Table 1)
- Lower mass auxiliaries fixed to the powertrain structure either directly (3) or by substantial cast brackets to achieve frequencies of local auxiliary modes outside of the range of excitation by the first harmonic of the firing order of the engine (Table 1).

Table 1 Ricardo target frequency guidelines for powertrain structural resonances

(f_{Fmax} = frequency associated with firing frequency of engine at maximum speed, e.g. for a 4 cylinder engine with a 6000 rev/min maximum speed f_{Fmax} = 200 Hz)

	Minimum Requirement	Target for Class Leading Refinement
Powertrain bending	$>f_{Fmax}$	$>\sqrt{2}\cdot f_{Fmax}$
Powertrain Mount Brackets	$>2\cdot f_{Fmax}$	$>2\sqrt{2}\cdot f_{Fmax}$
Auxiliaries	$> f_{Fmax}$	>>Powertrain bending frequency

Forced finite element models have also been used successfully in the minimisation of the noise radiation properties of new powertrain designs. This has resulted in cylinder blocks with reduced modal densities relative to weight, principally achieved by attention to detail in the main bearing to crankcase wall interface and through crankcase surface shaping and ribbing (7). The deep skirted crankcase design still dominates for cast iron cylinder blocks, with low mass bearing beams or ladderframes being applied to premium engines (9, 10). Substantial bottom end stiffening is a pre-requisite for good refinement from a new engine with an aluminium cylinder block (3, 7, 11).

2.3 Combustion

The degree of improvement made in controlling combustion noise varies between combustion systems:

- Combustion noise is generally well subdued in gasoline engines, having a small influence on overall levels and a larger influence on sound quality. At the start of the decade typical combustion noise targets for gasoline engines were centred on achieving rates of pressure rise of less than 3bar/°crank and maximum cylinder pressures of less than 70 bar (12). These targets are gradually reducing with values as low as 2.5bar/°crank and 65 bar P_{max} now being achieved in powertrains with acceptable performance and driveability, assisted by the use of variable geometry intakes and variable valve timing.

- Indirect injection (IDI) diesel engine combustion noise had, until recently, improved gradually with time but increasingly stringent gaseous emissions legislation has now resulted in a potential degradation as the heater plug has been either relocated downstream of the fuel injection spray, or retracted from the pre-chamber, to minimise the disturbance to the swirl in advance of the fuel injector. This has resulted in a reduction in turbulence, increasing the ignition delay period and hence noise. This can be addressed by detailed revision of the chamber geometry and/or re-specification of the FIE to reduce the initial injection rate and mass of pre-mixed fuel.

- Combustion noise from high speed direct injection (HSDI) diesel engines (13) has improved considerably over recent years to the point where noise levels from these engines are nearly competitive with IDI engines (Figure 1). This has been achieved by the use of higher fuel injection pressures (an increase in peak values from 700 bar to 1500 bar at the injector over the past ten years) from electronically controlled rotary pumps and through the use of two stage injectors (14).

2.4 Components

One of the principal relevant features of contemporary powertrain design is the emergence of components with a principal function of reducing noise and vibration. Typical examples include:

- Two stage fuel injectors for light load combustion noise reduction in HSDI engines (14)
- Dual mass flywheels for transmission rattle and driveline refinement improvement (15)
- Crankshaft torsional vibration dampers fitted to smaller engines with a principal noise and vibration reducing function (3)
- Dual mode (torsion and bending) crankshaft vibration dampers
- Combined acoustic and styling shields over the cylinder head and manifold(s) to reduce noise and improve the underhood appearance
- Sophisticated chain systems utilising novel chainwheels, damping rings and tensioners.

2.5 Detail design

Specific noise and vibration enhancing features can be found in the design and specification of most new powertrains. Typical examples include:

- Optimised bearing clearances
- Component isolation, especially top and front covers
- Use of damped plastics and laminated materials for covers and shields
- Single belt auxiliary drive systems
- Optimised mounting arrangements for auxiliary systems
- Optimally located powertrain mount attachment locations

3. CONTEMPORARY POWERTRAIN REFINEMENT ISSUES

The improving standard of powertrain refinement has led to a consolidation of the principal issues which need to be addressed in order to achieve further benefits. These issues are listed below and strategies for addressing them are proposed in the following sections of this paper.

Following examination of the mean surface noise source breakdown for contemporary gasoline and diesel engines tested at Ricardo, Figure 6, and from consideration of the potential for improvement for each surface source, several key components emerge:

- Noise radiation from **oil pans** has not reduced as a result of the swing away from pressed steel pans to structural aluminium ones. Whilst considerable effort is put into the optimisation of the crankcase, with a partial aim of reducing excitation into the oil pan, insufficient attention is often paid to the design of the oil pan itself.

- **Diesel engine fuel injection systems** remain a high noise source and have a particularly detrimental effect on sound quality due to their highly impulsive noise characteristic.

- The increasing complexity and size of **intake systems**, particularly in the use of multiple resonators and multi-mode plastic manifolds, has led to an increasing contribution from shell noise radiated by these components, particularly on gasoline engines. The relatively poor structural attenuation and rigidity of plastic, together with the need to use near production tooling before meaningful noise measurements can be made, has hampered noise reduction programmes to date, with isolation and shielding being the only consistently reliable noise reduction strategies.

- Direct noise radiation from the **cylinder block** (which is slightly underestimated by the sound intensity measurement technique used to generate Figure 6) is approaching insignificance in gasoline engines but remains an issue for diesel engines, where a significant proportion (which can be of the order of 50%) of this noise is in response to structural excitation by the fuel injection pump.

- The increase in standard fitment of air conditioning and power assisted steering, together with increased engine electrical loading, has led to the emergence of more complex **accessory drive** systems with consequential higher noise levels. Despite improvements in design specifically aimed at noise reduction (16), fan noise from **alternators** remains a significant noise source, particularly on gasoline engines at high speed (Figure 7).

- The remaining high surface noise sources shown in Figure 6 comprise the **exhaust manifold heat shield**, **front cover** and **camshaft cover**. High levels of noise radiation from these components are usually due to a combination of poor attachment to the engine structure, large surface areas and the use of materials with a relatively high noise radiation efficiency.

Figure 8 shows an internal noise source breakdown from a gasoline engine at full and part load conditions. Due to poor structural attenuation, this engine actually had an unusually high contribution from combustion noise, nevertheless it is clear that **mechanical noise sources** dominate, particularly at high frequencies. Today, the most commonly encountered major mechanical noise sources are the **chain drive** systems which are used increasingly for the engine timing and balancer shaft drives.

Powertrain sound quality has become an increasingly significant issue, the two most frequently encountered problems being **diesel engine part load combustion noise** (17) and gasoline engine structureborne **roughness** associated with the generation of modulating ½-order noise components from interactions between the engine structure and combustion in each cylinder (18, 19). This problem is particularly significant when the modulation coincides with engine bracket resonances between 250 and 750 Hz as shown in Figure 9.

4. FUTURE INFLUENCES ON POWERTRAIN REFINEMENT

In anticipating future powertrain noise and vibration reduction technologies it is firstly necessary to speculate about the influences affecting the direction taken in this area.

It is clear that there will be no relaxation of the customer demands for continuously improving vehicle refinement and hence significant competitive advantage will be gained from the development of refined powertrains. This advantage will be most important in the diesel powered vehicle sector, especially if manufacturers are required to expand sales volumes in order to meet tough corporate average fleet economy (CAFE) targets, hence further convergence towards the standards of refinement offered by gasoline powered vehicles is expected.

Currently, there is a lack of direction regarding future exterior noise legislation with respect to both the test procedure and the noise limits. However, it is clear that there will not be a relaxation of the current limits and it is hence inevitable that, in the majority of cases, base powertrain noise levels will be required to reduce further, unless the limits are so stringent that engine bay encapsulation becomes the only means of achievement.

The external influences on powertrain refinement will continue to be dominated by emissions, fuel economy, performance, cost, weight and packaging considerations. The most significant factors being summarised as follows:

- For **gasoline** engines, the expected transition towards **direct injection** (20) will present new refinement challenges previously confined to diesel engines, particularly if the capability of this type of combustion system to function largely unthrottled at part load is exploited to the full, namely:

    - Increased part load and idle combustion noise from an increase in burn rate
    - Increased flywheel velocity fluctuation at low speed and idle from an increase in maximum cylinder pressure. As a result this may well result in wider spread use of dual mass flywheels (15, 21) in gasoline powered vehicles
    - Reduced idle stability and increased idle shaking forces from increased maximum cylinder pressure and piston mass
    - Increased injector noise - a notable feature of the direct injection gasoline engines already in production.

- For **diesel** engines, the next generation of electronically controlled fuel injection systems, using either **unit injector** or **common rail** hardware, will offer the immediate refinement benefits of:
  - Reduced fuel system noise
  - Reduced combustion noise through rate shaping and increased injection pressure

  The greater EGR tolerance of emerging 4-valve HSDI engines (13) will also benefit part load combustion noise. However, the trend towards high levels of low speed torque back up, provided by variable nozzle turbochargers, will increase torque recoil at low speed, providing further challenges to maintaining installed driveline refinement.

  The demands for reduced cost and weight will lead to increased use of plastic intake (integrated) manifolds and aluminium cylinder blocks, even including diesel engines (7), both normally associated with degradation in powertrain refinement.

- The need to sell high volumes of low CO_2 emitting vehicles (to meet CAFE standards) will result in refined lightweight vehicles, with compact, light weight, low fuel consumption, refined powertrains. These packaging and economy requirements are encouraging the development of circa 1.0ℓ 3-cylinder direct injection diesel engines, which will require considerable refinement effort, particularly in the area of torque recoil induced low frequency vibration (22).

- Vehicle styling and crash performance constraints will continue to demand compact powertrains, leading to a predominance of chain driven timing drives and compact accessory drives, particularly on 4 and 5 cylinder engines.

5. FUTURE DESIGN TRENDS FOR POWERTRAIN REFINEMENT

The effects described above should, together with improvements in design and development technology, influence the design and specification of future powertrains considerably.

The increased use of aluminium should lead to increased focus on the **powertrain structure**. Recommended structural arrangements for refined powertrains comprise:

- Ladderframe or bedplate bottom end stiffening for reduced modal density (Figure 10)
- Rear of cylinder block and ladderframe or bedplate geometrically matched to bellhousing (8) with sufficient stiffening to achieve target powertrain bending frequency (Table 1)
- Relatively small steel or laminated steel oil pan attached to ladderframe to achieve low noise radiation (10).

The increase in computing power and experience should extend CAE models beyond reliable, time effective, global (low frequency) and local (high frequency) mode prediction to integrated noise and vibration prediction, routinely incorporating mechanical noise sources. These models should be capable of resolving the underlying interaction between combustion and the powertrain structure which leads to excessive ½-order generation, and hence ‘roughness’.

Secondary **balancer shafts** (23) will be required for larger (>2.0ℓ) gasoline and diesel 4 cylinder engines in order to achieve competitive levels of refinement (Figure 10), unless the combined forces of CAFE legislation and inexpensive electrically driven active control hardware intervene. Primary balance shafts are likely to feature on the 1.0ℓ 3-cylinder direct injection diesel engines in order to ease powertrain mounting.

Regarding **combustion noise** it is expected that significant benefits will accrue from the capability of future direct injection fuel systems to have much greater control over combustion, both cylinder by cylinder and cycle by cycle. With appropriate sensor and control technology, this should allow dynamic cylinder 'balancing' with the following benefits:

- Reduced, or even controlled, cylinder to cylinder variation in combustion characteristics should reduce roughness on gasoline engines and improve the characteristic 'knocking' noise of HSDI engines during part load transient operation.

- Diesel engine part load combustion noise could be further reduced by extending gasoline engine knock sensor technology to diesel engines. The sensors would be used to detect the onset of diesel 'knock' and the fuelling to each cylinder adjusted accordingly.

Throttling of HSDI engines may also be used to improve both idle and shut down noise and shake.

Two key areas for improvement in **mechanical noise** are immediately apparent:

The elimination of fan noise by the use of **liquid cooled alternators** (24) should significantly reduce high speed (>3000 rev/min) noise levels (Figure 7). A logical progression would be to integrate the alternator into the cylinder block thus reducing its contribution to structural vibration (Figure 4). Furthermore, the higher power capability of liquid cooled alternators could lead to the wider use of electrically powered steering pumps, and even electrically powered air conditioning systems. Adoption of these systems would allow relocation of these components away from the powertrain with a consequentially much simplified accessory drive system, resulting in reduced powertrain noise and vibration. A further development is the pseudo-hybrid flywheel generator system (25) which would give the additional benefits of vibration reduction through elimination of the starter motor and the potential to significantly improve driveline refinement by actively reducing flywheel velocity fluctuation.

Cost effective reductions in timing **chain noise** are required, especially since the move to dual mass flywheels has greatly increased low speed crankshaft nose velocity fluctuations, placing more emphasis on chain system durability. This responsibility will largely fall to the chain suppliers but should be supported by the development of better dynamic models of timing chains, capable of being used for high frequency force and hence noise prediction.

Of the remaining possible pathways to future improvements in powertrain refinement, the most significant would appear to be in the use of acoustically optimised **shields**. Cosmetic cylinder head shields are frequently used today for improving the appearance of engines and to a limited extent in reducing noise. With the exception of the shields used for most HSDI engines, these shields have only a limited acoustic effect. Future engine designs initiated with a requirement for acoustic shield fitment should allow the use of shields with absorptive backing, better sealing and a larger encompassing area. Beyond general cylinder head sourced noise reduction, this should offer the following further benefits when considered in the light of future engine design:

- Extension over plastic intake manifolds, effectively reducing noise from a high noise component with limited noise reduction potential
- Absorption of noise from fuel injection system. A feature of future gasoline and diesel (common rail and EUI) fuel injection systems is that the fuel delivery system is located either in or close to the cylinder head.

6. CONCLUSIONS

1. Substantial improvements in powertrain refinement have been achieved in recent years. Pressure for further improvements will continue into the long term, against a background of the need for lighter, more fuel efficient, vehicles. Several key target areas have been identified as requiring attention if significant improvements are to be achieved in the future.

2. The next generation of electronically controlled diesel fuel injection systems offer many opportunities for improving powertrain refinement through reduced fuel system noise and improved control of combustion noise. It is therefore important that the potential for noise minimisation is recognised throughout the combustion system development process

3. Economy and emissions optimised gasoline direct injection technologies may well transfer some of the current diesel engine refinement challenges to gasoline engines.

4. Future powertrain designs will benefit considerably from integrated structural and mechanical optimisation coupled with the emergence of components and subsystems with inherent low noise and vibration.

7. ACKNOWLEDGEMENTS

The authors would like to thank the Directors of Ricardo Consulting Engineers Ltd. for permission to publish paper. They would also like to acknowledge the assistance of Polly Courtney in producing the supporting information and data.

8. REFERENCES

1 ANDERTON, D. Trends in I. C. Engine Noise Generation and its Effects on Vehicle Noise.
ATA 2nd International Conference on Vehicle Comfort. 1992, Bologna.

2 WREN, C.S. and JOHNSON, O. Latest Simulation Techniques Applied to Intake and Exhaust System Design,
ISATA paper 94ME044

3 MENNE R., RUDDERHAM T., LATZ F., BROHMER H. Neue Vierventilmotoren fur den neuen Fiesta.
MTZ 56, July/Aug. 1995

4 KUIPERS J., POGGI M., POWELL N. Vehicle Interior Noise and Vibration Prediction Using the ADAMS Simulation Tool.
ATA, Paper No. 97A3003

5 POWELL N., CROWTHER P., KUIPERS J., POGGI M. Optimisation of Powertrain Mounting Systems using Multi-Body Dynamic Simulation.
EAEC Conference, July 1997

6 MO C., BEAUMOUNT A., POWELL N. Active Control of Driveability.
SAE 960046

7 ABTHOFF J., GELSE W., LANG J., SCHNARRENBERGER K. Lightweight Crankcase for a Turbocharged Diesel Engine with Direct Injection.
ATZ/MTZ, 1996. Materials in Automotive Eng.

8 SHIBUYA H., ISHIHAMA M., KUBOZUKA T. Improved Passenger Compartment Sound Quality of a Shell-Shaped Power Plant Structure.
IMechE, C420/034/90

9 OGAWA N., HATASHI Y., HAYAKAWA Y., KUBOZUKA T. Engine Noise Reduction by Suppression of Vibration of Main Bearing Bulkheads and Caps.
Noise Control Eng. Journal, Jan/Feb. 1988

10 BREITWIESER K., BEDNAREK G., SNITIL B. The New 1.8 and 2.0l Four Valve Engines for the New Opel Vectra.
MTZ, Sept. 1995

11 HILJEMARK S.L, KNIGHT K., SHILLINGTON S.A.C. The Development of an Innovative New Powertrain for the Next Generation of Rover Cars.
I.Mech.E. Autotech '89 Seminar on High Tech - New Engine, NEC, Birmingham, 1989

12 DE BOER C., JOHNS R., GRIGG D., TRAIN B., DENBRATT I., LINNA J. Refinement with Performance and Economy for Four-Valve Automotive Engines. IMechE C394/053/90

13 MONAGHAN M. High Speed Direct Injection Diesels Present and Future Perspectives. IMechE Lean Burn Combustion Engines Seminar, Dec. 1996

14 PETERS A., PUTZ W. A New Turbo Diesel Engine with Direct Injection for the Mercedes-Benz E-Class.
MTZ, July/Aug. 1995

15 SEBULKE A. The Two-Mass Flywheel - A Torsional Vibration Damper for the Power Train of Passenger Cars - State of the Art and Further Technical Development.
SAE 870394

16 KIRBERG U. The Compact Alternator - A New Alternator from Bosch. IMechE seminar on Vehicle Electric Power Supply,
Autotech 1991

17 PETERS A., PUTZ W. The New Diesel Engine with Common-Rail-Direct Injection for the Mercedes-Benz A-Class. Part 2: Combustion, Injection and Engine Management. Sonderausgabe
ATZ/MTZ 1997

18 DIXON J., RHODES D., PHILLIPS A. The Generation of Engine Half Orders by Structural Deformation.
IMechE C487/032/94

19 TONOSAKI H., SUMMERS T., YAMASHITA H., NAKADA T. Investigations into the Excitation of Low Frequency Half Order Vibrations in a Diesel Passenger Car Powertrain.
SAE 931304

20 JACKSON N., STOKES J., WHITAKER P., LAKE T. Stratified and Homogeneous Charge Operation for the Direct Injection Gasoline Engine - High Power with Low Fuel Consumption and Emissions.
SAE 970543

21 FITZPATRICK-ELLIS J., GOODWIN C. Unique Solutions for Driveline Noise Attenuation and Extended Clutch Life.
SAE 930909

22 DEACON M., HORROCKS R., WATTS M. The Route Towards a 3 litre/100km Passenger Car.
IMechE C524/154/97

23 BREITWIESER K., HOFMANN M., JACOBS H., VETTER U. The New 2.2l Four Valve Engine in the Opel Sintra.
MTZ, Sept. 1996

24 SCHIMANN A., SAILER W., DENTELER J. Kompakter Flussigkeitsgekuhlter Generator.
6. Aachener Kolloqium, Fahrzeug- und Motorentechnik, 1997

25 ZEYEN K., PELS T. Ein Computergestutztes Integriertes Starter- Alternator- Dampfer-System fur Verbrennungsmotoren.
6. Aachener Kolloqium, Fahrzeug- und Motorentechnik, 1997

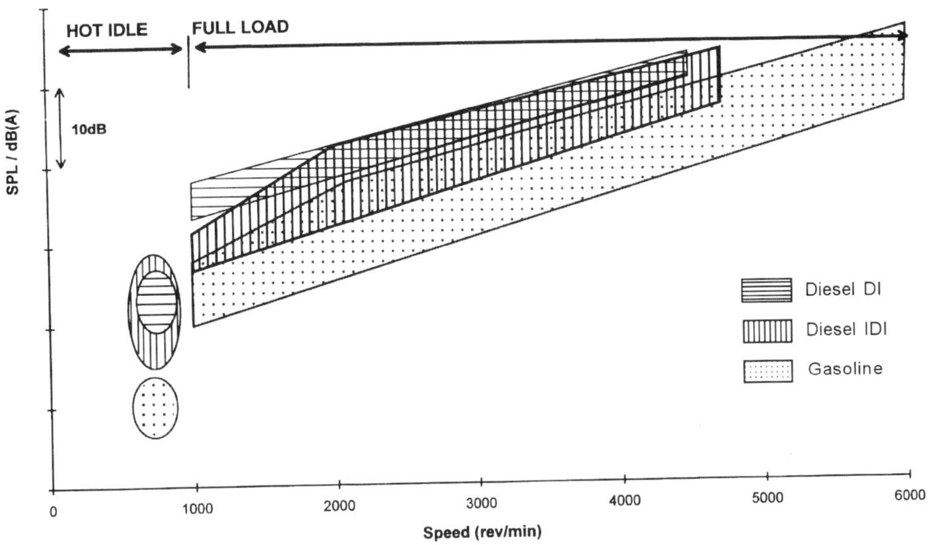

Figure 1 - Contemporary powertrain sound pressure levels (4 microphone average)

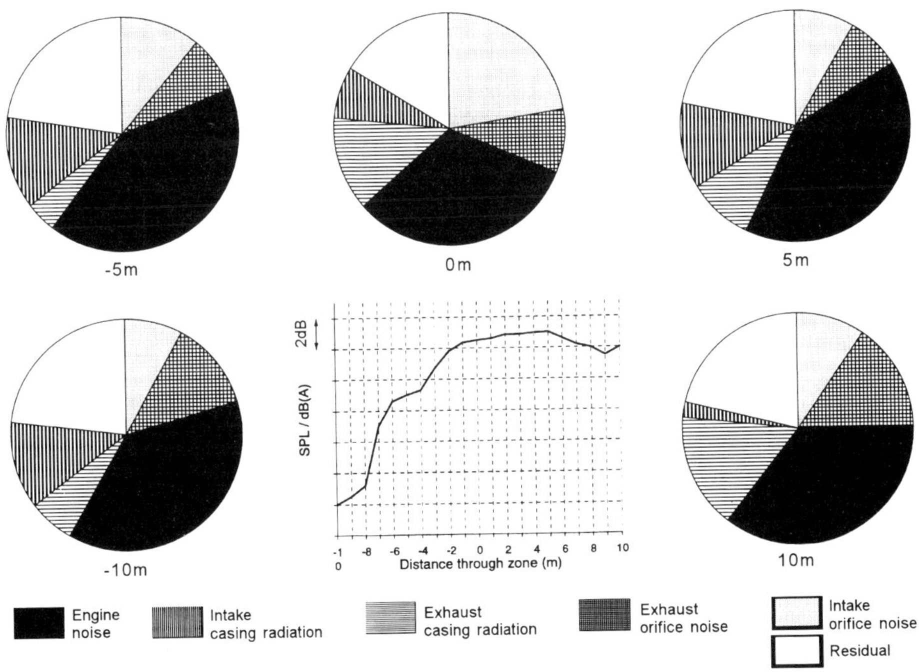

Figure 2 - Typical noise source contribution to pass-by test

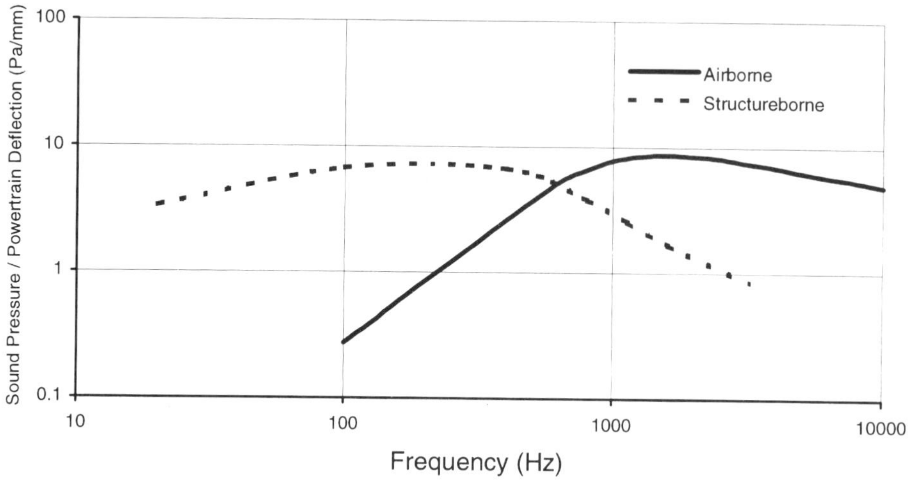

Figure 3 - Predicted vehicle interior airborne / structureborne noise response from simple model using typical measured transmission loss factors and frequency response functions

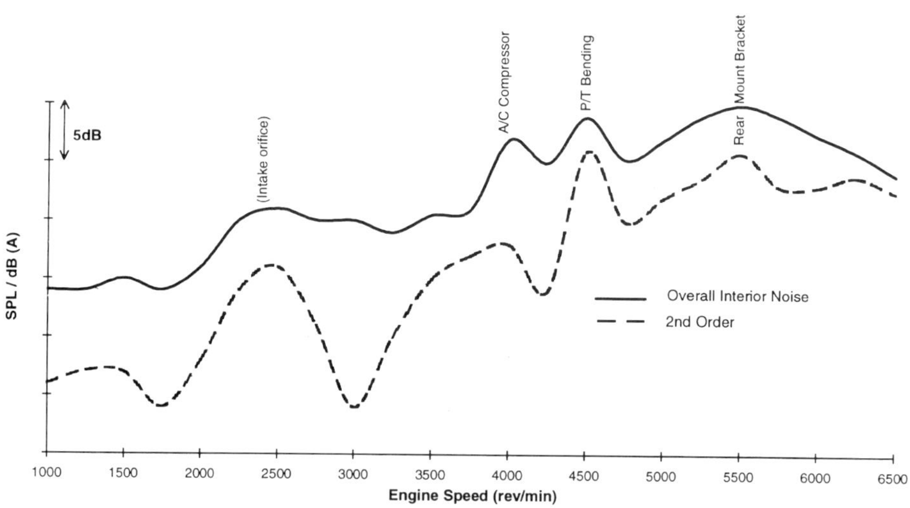

Figure 4 - Example contribution of powertrain structural resonances to vehicle interior 'booming' noise

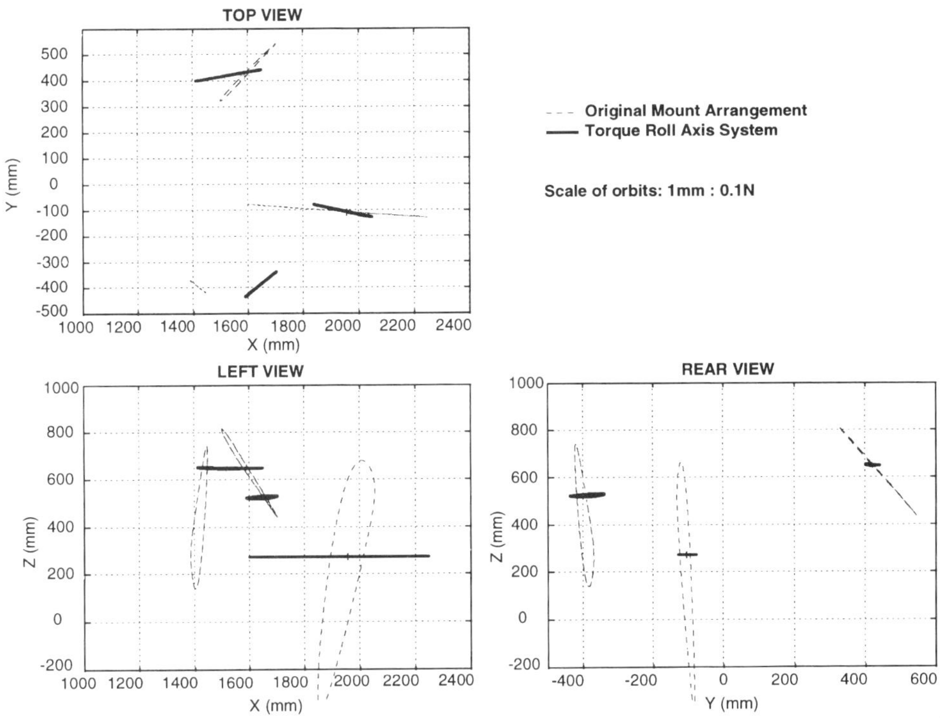

Figure 5 - Engine bay view comparing effect of mounting position on mount forces at idle

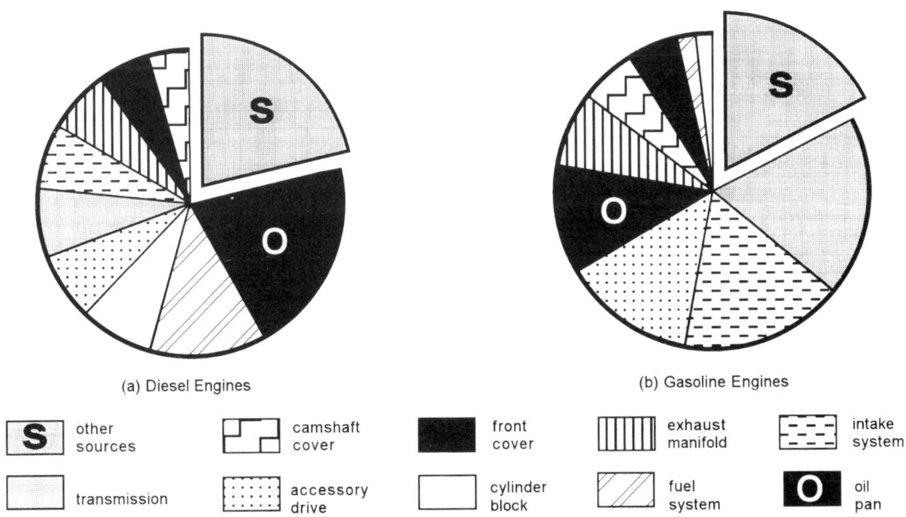

Figure 6 - Mean surface noise radiation ranking for contemporary powertrains

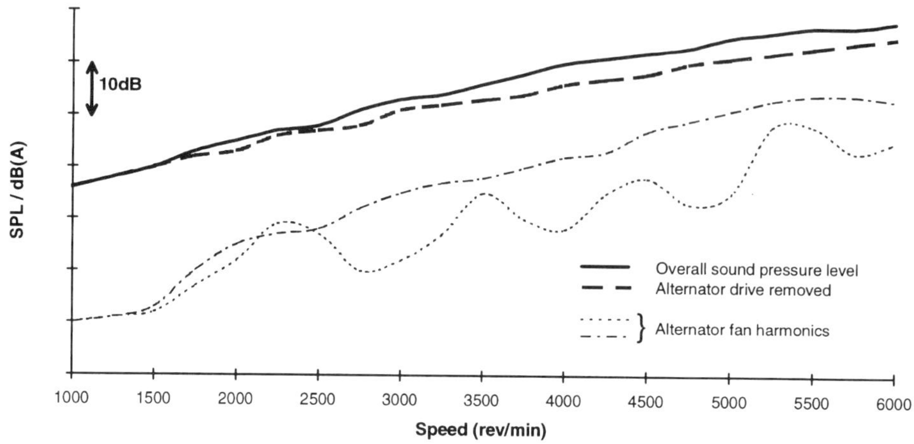

Figure 7 - Contribution of alternator to powertrain noise (0% load)

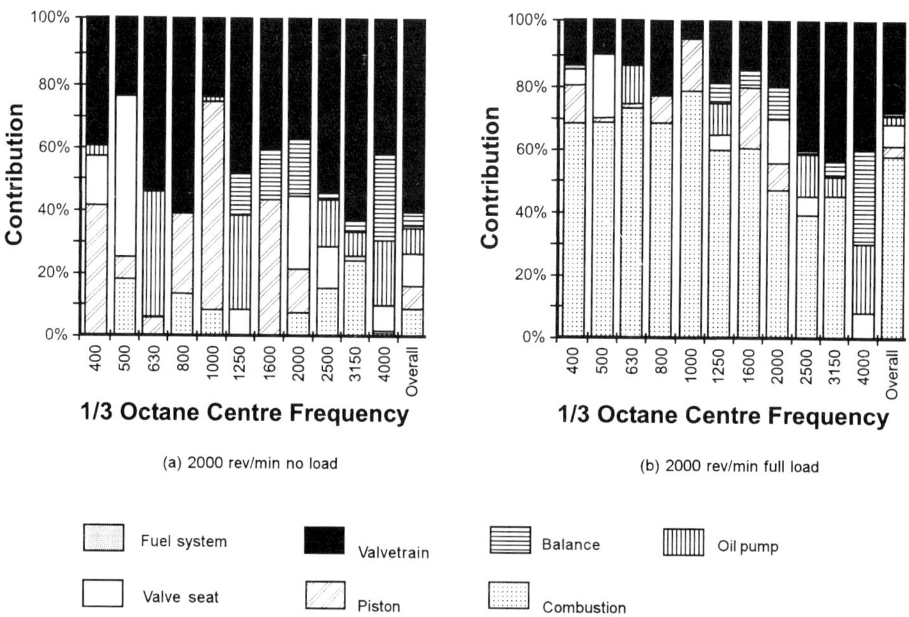

Figure 8 - Example powertrain internal noise source breakdown

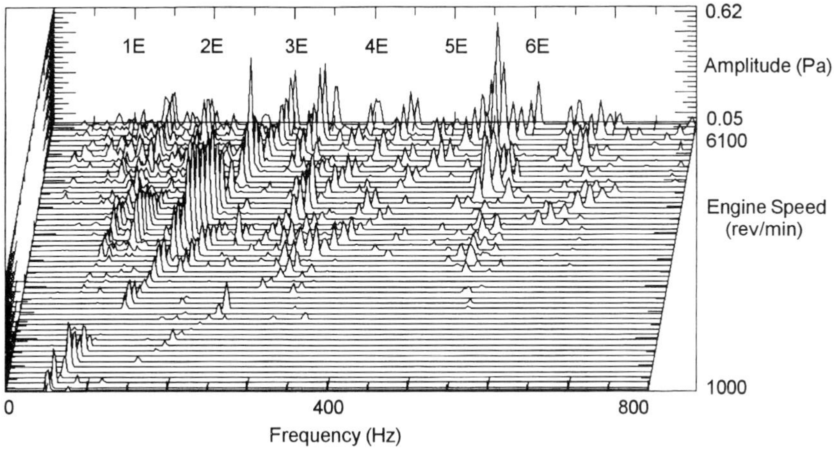

Figure 9 - Vehicle interior noise recording of 'roughness' during acceleration

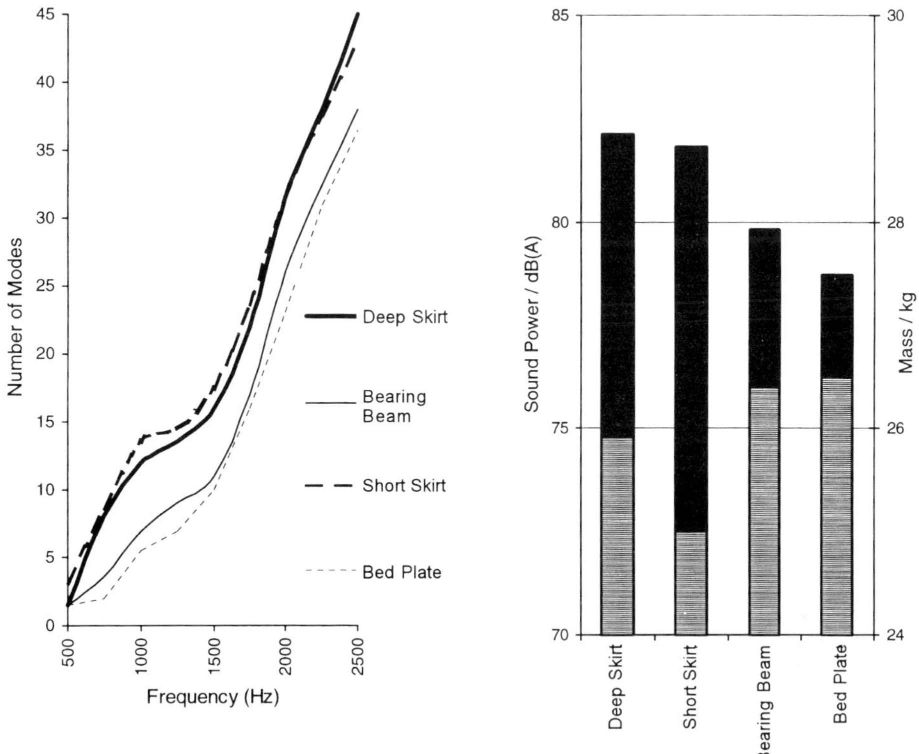

Figure 10 - Effect of cylinder block bottom end structure on modal density sound power level and mass

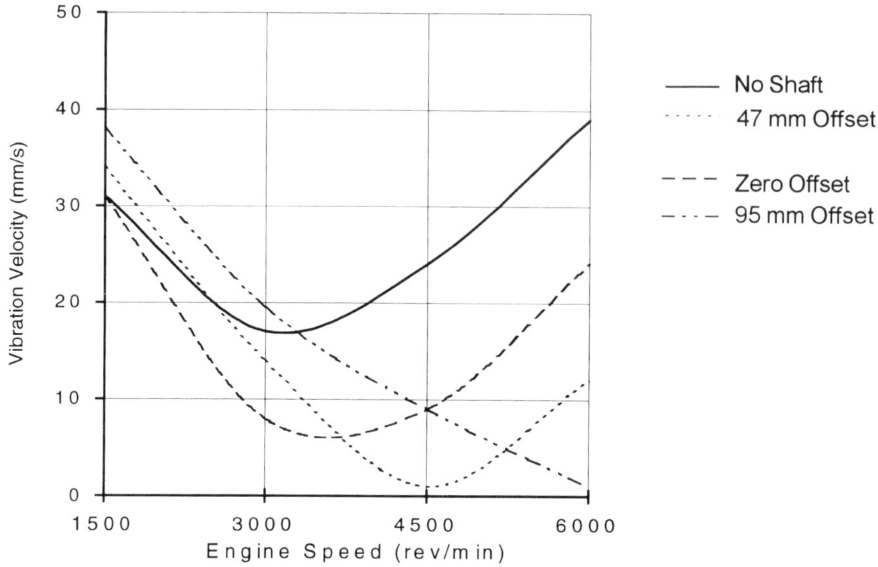

Figure 11 - Influence of secondary balancer shaft system (including vertical offset) on powertrain vibration level at roll restrictor

C521/028/98

Fuel injection pump drivetrain dynamics and their influence on pump drive torque

G P B BALFOUR BSc, MSc, MSAE and **C D YOUNG** BTech
Lucas Diesel Systems, Gillingham, UK

SYNOPSIS

One consequence of increasingly stringent emissions legislation for diesel engines is the increase in injection pressures demanded of fuel injection equipment (FIE). For conventional pump-pipe-nozzle systems, these higher pressures have a concomitant effect on the peak pump drive torque, which in turn increases the loading on the pump drive and pump bearings. The power consumption of the pump also increases with the increase of injection pressure. The influence of the pump drivetrain dynamics assumes great importance in such cases, since excitation of any resonance conditions by the torque pulse associated with the injection event will further increase the peak drive torque. A consequence of higher injection pressures is the reduction of injection periods, which can result in injection periods at or close to the period of the fundamental natural frequency of the drivetrain. A thorough understanding of the drivetrain dynamics is necessary in order to determine the factors governing peak drive torque, and to assist in producing designs where the effects of resonance are minimised.
This paper examines the dynamic behaviour of a simple pump drive system for a rotary fuel injection pump fitted to a pump test rig. The test rig included a torque transducer in the drive to provide a direct measure of pump drive torque, and allowed pump power consumption to be calculated from integration of the torque signal.
A mathematical model of the system is developed to explain the observed behaviour on the rig. This model then forms the basis for a more complete pump drivetrain model, which will allow the prediction of peak drive torque in the vehicle situation.

1) INTRODUCTION

The continuing drive to reduce diesel engine exhaust emissions in order to meet stringent new legislative targets has resulted in increases in injection pressures required from fuel injection systems. For rotary pump based fuel injection systems any increase of injection pressure is usually accompanied by a decrease of injection period.
The drive to the fuel injection pump will be by either a toothed rubber belt or chain, depending upon the maximum drive torque required. The trend for higher injection pressures will result in higher drivetrain loads, which will require more widespread use of chain drives. These higher drivetrain loads will also increase the load on the pump bearings.
The results of rig tests to measure drive torque and power consumption for rotary type fuel injection systems over several years have shown that a simple static analysis of the forces required to generate pumping pressure does not always give an accurate value for peak drive torque, because the dynamic effects of the driveline are ignored. It is likely that as period of injection reduces and maximum pressure increases then the drivetrain resonances, which had hitherto been considered high enough, will assume greater importance.
A mathematical model which predicts the dynamic behaviour of the drivetrain is required for use as a design tool, not only to gain insight into the influence of the mass elastic elements of the system but also to understand the importance of the pressure waveform shape. The model can then be used to optimise the drivetrain components, within the constraints of the rigidity required to maintain timing integrity and space limitations on inertial components.
A model is developed in this paper, and is compared to rig tests. The model is developed using a methodology which will allow simple extension to higher numbers of degrees of freedom, and can cater for force inputs at any inertial element. The model can also be readily applied to the drive for high pressure pumps used in common rail injection systems.

2) RIG TESTS

2.1 Rig details

A high pressure rotary type fuel injection pump with spill end of injection was fitted to a special pump test rig, as part of an exercise to quantify the drive torque requirements and energy balance for the pump.
A schematic diagram showing the principal drivetrain components of the rig is presented in Fig. 1. The flywheel and drivewheel are fixed to one end of the main driveshaft, which runs in two pillow block bearings. The drivewheel is a poly-vee type driven by a 10 horsepower DC motor. Any axial misalignment of the driveshaft is taken up by using an intermediate shaft with flexible couplings, these couplings being very stiff torsionally but allowing some bending.
The torque transducer is a simple strain tube design with a flange at each end, as shown in Fig. 2. Two double gauge torsion rosette strain gauges are diametrically opposed, and are connected so as to form a four arm bridge measuring torsion only. The wiring from the strain gauges passes through the centre of the driveshaft assemblies, and is connected to a set of silver slip rings on the flywheel, and then taken to a full-bridge strain gauge amplifier. It is designed to work for torques up to ±400 N-m, and is very linear with no evidence of

hysteresis. The transducer flanges are rigidly attached to the input driveshaft and the drive hub of the fuel injection pump, which is mounted on a massive bedplate on the rig.

2.2 Rig test results

The system was run over a range of speed and load conditions, in order to establish the relationships between peak drive torque and peak pumping pressure and also between power consumption and pumping pressure.

The nature of the pumping mechanism results in the pump drive shaft undergoing windup during the pumping process, with a sudden release of the strain energy stored in the driveshaft upon termination of injection. The very rapid pressure collapse within the pump at the end of injection means that the pumping components are no longer restrained, and consequently exhibit damped oscillatory motion superimposed on the steady speed of rotation until the next pumping event occurs. This results in oscillation of the drive torque immediately following the pumping event. This behaviour is clearly illustrated in Fig. 3, which shows time histories of drive torque and pump end fuel line pressure (for one line only) at low delivery, low speed and also high delivery, high speed conditions for one revolution of the fuel pump. Analysis of these plots reveals that the frequency of oscillation is about 900 Hz in both cases, and it is also clear that the damping is higher at the high speed high load condition, where the rate at which the oscillations decay is much higher. Detailed examination of the damping over a wide range of tests indicated that damping is dependent mainly on fuel delivery, with damping increasing with increase of delivery.

A static force analysis of a radial plunger rotary fuel injection pump during the pumping process allows a relationship between peak drive torque and peak pumping pressure (as applied to the rollers and shoes) to be derived. It is known that the torque required to overcome friction in the bearings and other losses is small in comparison to the pumping torque, and it can be shown that a relationship of the form T = constant * A * R * P is appropriate. In this equation T is the torque in N-m, A is the area of the pumping plungers in m^2, R is the cam lift rate in m/degree, and P is the pumping pressure in N/m^2. The constant is dependent upon the particular pump configuration, but is usually found to be approximately 70.

The relationship between peak drive torque and peak pumping pressure measured on the rig would therefore be expected to be close to linear. A plot of these quantities taken from the rig tests is presented in Fig. 4, and it is clear that there is deviation from the expected behaviour. A certain degree of scatter would be expected, in particular because the pressure is actually measured outside the pump and is therefore not a true measure of pumping pressure and may include the effects of reflected waves at certain conditions. The deviations observed are much higher than those which could be attributed to experimental errors, and it was postulated that they are due to the dynamic behaviour of the driveline.

These data were re-plotted in Fig. 5 as the ratio of peak torque to peak pressure versus injection period, which illustrates the dynamic effects when the injection period approaches approximately 1.1 milliseconds. At longer injection periods the ratio tends towards a constant value, as predicted by static analysis. The significance of the 1.1 millisecond period is that it corresponds to the period of the oscillatory motion observed in Fig. 3.

In order to better understand the dynamic effects, a mathematical model of the rig was developed. From the outset it was formulated in such a manner that extension to model the drivelines encountered in vehicles would be a straightforward process.

3) MATHEMATICAL MODELLING

3.1 Simple dynamic model of the pump on the rig

The fundamental premise in deriving a dynamic model is to disregard the steady speed component of angular displacement and simply consider angular displacements relative to a ground point on the rotating input shaft.

Calculations of inertias and stiffnesses of the drive system shown in Fig. 1 indicated that a lumped parameter model would be appropriate. The very high inertias and stiffnesses of the drivetrain to the left of the torque transducer permit the assumption that the torque transducer is attached to ground, and the system may therefore be considered to be a simple two degree of freedom system as shown in Fig. 6. This assumption was tested and shown to be valid by making large changes to the inertias and stiffnesses of some drive system components "upstream" of the torque transducer on the experimental test rig, which had no measurable effect on the measured dynamic behaviour of the pump.

The methodology adopted to solve this problem was chosen with a view to extending the analysis to systems with more degrees of freedom, as would be the case for a typical engine pump-camshaft-ancillary drivetrain system.

The system to be modelled is expressed in matrix form, thus allowing simple and rapid determination of the eigenvalues and eigenvectors for the system with damping neglected. This is a valid simplification, since it is known both from rig tests and experience that the damping is relatively light in the case under consideration, and is a necessary simplification to allow the equations to be expressed in terms of principal co-ordinates. The advantage of working in principal co-ordinates is that it allows the equations of motion to be de-coupled, which greatly simplifies the solution for non-harmonic forcing. Indeed, for more than two degrees of freedom, solution for the forced response of damped systems becomes extremely unwieldy.

The matrix formulation of the equations of motion is transformed into principal co-ordinates by pre-multiplying throughout by the transpose of the modal matrix (the matrix of eigenvectors) and then post-multiplying throughout by the modal matrix. One further simplification is necessary, since the generalised damping matrix will still have off-diagonal terms indicating coupled equations. When considering the response of any one principal mode of vibration, the contributions from other modes arising from the off-diagonal coupling terms can be shown to be very small if the damping is light, as is the case here, and so these off-diagonal terms may be set to zero to ensure de-coupling. This topic is discussed in detail in reference 1, pages 503-518, and assurance criteria are presented which can be used to establish if damping is light enough for the above approximations to be valid.

The uncoupled equations can then be expressed in receptance notation, and in particular in impulsive receptance form. These are the responses in each principal co-ordinate of the system to unit impulsive loading, and will be made up of both direct and cross receptance terms. The use of receptance methods and in particular impulsive receptances are discussed in detail in reference 1.

The convolution integral method (Duhamel's Integral) can predict system response to arbitrary inputs, and the method is particularly suitable for impulsive type loadings such as that applied to the pumping components of a fuel injection pump under pumping conditions, where the pumping period is much shorter than the period between pumping events. The system response in each of the principal co-ordinates is computed for a torque pulse applied to

mass 1, this torque pulse resulting directly from the pumping process. The time response of each mass can then be obtained in the original co-ordinate system by transforming back from principal co-ordinates. The drive torque measured in the rig tests is equivalent to the angular twist of the torque tube multiplied by its stiffness, k_2, and may be readily computed once the forced response to the torque loading arising from pumping is known.
A more complete mathematical treatment of this problem is contained in the appendix.

3.2 Results from model of pump on rig

The mathematical formulation described in the above section was realised using Mathcad ™ technical calculation software. The values for the mass and stiffness elements were determined mathematically either directly from component drawings or by using calculation features in the Lucas Diesel CAD system. The values of damping are much more difficult to estimate. The damping coefficient b_2 is likely to be small, being simply the inherent internal damping of the torque tube. The damping coefficient b_1 is controlled by the frictional forces acting on the driveshaft in it's bearings and also the viscous damping applied to pumping components as they oscillate in the pump housing, which is filled with diesel oil, as well as any inherent material damping.
Initial trials with the model were carried out, and these showed that the simple two degree of freedom lumped parameter model yielded a fundamental natural frequency of 902 Hz, which is very close to the value observed during rig tests. Test runs were then carried out to determine the forced response to pumping torque input waveforms, and the drive torque measured. These torque inputs were derived directly form experimental pumping pressure waveforms, although it would have been possible to generate these using one of the highly refined Lucas Diesel Systems FIE simulation programs. The effects of damping on the predicted response to typical pumping torque inputs were compared to rig test results in order to determine coefficients to give the best fit to experimental results. It was concluded that the damping coefficient b_2 is much less than coefficient b_1 and can be set to zero for practical purposes. The coefficient b_1 is known from rig tests to vary mainly with fuelling quantity, and suitable values were deduced to give a good match between rig tests and predicted results over a range of speed and load conditions.
Comparisons of predicted and measured responses at both low speed low delivery and high speed high delivery conditions are presented in Fig. 7. Good agreement is evident in terms of amplitude response, frequency of oscillation and decay rates. It should be pointed out that the pumping pressure profiles used in these predictions were inferred from the experimental data, and as such may deviate from the true profiles. A better result would be obtained either by measuring pumping pressure directly, which in reality is very difficult since the measurement is required in a rotating component, or preferably by using a well developed FIE simulation. A second potential source of error is that the model only predicts one cycle (due to computation time considerations) and therefore contains no contributions from previous cycles which may not have completely died away. It is not thought that this source of error is significant, however.
Once confidence in the model had been established, it was used to investigate the effect of pulse width on peak drive torque. The model was repeatedly run with the pulse width (injection period) increasing in small steps, and from these data a plot of the ratio of peak drive torque to peak pumping pressure (normalised torque) against pulse width could be derived. In order to simplify the exercise, the torque pulse shape was kept constant, and only

the period was changed No amplitude variation was necessary, since the normalised output would not change for a linear model. Triangular pulse shapes with a rise time to fall time ratio of either 3:1 (fast termination) or 2:1 (slow termination) were used for expediency, and the results for both fast and slow terminations with two different damping coefficients are presented in Fig. 8. It can be seen that the maximum value of normalised torque depends upon both the pulse shape and the damping coefficient b_1.

It is evident from Fig. 8 that the response curve shapes are of a similar form to the experimental data presented in Fig. 5, rising to an initial peak at about 1 msec pulse width and then exhibiting decaying oscillation before settling to a value of about 0.24, as would be predicted by the static force balance equation. It must be borne in mind, however, that the approximate curve presented in Fig. 5 is a composite derived from the different pulse shapes and amplitudes which are generated by the fuel pump over the range of speeds and deliveries tested, and subject to a varying damping coefficient.

The effects of pulse shape and damping coefficient were further investigated, using half sine and rectangular pulse shapes, and these results together with those using triangular pulses are summarised in table 1.

	b_1 = 0.15 N-m-s/rad		b_1 = 0.5 N-m-s/rad	
	Maximum ratio of peaks	Pulse Width (msec)	Maximum ratio of peaks	Pulse Width (msec)
3:1 triangle	0.32	0.685	0.298	0.685
2:1 triangle	0.3565	0.792	0.33	0.792
Half sine	0.424	0.879	0.4014	0.879
Rectangular	0.463	0.586	0.431	0.586

Table 1 : Effect of pulse shape on maximum ratio of peak torque to peak pressure

The slow termination triangular pulse shape gave a higher maximum normalised torque than the fast termination, and this maximum occurred at a higher value of pulse width. A half sine pulse gave a higher peak value than either of the triangular pulses, for a given value of damping, at a pulse width equal to about 0.8 times the period of the fundamental natural frequency of the system.

A rectangular pulse shape was found to give the highest maximum value for normalised torque of all the pulse shapes tested. The response curve to this pulse shape differed from the others investigated, with the normalised torque increasing up to a maximum at a pulse width close to one half of the period of the fundamental natural frequency, and thereafter remaining flat with increasing pulse width.

It is evident that the maximum normalised torque increases as the steepness of the rising portion of the pumping pressure waveform increases, and that the pulse width corresponding to the maximum normalised torque is different for different pulse shapes.

The effect of damping is as expected, with increase of damping reducing the maximum value of the ratio of peak torque to peak pumping pressure, and also resulting in a more rapid settling to a constant ratio as pulse width increases. The damping coefficient had no apparent influence on the value of pulse width corresponding to the maximum normalised torque.

The maximum normalised torque is therefore a function of :-

1) The steepness of the rising portion of the pumping pressure waveform, which increases with increasing speed for rotary type fuel injection pumps.

2) The pulse duration, which increases with increasing delivery at a fixed speed and decreases with increasing speed at fixed delivery.

3) The amplitude of the pressure waveform, which increases with both speed and delivery.

4) The damping coefficient, which has been observed to vary mainly with delivery. The damping increases with increase of delivery.

These results are significant if changes to the mass-elastic parameters are to be made in order to shift the natural frequencies as a means of eliminating unacceptable torque amplification, and it is clear that the full load rated speed condition will not necessarily correspond to the highest peak drive torque, even though it is the condition with the highest pumping pressure.

3.3 Extension of model concepts to engine situation

The use of receptance methods advocated in section 3.1 allows a straightforward extension to higher numbers of degrees of freedom, as is the case when a complete vehicle drivetrain is considered. Impulsive or other types of loading can be applied to other dynamic elements with ease once the receptances have been determined in the principal co-ordinates, and the responses summed prior to transformation back into the initial co-ordinate system.
The difficulties of calculating accurate damping coefficients from design data are well known, and it is likely that the use of empirical values for generic pump types is the best approach..
The inclusion of belt or chain tensioner devices introduces non-linearities. Models of hydraulic chain tensioners, for example, do exist and can be incorporated into the model, but due regard needs to be paid to non-linear effects, and in particular the superposition of responses to each forcing function if more than one is modelled.

4) SUMMARY

The trend for increasing injection pressures has necessitated a re-examination of fuel injection pump drive torque issues, in particular the peak drive torque and the consequent increase in fuel pump bearing load.
Experimental work on a purpose built fuel injection pump test rig has shown that the peak drive torque is not a simple function of peak pumping pressure, as would be expected from static force analysis. The dynamics of the drivetrain have been shown to have a very significant effect.
A lumped parameter model of the fuel pump on the rig was developed to gain insight into the dynamic behaviour. This model was formulated with a view to extending the theory to model complete in-vehicle pump drivetrains. The basis for the model was the use of impulsive receptances, expressed in terms of principal co-ordinates.
The model results were compared with the rig tests, and suitable damping coefficients were extracted. The model results compared very well with the rig tests, both in terms of natural

frequency and amplitude response, over a range of test conditions.
The effect of the shape and duration of the pumping pressure waveform on peak drive torque was investigated using the model. The results were presented as the ratio of peak drive torque to peak pumping pressure (normalised torque) as a function of pulse width. It was found that the maximum value of normalised torque was dependent upon

1) The steepness of the rising portion of the pumping pressure waveform, which increases with increasing speed for rotary type fuel injection pumps.

2) The pulse duration, which increases with increasing delivery at a fixed speed and decreases with increasing speed at fixed delivery.

3) The amplitude of the pressure waveform, which increases with both speed and delivery.

4) The damping coefficient, which has been observed to vary mainly with delivery. The damping increases with increase of delivery.

The maximum drive torque does not necessarily occur at the same speed and load as the maximum pumping pressure, which usually occurs at full load rated speed.
In a vehicle application, it may be necessary to change the mass-elastic parameters in order to shift the natural frequencies thereby eliminating unacceptable torque amplification. However, the scope for changing the stiffnesses and inertias are limited, since stiff elements are required to preserve injection timing integrity, and there are size constraints on inertial components. A thorough understanding of the system dynamics and the pumping pressure waveform are essential to ensure optimal design.

REFERENCES

1) R E D Bishop and D C Johnson
The Mechanics Of Vibration
Cambridge University Press, 1979
ISBN 0 521 04258 5

ACKNOWLEDGEMENTS

The authors would like to thank the directors of Lucas Varity Diesel Systems for their permission to publish this paper, and their many colleagues for helpful advice and suggestions.

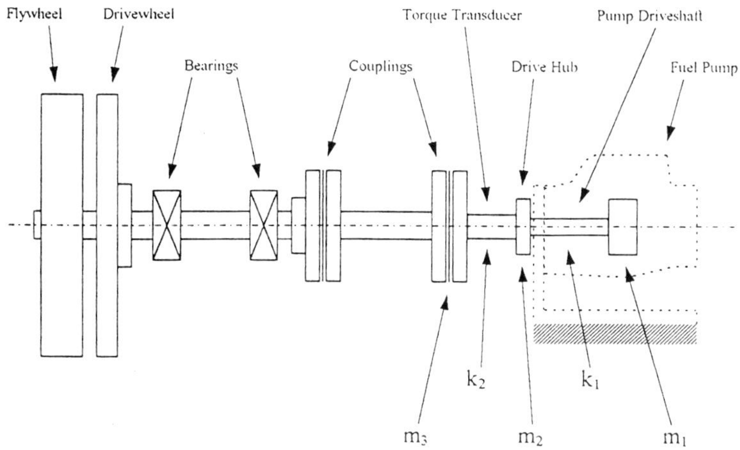

Fig. 1 Schematic diagram of the rig driveline

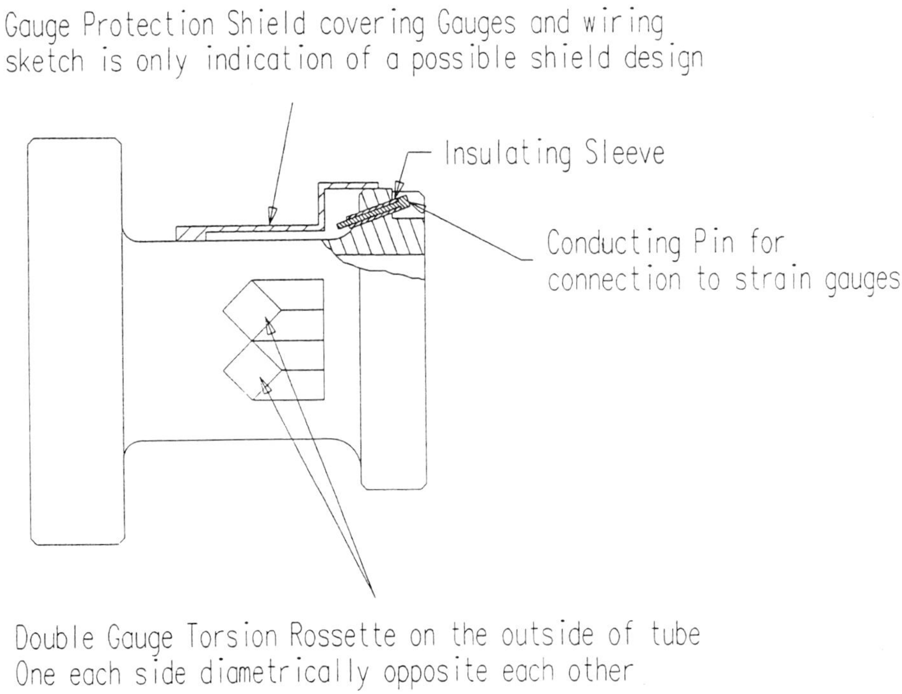

Fig. 2 Torque transducer construction

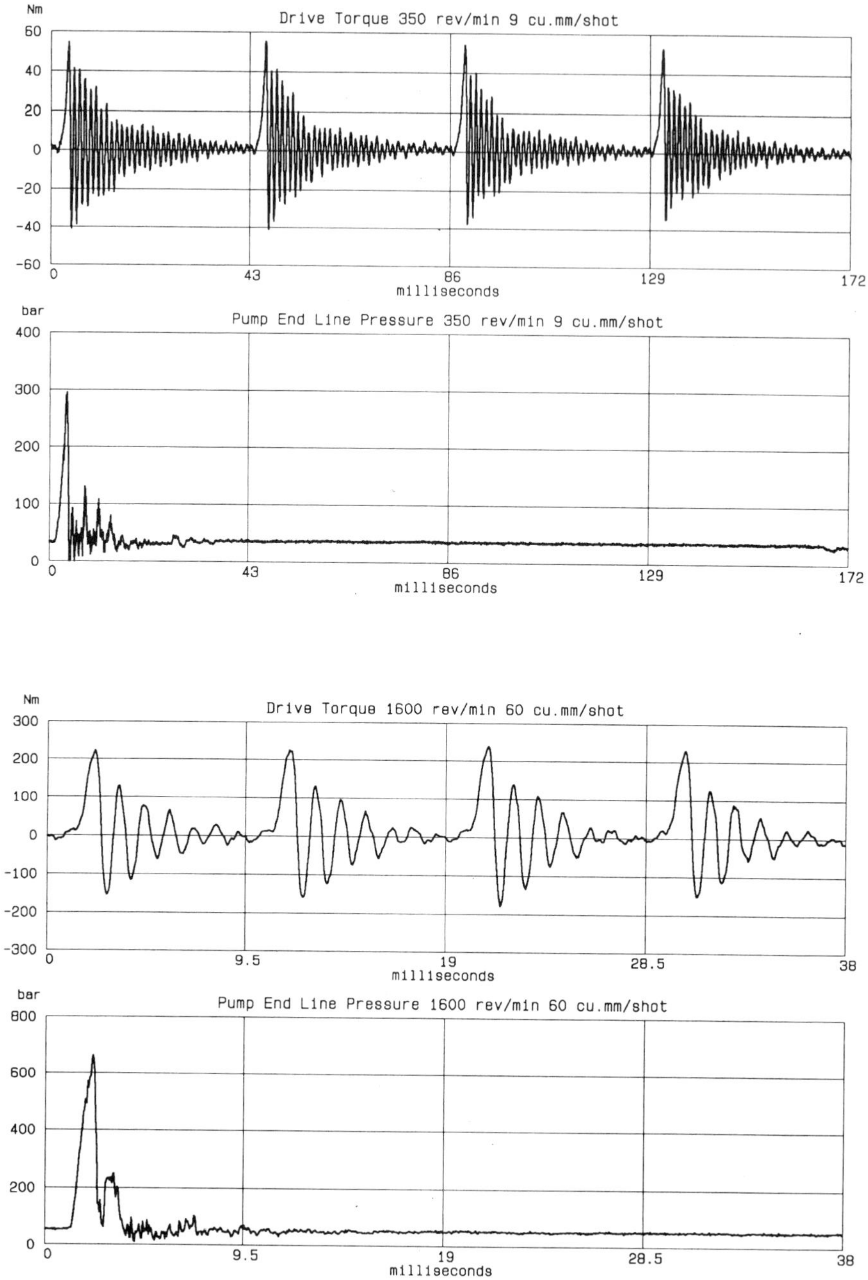

Fig. 3 **Typical drive torque and pump end line pressure time histories**

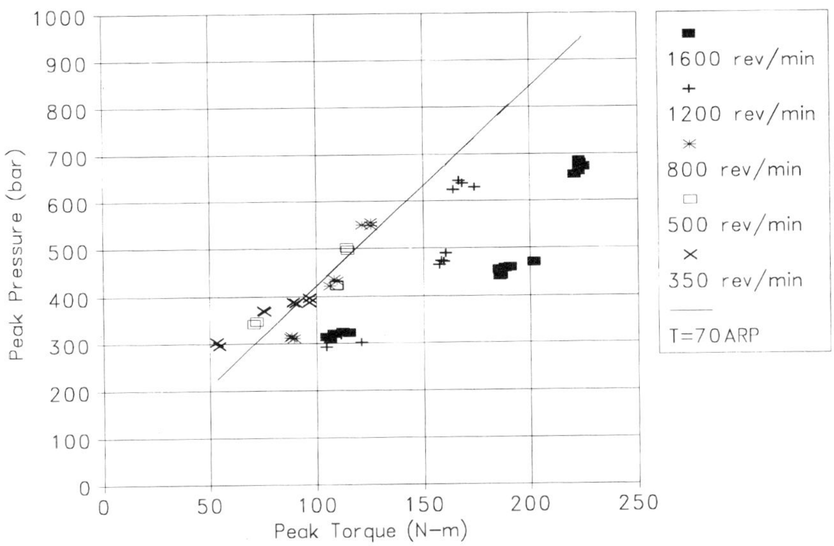

Fig. 4 **Peak pressure versus peak torque for a range of speeds and loads**

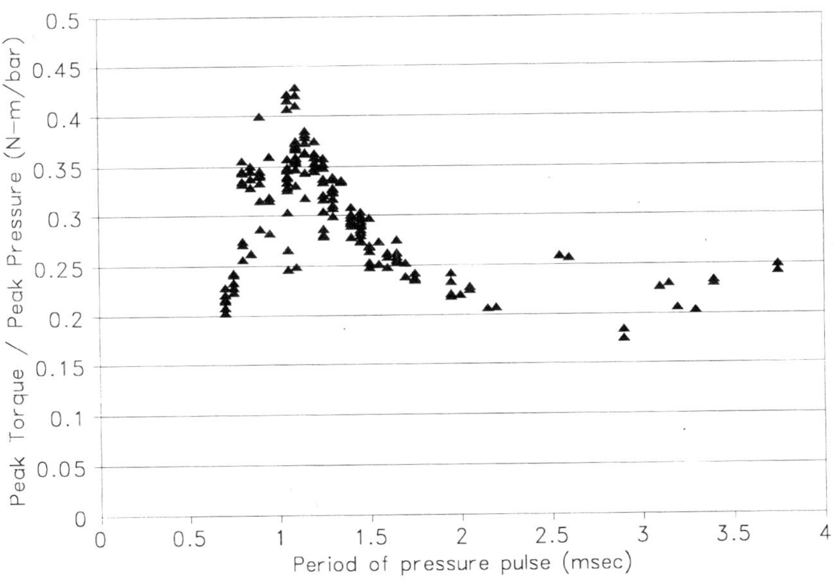

Fig. 5 **Ratio of peak torque to peak pressure versus period of injection**

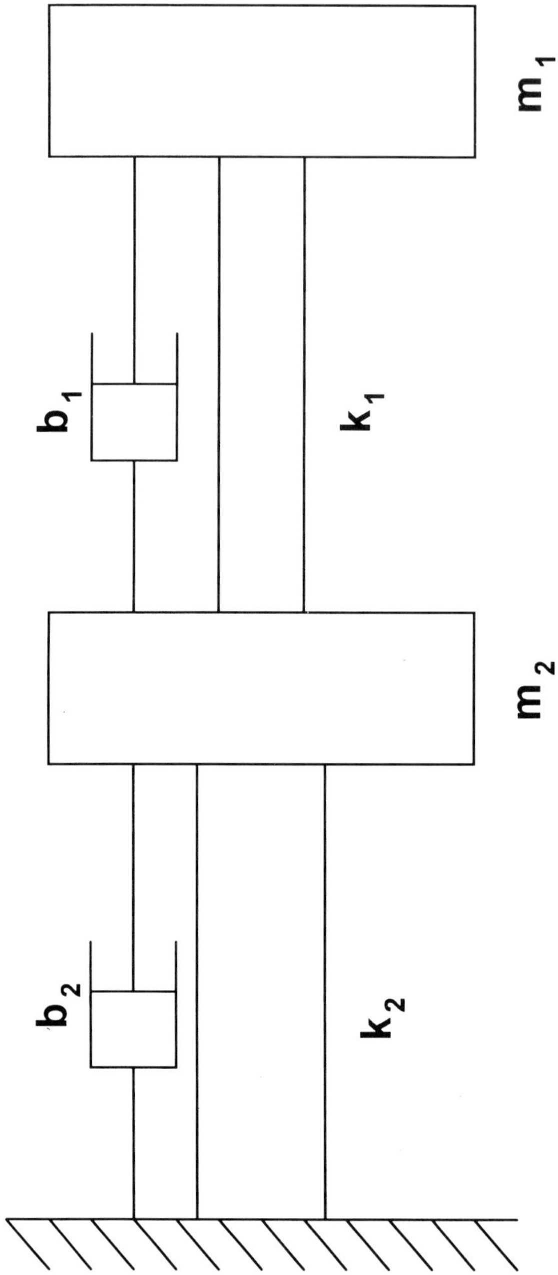

Fig. 6 Schematic diagram of two degree of freedom system

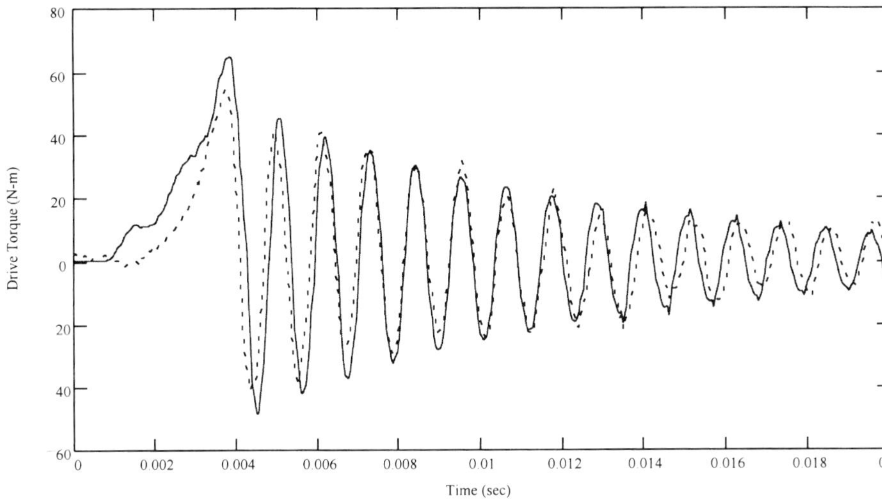

350 revs/min, 9 mm³ delivery

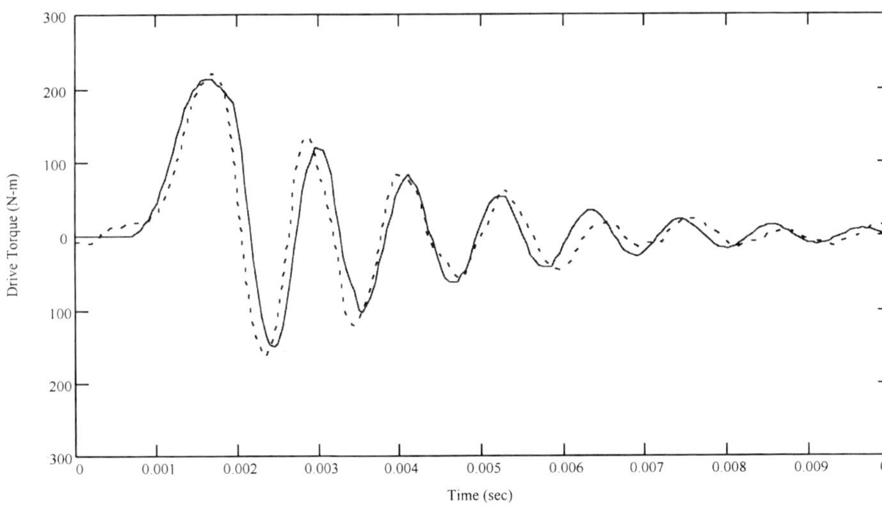

1600 revs/min, 60 mm³ delivery

Fig. 7 Comparison of predicted (solid line) and measured (dashed line) drive torques

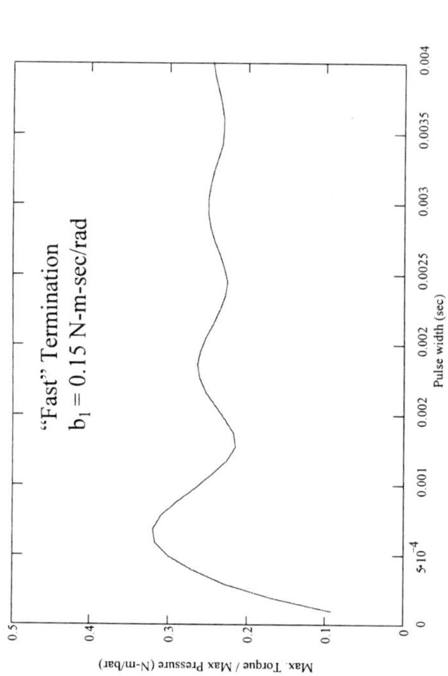
"Fast" Termination
b_l = 0.15 N-m-sec/rad
Max. Torque / Max Pressure (N-m/bar)
Pulse width (sec)

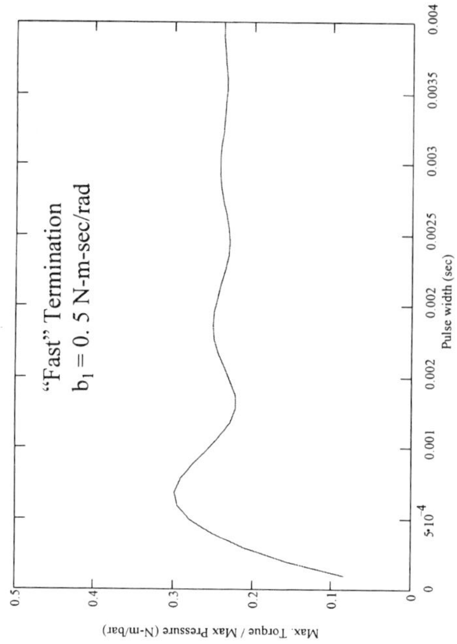
"Fast" Termination
b_l = 0. 5 N-m-sec/rad
Max. Torque / Max Pressure (N-m/bar)
Pulse width (sec)

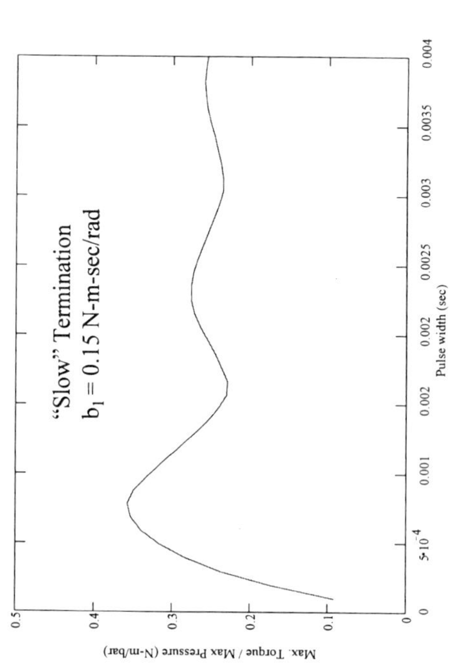
"Slow" Termination
b_l = 0.15 N-m-sec/rad
Max. Torque / Max Pressure (N-m/bar)
Pulse width (sec)

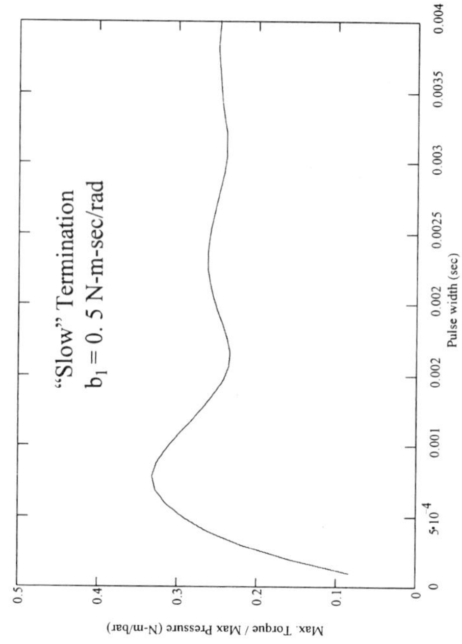
"Slow" Termination
b_l = 0. 5 N-m-sec/rad
Max. Torque / Max Pressure (N-m/bar)
Pulse width (sec)

APPENDIX

Notation

ω	circular frequency, radians/sec
ω_k	undamped natural frequency of the kth mode, radians/sec
m_i	mass moment of inertia, kg-m^2
k_i	rotational stiffness, N-m/rad
b_i	damping coefficient, N-m-s/rad
q_i	angular displacement co-ordinate for inertia m_i, radians
f_i	forcing function acting upon inertia m_i
[M]	inertia matrix
[K]	stiffness matrix
[B]	damping matrix transient forcing function
{f}	force vector
{q}	angular displacement vector
[EV]	matrix of eigenvectors (modal matrix)
$\{\lambda\}$	eigenvalues
$[\mathbf{M_p}]$	generalised inertia matrix
$[\mathbf{K_p}]$	generalised stiffness matrix
$[\mathbf{B_p}]$	generalised damping matrix
{p}	principal co-ordinate displacement vector
[I]	unit matrix
α	receptance
I	unit impulse
$F(t)$	transient (pulse) forcing function due to pumping process

Mathematical analysis of problem

The equations of motion for the two degree of freedom system under consideration are

$$m_1\ddot{q}_1 + b_1(\dot{q}_1 - \dot{q}_2) + k_1(q_1 - q_2) = f_1$$
$$m_2\ddot{q}_2 + b_2(\dot{q}_2 - \dot{q}_1) + k_1(q_2 - q_1) + k_2q_2 = 0$$

Expressed in matrix notation, these equations may be written

$$[\mathbf{M}]\{\ddot{\mathbf{q}}\} + [\mathbf{B}]\{\dot{\mathbf{q}}\} + [\mathbf{K}]\{\mathbf{q}\} = \{\mathbf{f}\}$$

Solve for natural motion by setting $\{\mathbf{f}\} = 0$. Assuming a solution of the form $\{\mathbf{q}\} = \{\mathbf{Q}\}\, e^{j\omega t}$ and neglecting damping terms, which are light, allows solution for the eigenvalues. The matrix solution is therefore given by

$$[\mathbf{K}]\{\mathbf{Q}\} - \omega^2[\mathbf{M}]\{\mathbf{Q}\} = 0$$

Pre-multiply throughout by the inverse of the mass matrix gives

$$[\mathbf{M}]^{-1}[\mathbf{K}] - \omega^2[\mathbf{I}] = 0$$

which can be readily solved to yield the eigenvectors $\{\lambda\}$ and the eigenvector matrix **[EV]**, also called the modal matrix.
In order to significantly reduce the complexity of the mathematics, particularly for problems with higher numbers of degrees of freedom, the use of principal co-ordinates allows de-coupling of the equations of motion provided that the damping terms are considered to be light enough. It can be shown that the relationship between the initial co-ordinate system and the principal co-ordinates is given by

$$\{\mathbf{q}\} = [\mathbf{EV}]\{\mathbf{p}\}$$

Consider the original equations of motion

$$[\mathbf{M}]\{\ddot{\mathbf{q}}\} + [\mathbf{B}]\{\dot{\mathbf{q}}\} + [\mathbf{K}]\{\mathbf{q}\} = \{\mathbf{f}\}$$

The forcing function vector is of the form $\{\mathbf{f}\} = \{\mathbf{F}\}\, e^{j\omega t}$ and the solution will be of the form $\{\mathbf{q}\} = \{\mathbf{Q}\}\, e^{j\omega t}$. Hence

$$[\mathbf{K}]\{\mathbf{q}\} - \omega^2[\mathbf{M}]\{\mathbf{q}\} + j\omega[\mathbf{B}]\{\mathbf{q}\} = \{\mathbf{f}\}$$

Substitute $\{\mathbf{q}\} = [\mathbf{EV}]\{\mathbf{p}\}$ throughout gives

$$[\mathbf{K}][\mathbf{EV}]\{\mathbf{p}\} - \omega^2[\mathbf{M}][\mathbf{EV}]\{\mathbf{p}\} + j\omega[\mathbf{B}][\mathbf{EV}]\{\mathbf{p}\} = \{\mathbf{f}\}$$

Pre-multiply throughout by the transpose of the modal matrix yields

$$[\mathbf{EV}]^{\mathbf{T}}[\mathbf{K}][\mathbf{EV}]\{\mathbf{p}\} - \omega^2[\mathbf{EV}]^{\mathbf{T}}[\mathbf{M}][\mathbf{EV}]\{\mathbf{p}\} + j\omega[\mathbf{EV}]^{\mathbf{T}}[\mathbf{B}][\mathbf{EV}]\{\mathbf{p}\} = [\mathbf{EV}]^{\mathbf{T}}\{\mathbf{f}\}$$

Define the generalised mass, stiffness and damping matrices as follows

$$[\mathbf{M_p}] = [\mathbf{EV}]^{\mathbf{T}}[\mathbf{M}][\mathbf{EV}]$$

$$[\mathbf{K_p}] = [\mathbf{EV}]^{\mathbf{T}}[\mathbf{K}][\mathbf{EV}]$$

$$[\mathbf{B_p}] = [\mathbf{EV}]^{\mathbf{T}}[\mathbf{B}][\mathbf{EV}]$$

Hence the equations of motion may be written

$$[\mathbf{K_p}]\{\mathbf{p}\} - \omega^2[\mathbf{M_p}]\{\mathbf{p}\} + j\omega[\mathbf{B_p}]\{\mathbf{p}\} = [\mathbf{EV}]^{\mathbf{T}}\{\mathbf{f}\}$$

The generalised mass and stiffness matrices are diagonal, as a result of the orthogonality properties of the eigenvectors. However, the generalised damping matrix retains off-diagonal

terms, which results in modal coupling. If the damping is light enough, however, it can be shown that the contributions to the response in any principal mode of terms arising from the coupling of other modes is insignificant. The off-diagonal terms may therefore be set to zero, thereby resulting in de-coupled equations of motion.
Hence, if the force vector **{f}** is defined, then the response in the principal co-ordinates is given by

$$\{\mathbf{p}\} = \left[[\mathbf{K_p}] - \omega^2[\mathbf{M_p}] + j\omega[\mathbf{B_p}]\right]^{-1}[\mathbf{EV}]^{\mathbf{T}}\{\mathbf{f}\}$$

The response in the original co-ordinate system can the be computed using the relationship between co-ordinates, yielding

$$\{\mathbf{x}\} = [\mathbf{EV}]\left[[\mathbf{K_p}] - \omega^2[\mathbf{M_p}] + j\omega[\mathbf{B_p}]\right]^{-1}[\mathbf{EV}]^{\mathbf{T}}\{\mathbf{f}\}$$

The preceding analysis applies to the general case of forced response of multi-degree of freedom systems. The next stage of the analysis is to consider the response of the uncoupled equations to unit impulse loading applied to mass 1. This will allow the direct and cross impulsive receptance terms to be derived. The uncoupled equations for natural motion may be written

$$m_{p1,1}p_1 + b_{p1,1}p_1 + k_{p1,1}p_1 = 0$$
$$m_{p2,2}p_2 + b_{p2,2}p_2 + k_{p2,2}p_2 = 0$$

The mass, stiffness and damping terms are the diagonal terms from the generalised matrices. The natural motions in each principal co-ordinate are of the form

$$p_1 = \exp\left(\frac{-b_{p1,1} * t}{2 * m_{p1,1}}\right) * (G_1 * \cos(\omega_1 * t) + H_1 * \sin(\omega_1 * t))$$
$$p_2 = \exp\left(\frac{-b_{p2,2} * t}{2 * m_{p2,2}}\right) * (G_2 * \cos(\omega_2 * t) + H_2 * \sin(\omega_2 * t))$$

Reverting back into the original co-ordinate system, for an impulse applied at mass m_1 the initial displacement conditions are $x_1 = x_2 = 0$. The initial velocities of masses m_1 and m_2 will be (I/m_1) and 0 respectively, where I is a unit impulse. By substituting t = 0 into the uncoupled response equations they reduce to $p_1 = G_1$ and $p_2 = G_2$,, and providing that the modal matrix has a non-zero determinant (a necessary condition for this analysis) then it can be deduced that $G_1 = G_2 = 0$. The derivatives of p_1 and p_2 at t = 0 are $(H_1 * \omega_1)$ and $(H_2 * \omega_2)$ respectively, so knowing the initial velocities of m_1 and m_2 allows values of (H_1/I) and (H_2/I) to be deduced as follows.

$$\begin{Bmatrix} \dot{p}_1 \\ \dot{p}_2 \end{Bmatrix} = [\mathbf{EV}]^{-1} * \begin{Bmatrix} \dot{x}_1 \\ \dot{x}_2 \end{Bmatrix}$$

Hence

$$\begin{Bmatrix} H_1 * \omega_1 \\ H_2 * \omega_2 \end{Bmatrix} = [\mathbf{EV}]^{-1} * \begin{Bmatrix} \left(\frac{I}{m_1}\right) \\ 0 \end{Bmatrix}$$

which may be rewritten

$$\begin{Bmatrix} \left(\frac{H_1}{I}\right) \\ \left(\frac{H_2}{I}\right) \end{Bmatrix} = \begin{bmatrix} \left(\frac{1}{\omega_1}\right) & 0 \\ 0 & \left(\frac{1}{\omega_2}\right) \end{bmatrix} * [\mathbf{EV}]^{-1} * \begin{Bmatrix} \left(\frac{1}{m_1}\right) \\ 0 \end{Bmatrix}$$

The direct impulsive receptance $\alpha_{1,1}$ is defined to be the response of mass 1 (in the original co-ordinate system) to unit impulse applied at mass 1 and the cross receptance $\alpha_{2,1}$ is defined as the response of mass 2 to the impulse applied to mass 1. Combining the equations for natural motion in principal co-ordinates, the relationship between co-ordinates and the vector for the constants (H_1/I) and (H_2/I) yields

$$\begin{Bmatrix} \alpha_{1,1} \\ \alpha_{2,1} \end{Bmatrix} = [\mathbf{EV}] * \begin{Bmatrix} \exp\left(-\frac{b_{p1,1} * t}{2 * m_{p1,1}}\right) * \left(\frac{H_1}{I}\right) * \sin(\omega_1 * t) \\ \exp\left(-\frac{b_{p2,2} * t}{2 * m_{p2,2}}\right) * \left(\frac{H_2}{I}\right) * \sin(\omega_2 * t) \end{Bmatrix}$$

A similar process can be followed for the case of an impulse applied at mass 2 in order to fully define the problem. In the case under consideration, however, it is sufficient to limit the analysis to impulse excitation at mass 1 only.

The use of the convolution integral to compute the response to a transient forcing function $F(t)$ of duration τ applied to mass 1 completes the analysis. The responses in each (original) co-ordinate are given by

$$x_1(t) = \int_0^\infty F(\xi) * \alpha_{1,1}(t - \xi)d\xi$$

$$x_2(t) = \int_0^\infty F(\xi) * \alpha_{2,1}(t - \xi)d\xi$$

The above integrals are computed numerically at time increments chosen to adequately define the response. Too short an increment increases the computational burden, whilst too long an increment will reduce resolution. The upper integration limit of infinity does not present a problem, since the upper limit can be set to t for $0 \le t \le \tau$ or τ for $t > \tau$.

C521/030/98

A new approach for powertrain structure borne noise analysis

F ALBERTZ MSc, PhD
Ford Werke AG, Koeln, Germany

A new approach has been developed which provides a quantitative breakdown of the measured powertrain vibration into powertrain rigid motion, component modes and global powertrain compliance (e.g. powertrain bending), based on a kinematic analysis of the measured running mode shape data. This method clearly separates the relative influences of component from global modes, and thereby indicates which powertrain design changes will provide the greatest structure-borne noise improvements. Additionally, by applying the method to analytical data the correlation of an FE model can be very efficiently supported.
This paper describes the new approach and an example of a complete structure-borne noise analysis. The correlation of an FE-model of a powertrain model is also shown.

Notation

a, **A**		matrix, vector	*e*	Mode Number
$\overline{a}$		complex variable	*flexible*	flexible contribution
			rigid	contribution of rigid powertrain
T	Nm	Torque	*2EO*	2nd Engine order shaking
F	N	Force	*j*	location of force
φ_{ei}	$\sqrt{1/kg}$	transitional eigen-vector	*i*	location of displacement
γ_{ej}	$\sqrt{1/kg\,m^2}$	rotational eigen-vector	*ref.*	reference
$m_{P/T}$	kg	mass of powertrain	*comb*	combustion
$\Theta_{P/T}$	$kg\,m^2$	inertia tensor of powertrain	*x, y, z*	co-ordinates
ω	$1/s^2$	frequency of vibration	*meas.*	measured
D_e	-	critical viscous damping ratio	*k*	number of force locations
trans		rotation	*l*	number of torque locations
rot		measured	*m*	number of considered modes
FE		Finite Elements	*DOF*	Degree of Freedom

1 INTRODUCTION

A major component of the interior noise levels in vehicles is structure borne noise (50-500Hz) transmitted from the powertrain. The main noise paths are the engine mounts. These paths are determined by the vibration levels of the powertrain at the mount positions, the dynamic stiffness of the mounts and the noise transfer functions from the body fixing point of the mounts to the drivers ear. Sharp peaks in noise levels during engine acceleration (often called "booms") are particularly unpleasant and should be avoided. This leads to 3 possibilities to reduce the interior noise levels.

- reduce mount stiffnesses
- improve body noise transfer functions
- reduce vibration levels of the powertrain at the mount fixing points

The reduction of mount stiffnesses is often limited by trade-offs with other attributes like durability and drivability. Due to the fact that the noise transfer mechanisms are very complex, it is often very difficult to improve the structure borne noise transfer functions by design improvements [1].

A very efficient way to improve the low frequency interior noise is to reduce the powertrain vibration levels at the mount fixing points. Vibration peaks due to resonance effects should be reduced by increasing stiffness and reducing mass. Introducing absorbers can also provide improvements but in many cases it does not provide a robust solution for the life cycle of the vehicle.

The identification of the root causes for peaks in the vibration level curves vs. speed is a difficult task due to the interaction of complex mode shapes. One common approach is to visualise the running mode shapes via on screen animations. The root causes are often difficult to identify from such animations due to the superposition of the rigid powertrain motion and the various resonance effects, and the relative contributions from local bracket modes and global bending modes cannot be quantified using this technique. This makes it difficult to determine which design changes will provide the greatest improvements. To overcome this problem the application of kinematic analysis as a new approach to analyse running mode shapes was developed.

In the following section, methods to establish structure-borne noise data are considered. After that a theoretical background of the powertrain behaviour is discussed (see section 3). Section 4 contains a description of the kinematic analysis as a new approach for an advanced analysis of powertrain running mode shapes. Section 5 describes an example of a complete structure-borne noise analysis including the correlation of an FE-model.

2 ANALYSIS OF POWERTRAIN STRUCTURE BORNE NOISE

To estimate the powertrain structure-borne noise behaviour three methods are common practice [2]:

- Rig tests
- modal test
- FEM based modal analysis and frequency response calculation

To perform a rig test the powertrain (engine, transmission, accessories, driveline etc.) is assembled separately from the vehicle in a stiff frame. The vibration levels are measured with accelerometers on different locations on the powertrain (50-150 measurement points). All measured vibrations are phase-related to a reference signal e.g. a specific accelerometer or the cyclic irregularity on the flywheel. The rig test provides a representative measurement of vibration levels but the operational deflection shapes or running mode shapes are difficult to understand because they consist of the superposition of the rigid motion and all flexible mode shapes. Also the restricted number of measurement points makes the interpretation difficult.

The modal test does not provide representative vibration levels but it does serve to identify natural modes, their eigen-frequencies and damping. The modal test is never completely representative, because several modes are effected by the compliance of the crank-system. To establish these effects, the stiffnesses of the rotating hydrodynamic bearings of the crankshaft must be taken into consideration, which is not possible with the standard modal test set up of a not running engine [3].

The finite element approach is the only way to evaluate structure borne noise without prototypes. It is a very effective experimentation tool for design optimisation. Due to unknown model parameters like damping and joint stiffnesses the confidence level for predicting vibration levels is in many cases not sufficient. To overcome this problem the correlation with rig test data is required.

Because modal testing is not representative of a running powertrain this method should be restricted to subsystem investigations e.g. front-end accessory drive resonance effects. To analyse and optimise the powertrain behaviour the rig test and FE-based modal analysis and forced response calculation is preferred.

Using these methods separately leads in a real world development process often to the following problems:

- fully representative hardware for rig test comes too late to support the design
- the FE-model is not accurate enough to predict absolute levels

To overcome these problems a combined or hybrid approach of rig test and FE-analysis is required. Using a very early prototype or a comparable powertrain and an appropriate FE-model of this structure a correlation based on the frequency response calculation and the measured vibration levels is performed. Only the modes which really affect the vibration levels on the engine mounts are taken into consideration. The correlation leads to realistic model parameters for the damping and the joint stiffnesses. The FE-model is then updated based on the current and future design assumptions. This approach allows the prediction and optimisation of vibration levels with a high confidence level, early in the design stage.

To support this approach effectively it is necessary to understand the vibration mechanism which leads to critical vibration levels on the engine mounts. Therefore in the next section a theoretical background of the vibration behaviour of powertrains is discussed.

3 STRUCTURE BORNE NOISE BEHAVIOUR OF POWERTRAINS

3.1 Basic assumptions for the structure borne noise behaviour

The following considerations are valid for a powertrain in East-West installation for front wheel drive with an I4 engine without balance shafts. The mounting system is a 3 point torque roll axis system (cf. Fig 1).

Regarding the structure borne noise behaviour a powertrain can be considered as an linear, time invariant mechanical system. Non-linearity in the stiffness and the damping can be neglected if the analysis is restricted to a specific operational point. Time variance caused by the crank-system can also be ignored in most cases. With these simplifications the powertrain can be considered as a linear multiple input-multiple output (MIMO) system where the excitation forces and torques are the inputs and the vibration levels on the powertrain are the outputs. Due to the fact that the forces at the interfaces of the powertrain (i.e. mounts, exhaust, drive-shaft etc.) are very small in comparison to the excitations which come from the engine, the powertrain can be considered as an approximately free-free system regarding the vibration levels in the frequency range of 50-500Hz.

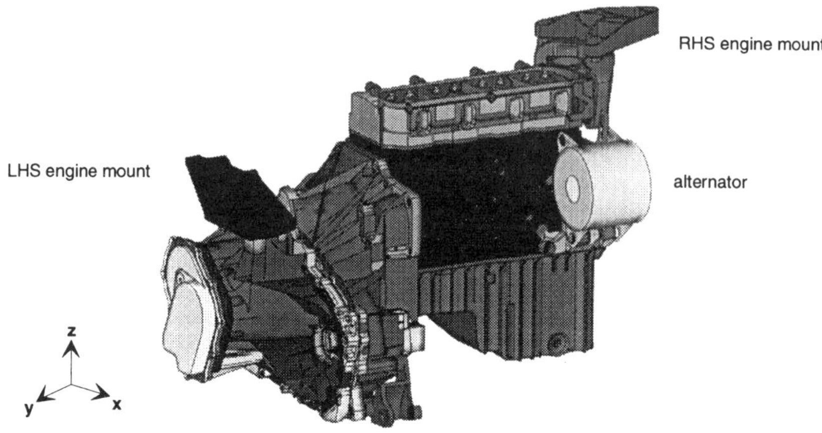

Figure 1: Example I4 powertrain

In this case the powertrain displacements can be split in two parts, the free body rigid motion and the flexible contribution

$$\overline{\mathbf{x}}_i = \overline{\mathbf{x}}_{i\,rigid} + \overline{\mathbf{x}}_{i\,flexible} \tag{1}$$

with the rigid contribution of the displacements

$$\overline{\mathbf{x}}_{i\,rigid} = -\frac{1}{\omega^2}\sum_{j=1}^{k}\left(\overline{\mathbf{F}}_j/m_{P/T} + \left(\Theta_{P/T}{}^{-1}\left(\overline{\mathbf{T}} + \mathbf{r}_j \times \overline{\mathbf{F}}_j\right)\right)\times \mathbf{r}_i\right). \tag{2}$$

Due to the assumed linearity the modal theory can be applied, to calculate the flexible contribution. Assuming a damping behaviour which is described by a speed-proportional viscous damping function, the displacement frequency response function on the location i may be inferred by the eigen-vector components of the translation φ_{ei} and the rotation γ_{ei}, the eigenfrequency ω_e, the critical viscous damping ratio D_e the forces F_j and the torques T_j as

$$\overline{\mathbf{x}}_{i\,flexible}(\omega) = \sum_{j=1}^{k}\overline{\mathbf{F}}_j\sum_{e=1}^{m}\frac{\varphi_{ei}\varphi_{ej}}{-\omega^2 + j2D_e\omega_e\omega + \omega_e^2} + \sum_{j=1}^{l}\overline{\mathbf{T}}_j\sum_{e=1}^{m}\frac{\varphi_{ei}\gamma_{ej}}{-\omega^2 + j2D_e\omega_e\omega + \omega_e^2}. \tag{3}$$

where m is the number of natural frequencies taken into consideration and l,k are the numbers of force and torque locations.

3.2 Powertrain excitation of I4 2EO

The main excitation of the powertrain relevant for the considered frequency range of 50-500Hz is caused by the engine shaking forces and the torque excitation which consist of combustion torque and inertia torque. The levels of an I4 engine are dominated by the 2nd engine order which is relevant for the interior noise levels. The higher orders and especially the order pattern influence the sound quality. The 2nd engine order shaking forces follow a parabolic characteristic. The equations

$$\overline{F}_{2EO\,z\,j} = F_{ref}\left(\frac{n}{n_{ref}}\right)^2 \tag{4}$$

$$\overline{T}_{2EO\,y\,j} = j\left(T_{ref}\left(\frac{n}{n_{ref}}\right)^2 - \overline{T}_{2EO\,Comb}(n)\right),\quad \overline{T}_{2EO\,Comb}(n) \approx T_{2EO\,Comb}(n) \tag{5}$$

shows the characteristic of the 2nd engine order excitation. The force in z direction caused by the shaking force of the crank-system only has a real contribution. The torque in y direction caused by the inertia torque of the crank-system and the combustion torque in anti-phase has approximately only an imaginary contribution [4].

3.3 Vibration characteristics of powertrains

This excitation leads to a rigid powertrain motion vs. speed which is approximately a real constant translation in the z-direction, a real constant rotation around the x-axis and an imaginary rotation around the y-axis. Appropriate to these motions the rigid contributions of the mount displacements in the y- and z-directions at the mounts are dominated by the constant z-translation and x-rotation. The x-displacements follow the imaginary y-rotation and shows for a full load run up the characteristic shape with a minimum in the mid speed range (cf. Fig. 2).

According to equation (2), the flexible contribution of the displacements is a superposition of the frequency response functions of several modes. Those modes, where the real force excitation dominates, show a peak in the imaginary contribution and a step in the real contribution. Modes mainly excited by the torque show resonance patterns which are vice versa. Additionally, modes whose peak frequencies lie above the engine speed range affect the displacement levels by sharply increasing displacements in the upper engine-speed range. In general 5 different types of flexible powertrain modes can be distinguished (cf. Table 1).

Table 1: Types of flexible powertrain modes

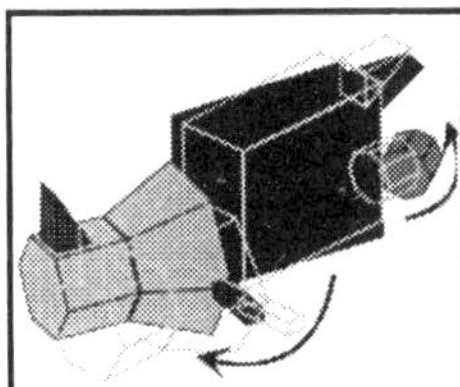	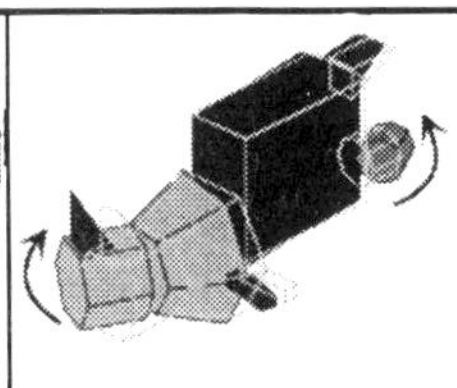	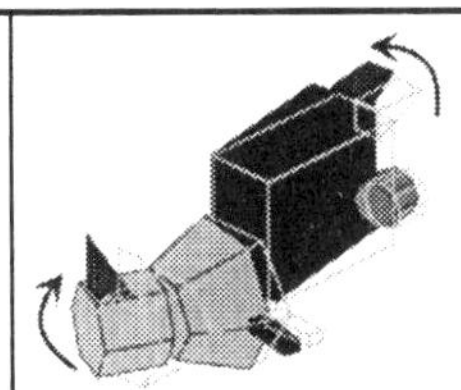	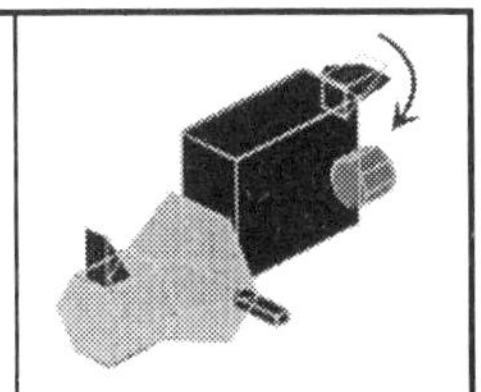
local mode of accessories	**combined component/-global powertrain mode**	**global powertrain mode**	**local mode of the mount bracket**
50-150Hz	150-200Hz	170-400Hz	250-500Hz
an accessory component with relatively high mass (>2kg) vibrates in anti-phase relative to the (nearly rigid) rest of the powertrain	an accessory component with relative high mass (>2kg) vibrates in anti-phase to a flexible power-train, mostly the trans-mission end	transmission vibrates in anti-phase to the engine	the mount vibrates vs. a nearly fixed rest of the powertrain due to the compliance of the mount bracket
alternator catalyst linkshaft	flywheel alternator starter	torsion vertical & lateral bending	especially RHS and LHS mounts

The vibration levels on the mounts result from a superposition of such modes. The identification of these mechanisms and their relationship to the vibration levels on the mounts is an important task enabling the optimisation of the powertrain structure borne noise. The current approach is to visualise the running mode shapes via on screen animations. The vibration mechanisms are often difficult to identify from such animations, due to the interaction of complex mode shapes. Only qualitative information can be gathered. To overcome this problem a new approach based on a kinematic analysis of the running mode shapes was developed.

4 KINEMATIC ANALYSIS OF POWERTRAIN MODE-SHAPES

The different types of powertrain modes (cf. Table 1) can be approximately described by the rigid motion of certain selected reference planes.

Assuming that the powertrain is longer than its height and width the first global powertrain modes are a torsion mode along the y-axis, a vertical bending mode around the x-axis and lateral bending mode around the z-axis. The simplification is to compare the powertrain with a simple one-dimensional structure like a beam, which has also these two bending modes and a torsion mode. The kinematics of these modes can be described by the sections of the beam. The 6 DOF motion of these sections allows the identification of the global behaviour. Relative differences in these kinematic parameters between the sections indicates bending or torsion in the structure. This approach assumes no relative compliance in the section, which is valid for a slim beam. For a powertrain this assumption is not valid. But based on experience the relative compliance of sections perpendicular to the crankshaft is relative small in comparison to the above mentioned global bending and torsion. Therefore approximated rigid body motions of powertrain sections can be used to describe the kinematics of global powertrain modes.

The same approach can applied to describe local component modes and combined modes if it is assumed that the component itself has relatively little flexibility. In this case the component motion can be described by an approximated rigid body motion of the component.
To establish these approximated rigid body motions it is assumed that the displacements are small in comparison to the dimensions of the powertrain which is the case for vibration levels in the frequency range from 50-500Hz (<0.2mm). Under this assumption a mean 6 DOF rigid body motion consisting of the complex vector of translation $\overline{\mathbf{x}}_{trans}$ and the complex vector of rotation $\overline{\mathbf{x}}_{rot}$ can be calculated by approximation using the equation (6).

$$\overline{\mathbf{x}}_{trans} + \overline{\mathbf{x}}_{rot} \times \mathbf{r}_i \approx \overline{\mathbf{x}}_{i\,meas} \tag{6}$$

At least 3 points which are not on one line are necessary for this approximation. Using more points increases the accuracy because local compliance or measurement inaccuracies are better averaged.

To establish the type of global mode (cf. Table 1) and to detect the main compliances, the kinematic parameters of planes perpendicular to the crankshaft are used. A good indicator for the global powertrain modes are differences of the rotational degrees of freedom for these planes. In vehicle co-ordinate system, the x-rotation describes vertical bending, the z- rotation lateral bending and the y- rotation describes the powertrain torsion. For this analysis the real and imaginary contribution has to be analysed separately because the sign of the displacement has to be taken into consideration. The same approach can be applied to identify component modes. In this case a reference plane defined by locations on the component instead of powertrain sections are used.

A further application is to establish the quantitative breakdown of engine mount vibration levels. The approximated rigid body motion of the reference plane is geometrically extrapolated to give the theoretical displacements which would exist given a perfectly rigid coupling of the engine mount location with the reference plane.

Applying this procedure to 2-3 relevant planes along the powertrain provides a clear quantitative breakdown of the contributions of the various sections of the structure to the actual displacements at the engine mounts

5 APPLICATION TO AN I4 FWD POWERTRAIN

The application of the kinematic analysis is shown in this section. The example is a powertrain with an I4 1.3l engine with manual transmission in a base front end accessory drive configuration, i. e. having only an alternator and no AC-compressor or power assisted steering pump. It is a powertrain in EAST-WEST installation for a front wheel drive vehicle. The mounting system is a 3 point torque roll axis system (cf. Fig 1). The powertrain does not represent a current production powertrain or a prototype. It is only used for principle investigations.

5.1 Judgement and risk assessment of structure borne noise

The first step is to compare the measured vibration levels from the rig test with the vibration levels of a perfectly rigid powertrain. Using equations 2 ,4 and 5 the rigid powertrain vibration levels can be easily calculated. This comparison shows the potential to improve the powertrain by stiffening or mass reductions.

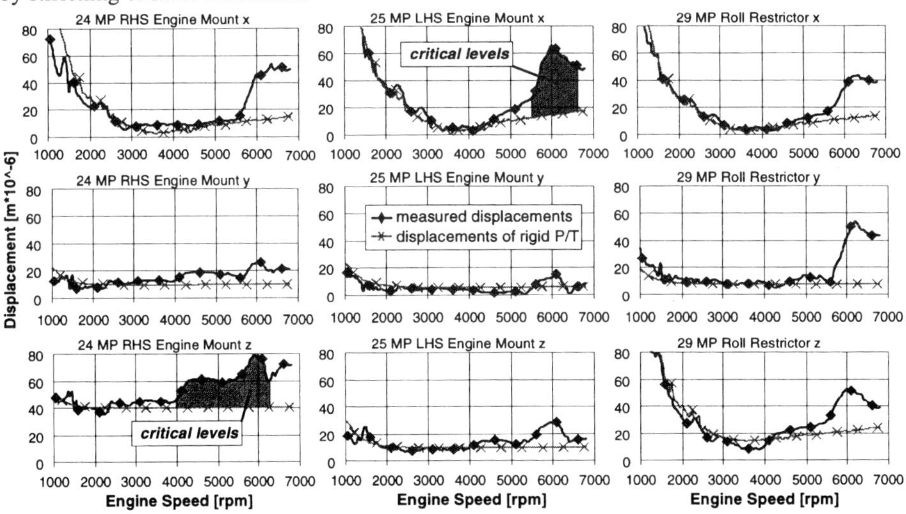

Figure 2: Measured displacement vs. displacements of a rigid powertrain

Figure 2 shows the displacement levels on the three mounts in comparison to the displacements of a rigid powertrain. There is obviously big great potential for improving the dynamic stiffness of the example powertrain. Significant deviations above 5000rpm indicate resonance effects. The results of a noise analysis (NPA) shows that especially the high levels on the LHS mount in x direction and on the RHS mount in z-direction the are critical for the interior noise.

5.2 Estimation of kinematic parameters

The next step is to define sections or reference planes for the powertrain. Figure 3 shows the front-end section of the engine as an example. To assess the quality of the measured data and to evaluate whether the assumption of low relative compliances in the section and components is valid, an easy check is performed. The measured displacements of the measurement points which define the section, are compared with calculated displacements using the approximated 6DoF rigid body motion of the section. This comparison is also shown in Figure 3. An indicator for the measurement quality is the correspondence in the low speed range. The small deviations in the high speed range show the low relative compliance of the sections.

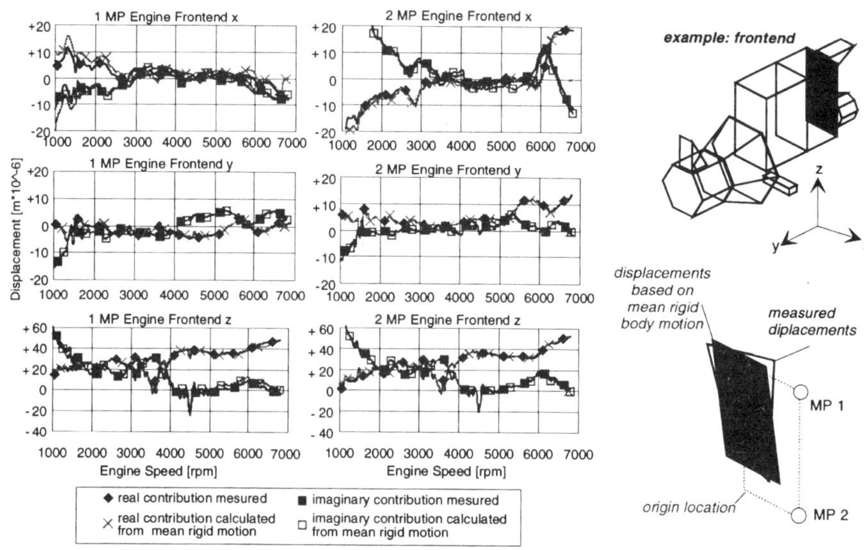

Figure 3: Comparison of displacement calculated from the mean rigid body motion with the measured displacements

5.3 Compliance path analysis

These kinematic parameters are used in the quantitative breakdown of engine mount vibration levels. The task is to identify which areas of the powertrain structure (including mount brackets) show significant flexibilities. This analysis is shown in figure 4 for the previously established critical mounts, the LHS mount in x-direction and the RHS mount in z-direction. Also the selected sections and reference planes are shown.

The analysis shows for RHS mount that the main flexibility in the critical speed range above 5000 rpm is between the engine front-end and the measurement point on the mount. This indicates that as a first priority for design improvements, the mount bracket and its fixing to the engine front-end should be stiffened.

The situation on the LHS mount is completely different. The main level increase occurs between the rigid displacements and the clutch housing section. There is nearly no flexibility between the clutch housing and transmission sections. The transmission between the sections "transmission" and "LHS base" shows some compliance but in comparison to the difference between the rigid displacements and the clutch housing section it is relatively small. Also the compliance in the mount structure, i.e. between the LHS mount base plane and measurement point has only a minor contribution. Obviously a global powertrain mode is the major root cause of the high levels on the LHS mount in x-direction.

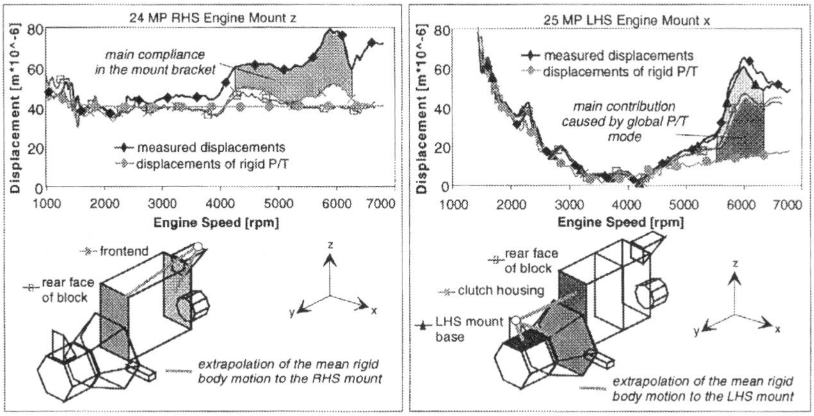

Figure 4: Analysis of global powertrain modes

5.4 Powertrain global mode analysis.

The next step is to identify the type of global mode which is causing the high vibration levels on the clutch housing section to give more specific directions for design improvements. A powertrain bending analysis as described in section 4 shows significant differences in the y rotations at 6000 rpm which indicate a powertrain torsion mode (cf. fig 5). The main compliance is in the engine, and the interface between engine and transmission has also a contribution. The imaginary contribution shows the same characteristic as the LHS engine mount vibration in the x-direction in the high speed range (cf. fig. 4). This indicates that this global mode is responsible for the steep increase of vibration levels at the LHS mount in x-direction.

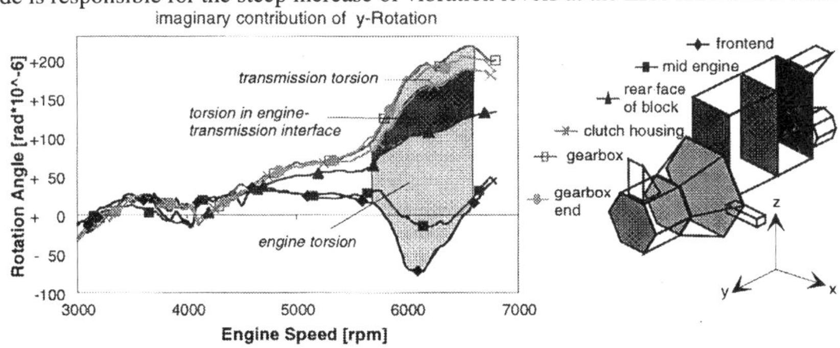

Figure 5: Analysis of global powertrain modes

5.5 Evaluation of component influence

Experience has shown that global powertrain modes can be strongly influenced by component flexibilities. The evaluation of such potential influences is the next task. The most critical component on this powertrain is the alternator due to its weight and its distance from the centre of gravity. For this purpose the mean rigid body motion of the alternator and the section of the engine front-end are compared (see Figure 6). The chart of y-rotations shows that the alternator moves in phase with the front-end with 80% higher amplitudes. The rotations around the z-axis indicate a very strong local mode at 190 Hz. The alternator amplifies the global powertrain torsion at 200 Hz which could be improved via a stiffer alternator fixing. The local alternator mode at 190Hz has only a minor influence on the rest of the powertrain.

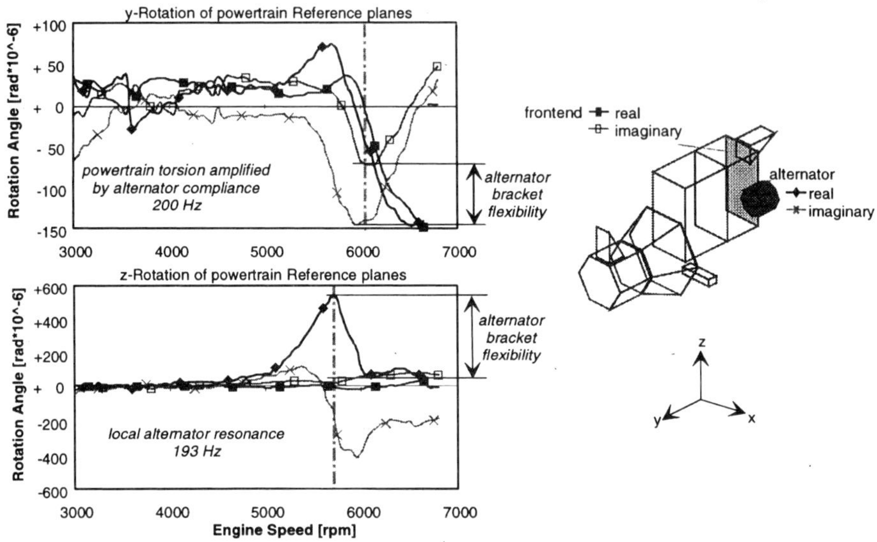

Figure 6: Influence of the alternator to powertrain torsion and bending

This detailed analysis based on the mean rigid body motion of powertrain sections and components leads to the following conclusions:

- the global powertrain torsion is the main root cause of the high vibration levels in x-direction on the LHS engine mount
- torsion within the engine is the major contributor to the total powertrain torsion
- the powertrain torsion is not caused by the alternator, but the flexibilities in the alternator fixing leads to an amplification.

The recommended design improvements are to increase the torsional stiffnesses of the engine and the interface to the transmission. Additionally the stiffness of the alternator bracket should be increased.

5.6 Correlation of an FE model

FE-analysis is a very effective tool for developing design improvements. An accurate calculation of the levels requires an FE-model which is correlated to rig test results. Instead of simply comparing vibrations at single locations it is of great advantage if the mean rigid body motions of the powertrain sections are used. This results in a much more efficient correlation of the global powertrain modes. In this way the measured and analysed rig test data are used to adjust the joint stiffnesses and, if required, to correct the geometry in the FE model. The result of the correlation of the investigated powertrain (cf. Fig. 1) is shown in Fig. 7. The charts show a comparison of the calculated 2nd engine order with test results. A good correspondence is visible for the mean rigid body rotations of the engine front-end which indicates the global powertrain behaviour. Also the displacements at the engine mounts show only small differences. This indicates good FE-modelling of the engine mount brackets.

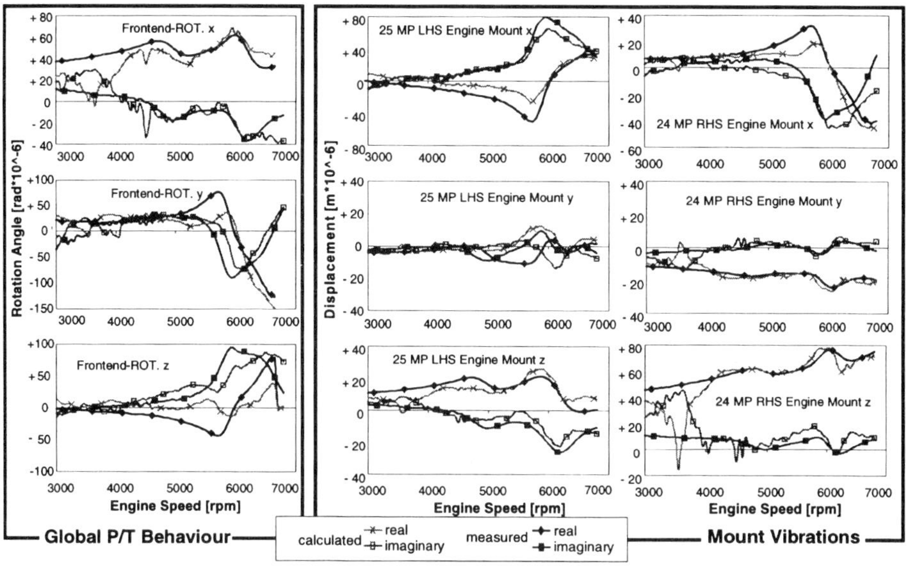

Figure 7: Comparison between FE-data and test results

6 CONCLUSIONS

Measured or calculated running mode shape data allows a very advanced analysis of the dynamic behaviour. In a first step the kinematic analysis provides a quantitative breakdown of the mount vibrations into the contributions of the transmission or engine and the mount brackets. The global powertrain modes and the component modes are analysed using the same basic approach. The application of both methods gives a very good and detailed understanding of the vibration mechanisms. The analysis is not restricted to qualitative information. Quantitative information about the contributions and the intensity of global modes can be gathered.

Based on these results specific design improvements which provide the greatest structure-borne noise improvements can be efficiently developed using correlated, accurate FE models. The un-reliable and time consuming alternative of "trial and error" tests using cobbled parts is no longer required.

REFERENCES

[1] *Schröder, E., Laufs J.*: Neue Konzepte zur Geräusch- und Schwingungsoptimierung von Antriebssträngen, 3. Aachener Kolloquium Fahrzeug- und Motorentechnik, 1991

[2] *Brügmann, G., Heitmann, F.*: Methoden der Strukturdynamik bei der Entwicklung von Motor-Getriebe Verbänden, ATZ 95 (1993), S. 582-590

[3] *Goto, Y., Kazuo, Y.,*: Experimental Analysis of Power Plant Vibrations by Shaker Test, SAE Paper 911069, 1991

[4] *Bharj, T.*: Engine Excitation Forcing Functions For Vehicle System Models, Advanced Engineering-Report, Ford Motor Company Ltd, 1989

C521/032/98

Power unit low-frequency airborne noise

J DIXON
Institute of Sound and Vibration Research, University of Southampton, UK
A V PHILLIPS
Ford Motor Company, UK

Synopsis

The quantification of low frequency powertrain airborne noise presents a number of problems, some of which are discussed in this paper. A simple methodology for overcoming these issues is presented and the possibility of this low frequency noise contributing to either interior or exterior vehicle noise is investigated.

1. INTRODUCTION

For many years it has been generally accepted that power unit noise enters the vehicle's passenger compartment via structural paths at low frequency and airborne paths at high frequency, with similar levels of contribution occurring at around 500 Hz (Ref 1). However, there is increasing suspicion that power unit airborne noise can be a significant contributor to "Boom" frequencies at as low as 100 Hz. Due to its inherent 2nd order imbalance, the four cylinder, four stroke engine clearly suffers from low frequency excitation and therefore its specific case forms the basis of this paper.

Due to the ever increasing performance of engine mounting systems as well as intake and exhaust silencers, the contribution dominance from these sources is lessening and therefore ignoring the 2nd order airborne contribution from the base engine is being questioned. In a similar way the ever lowering of vehicle exterior noise levels has meant that the traditional dominance of the intake and exhaust noise at 2nd order frequencies is being reviewed and the possibility of a base engine radiation contribution being investigated.

Although modern Noise Path Analysis techniques should allow an accurate measure of airborne contribution to be achieved, quantifying the actual source strength of a power unit at these low frequencies presents a number of problems.

Measuring 2nd order radiated noise from a 4 cylinder engine at 1 metre in a semi-anechoic chamber has two inherent shortfalls namely the installation and the chamber's acoustic performance. Although both the intake and the exhaust gases are traditionally piped away to remote systems, both this piping and the production intake and exhaust components used will often radiate high levels of 2nd order shell noise. In the case of the production components, it may be argued that this is part of the actual engine noise and should therefore be measured. However, this would compromise the traditional target cascading philosophy and could result in double counting of contributions, as the intake and exhaust system shell noise is quantified and targeted independently. Unfortunately the traditional test chamber Ad Hoc installation attenuation method of lead cladding is inappropriate at such low frequencies, with amplification just as likely as attenuation. It is therefore often very difficult to state categorically that an engine test cell is devoid of all low frequency intake and exhaust noise.

The acoustic performance of an engine test chamber is principally controlled by the absorption properties of the wall linings and the physical dimensions of the chamber itself. Although a chamber of around 80 cu metres lined with traditional perforate tiles is perfectly adequate for measurements down to around 300 Hz, to cover measurements over the 2nd order frequency range (down to 33 Hz) a chamber with appropriate low frequency absorption and a volume in excess of 2000 cu metres is desirable.

A combination of these installation and chamber acoustic issues are probably the two principle reasons why test bed SPL data at 2nd order generally appear to be of an almost random nature with speed rather than the steadily increasing characteristic as suggested by theory and vibration measurement. For an example Figure 1 compares some recently measured 2nd order radiated noise from a number of 4 cylinder engines when installed in nominally semi-anechoic test chambers.

In an attempt to reliably measure the 2nd order radiated noise of the four cylinder engine an experiment was conceived where both chamber acoustics and the installation issues were minimised.

2. THE MOTORED ENGINE EXPERIMENT

2.1 Description of the Rig

A 1.1 litre, 4 cylinder engine and transverse gearbox was installed on a 30 m × 30 m open air concrete pad which provided an environment approaching free-field. The powertrain was then connected to a motoring rig. To remove all possible pipework shell noise radiation, the camshaft of the engine was removed, thus ensuring no airflow took place. All intake and exhaust systems were removed bar the manifolds which for noise radiation purposes are generally considered as part of the power unit.

In this build, it was assumed that when motored any 1 metre 2nd order noise measured, would be directly radiated from the power unit. The generation of this 2nd order being the inherent imbalance of the crank train and the 2nd order torque fluctuations due to this imbalance and the small compression and expansion of the trapped air.

2.2 Noise Measurement

The engine was motored over the speed range and noise at the standard SAE 1 metre microphone positions was measured, together with vibration at selected points on the power unit.

Both the 2nd order noise and vibration levels showed fairly smooth characteristics with speed as is shown in the vibration plot of figure 2, and therefore to reduce data manipulation the remainder of the study was carried out at selected speeds over the operational range.

Figure 3 compares the 2nd order noise for the four SAE microphone positions. The two sides and the front of the engine exhibited very similar characteristics over the speed range, with a clear increase in noise with speed of close to 40 dB/decade, ie Noise $\propto$ (speed)^2. Although the left hand side was marginally noisier than the right hand side, the greatest recorded difference at 3000 rev/min was just over 1 dB or within the experimental errors expected in such a test. The noise to the front was on average 1 dB lower than the noise to the sides.

The noise to the top however had a very different character, being some 6 dB quieter than the sides at 2000 rev/min rising rapidly to around 4 dB noisier at 3000 and 4000 rev/min and then returning to a similar level as the sides by 5000 rev/min. Due to the recognised dominance of the vertical vibration resulting from the 2nd order crank train imbalance, it is believed that the noise radiated from the top of the engine should be the most uniform with noise rising steadily with increasing speed. It is therefore hypothesised that this irregular characteristic seen to the top is due to ground reflection effects off the concrete test floor.

2.3 Vibration Measurements

Vibration of the power unit was measured both vertically and horizontally at the locations shown in Diagram 1.

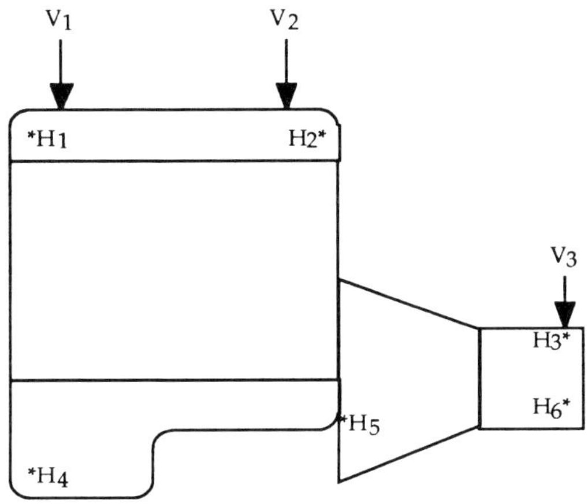

Diagram 1 - Vibration Measurement Locations

Although the level of noise radiated from the power unit could be highly influenced by the form of the relevant whole body mode of vibration, for this initial study the shape of the modes and thus phase were not considered and just average vibration levels were taken as a first estimate of radiated noise generation.

Figure 4 compares the vertical and horizontal average vibration of the power unit over the speed range. Although both average vibration levels increase regularly with speed, the horizontal vibration increases at a far greater rate with speed than the vertical vibration. As might be expected the vertical vibration increases at close to the theoretical rate for imbalance of 40 dB/decade, although a ramping up at high speed is in evidence, suggesting an influence from an early power unit bending resonance. The steeper slope seen on the horizontal vibration above 3000 rev/min may also be party due to an early power unit bending mode, which is supported by the fact that above this speed the vibration level at H_5 becomes greater than that at both H_4 and H_6 suggesting a horizontal bending mode. However, a steeper slope would be expected in the horizontal direction as the torque fluctuation increases away from the low speed point of cancellation between the inertia and trapped air force.

2.4 Effect of Distance on Noise

The opportunity was taken to verify that noise in the 2nd order frequency range propagates away from 1 metre according to the free field inverse square relationship. Noise was measured at 1 metre intervals out to 7.5 metres (The ISO drive by microphone distance) and as can be seen in Figure 5 the noise reduces at close to 20 dB/decade demonstrating that indeed the noise propagates according to the inverse square law and thus suggesting that acoustically the SAE 1 metre measurement procedure is still reliable at these low frequencies.

3. DISCUSSION

3.1 Comparison of Free Field Motored Measurements With Test Cell Data

If the test cell data shown in Figure 1 is compared to the respective measurements taken on the motored engine in free field, a clear difference in character can be seen. Figure 6 compares the motored engine results with the data base over the speed range for the left hand microphone. Whilst the motored engine noise is seen to regularly increase with speed, the data base has severe excursions from an underlying increasing characteristics. Although no attempts have been made to quantify the distribution statistically, it would appear that the excursions from the mean reduce with increasing speed (or frequency). This observation would be consistent with room acoustic issues, where the room size influences the measurement increasingly with reducing frequency.

The free field motoring tests themselves were not totally devoid of environmental distortion. As described in Section 2.2 the top noise measurements were believed to be severely influenced by the noise reflected off the concrete test site floor. By studying the tracked 2nd order noise to the sides of the motored engine, clear regions of noise cancellation were seen around 155 Hz which are believed to be due to interference between the direct path to the microphone and the reflected path via the floor.

3.2 Estimation of Radiating Efficiency

It has been shown that there is a direct relationship between average surface vibration of an engine and its noise radiation (Ref 2). At frequencies above 500 Hz and when considered in broad band terms (traditionally $^1/_3$ octaves) this relationship has been shown to be fairly robust (Ref 3).

When considering just a single low frequency, such as 2nd order, this relationship should be maintained although the radiating efficiency is probably no longer a relatively bland value, but is now more likely to be sensitive to the mode of vibration of the power unit.

Using the following relationship given in Reference 2 an estimation of the radiating efficiency of the motored engine was made over the considered frequency range.

$$10 \log \sigma_{rad}(f) = L_v(f) - 20 \log \frac{f}{4} - 10 \log \left[\frac{\pi \left(\left(\frac{S_{rad}}{\pi} \right)^{1/2} + 2 \right)^2}{S_{rad}} \right] - SPL(f) + 138$$

where

f	= 2nd order frequency under consideration (in Hz)
$\sigma_{rad}(f)$	= radiating efficiency at frequency f
$L_v(f)$	= average surface vibration acceleration at frequency f (rel 10 m/sec^2)
S_{rad}	= surface area of power unit (estimated at 1 sqm)
SPL (f)	= average 1 metre 2nd order sound pressure level at frequency f

Figure 7 compares this derived radiating efficiency with that predicted for an oscillating sphere of the same (1 metre) surface area (Ref 4). At the lower frequencies the slopes of the two radiating efficiencies are similar at around 40 dB/decade, but at higher frequencies the experimentally derived efficiency dramatically decreases. Examination of the vibration data used for deriving this efficiency reveals that the power unit vibrates as a solid free body up to around 3500 rev/min (116 Hz) and then increasingly distorts with increasing frequency suggesting influence from an early bending mode. As any such bending will be reflected in a lowering of the mean vibration radiating efficiency, the high frequency diverging characteristic seen in Figure 7 would appear plausible.

The apparent 4 dB difference seen in Figure 7 between the two radiating efficiencies in the whole body mode region is probably due to the simplification of the approach adopted. For the experimentally derived radiating efficiency a global mean vibration is compared to a global sound pressure in an attempt to minimise ground reflection issues, whereas for the theoretically derived efficiency, just the on-axis radiation is considered for the oscillating sphere.

3.3 Prediction of 2nd Order Noise From Vibration

To demonstrate the potential of quantifying power unit airborne noise by using a small number of vibration measurements, the top noise of the engine was predicted using the mean vibration of the 3 vertical measurements in combination with the theoretical radiating efficiency of an oscillating sphere. Figure 8 compares this predicted noise with that measured to the top in freefield. Although at first sight the comparison looks disappointing, some of the differences seen in the comparison have possible explanations. At 5000 revs/min the measured value is artificially low due to ground reflection cancellation at around 169 Hz to the top 1 m microphone, whilst the predicted noise is artificially high due to the onset of the power unit bending (see 3.2). At 2000 revs/min the measured noise is uncharacteristically low by some 10 dB, and although an acoustic environment issue is suspected, simple reflection theory fails to provide a hypothesis. The general under prediction in the mid speed range is probably again a reflection of the errors resulting from the use of a simplistic approach for the radiation efficiency of an oscillating sphere.

Although this prediction of noise does not totally agree with that measured, its character looks plausible, especially when compared with the characteristics seen in Figure 1. It is therefore believed that with a more appropriate estimate of radiating efficiency and a method for handling resonant conditions, a process for quantifying low frequency noise using vibration would appear to have great potential.

3.4 2nd Order Radiated Noise Contribution to Exterior and Interior Noise

Having established a reliable measure of power unit 2nd order radiated noise through both measurement and prediction, it is now possible to estimate the likely contribution this level of noise would make towards both exterior and interior overall noise levels.

3.4.1 Exterior Noise

Most small family saloons pass the microphone during driveby testing at around 4000 and 3000 rev/min in 2nd and 3rd gear respectively. Thus the stage data manipulation laid out below considers just these two speeds.

	2nd gear	3rd gear	units
Engine speed	4000	3000	rev/min
1 metre noise from motoring experiment (Figure 3)	83.5	78	dB
7.5 metre noise using inverse square law (Sec 2.4) (less 17.5 dB)	66	60.5	dB
Typical body attenuation (less ~ 5 dB)	61	55.5	dB
A weighting (16 and 20 dB respectively)	45	35.5	dBA
Average of the two gears	40		dBA

As for the foreseeable future, it is envisaged that vehicle exterior noise legislation will be around 70 dBA, it is suggested that this 2nd order radiated engine contribution at around 40 dBA would appear to be insignificant.

However, before totally dismissing this radiated noise contribution, the following points should be considered.

3.4.1.1 The above data manipulation uses the measured results from a motored engine which is devoid of any combustion forces. However, in general, above 3000 rev/min the 2nd order torque fluctuation reduce with increasing combustion force (Ref 5). The 2nd order vertical imbalance force is unaffected by combustion forces.

3.4.1.2 The manipulation uses data from a relatively small engine (1.1 litre). In general the larger the engine the greater the 2nd order noise radiated.

3.4.1.3 The nominal body attenuation used (~ 5dB) may be an over estimation. No attenuation, or even amplification is possible at these frequencies depending upon the vehicle and installation.

3.4.1.4 Due to the steepness of the A-weighting curve at these low frequencies and the increasing noise with engine speed, the engine radiated noise contribution rapidly increases with increasing speed. This effect is well demonstrated by the 9.5 dB difference between the 3000 and 4000 rev/min contribution compared to typical total vehicle differences of 4 - 5 dB.

Although the above points will all influence the low frequency radiated noise contribution from the engine, they are all either readily predictable, measurable or else accurately estimateable.

3.4.2 Interior Noise

Using a similar data manipulation process as that for exterior noise the following interior noise contributions may be derived:-

	2000 rev/min	3000 rev/min	4000 rev/min	5000 rev/min
1 metre noise from motored experiment (Figure 3) in dB	72	78	83.5	87
Less body attenuation (Ref 6) in dB	60	64	65.5	67
Typical vehicle interior noise level in dB	84	80	80	80
Δ between typical interior noise and engine contribution in dB	24	16	14.5	13

Although the above manipulation only used a single 1 metre measurement and a single body attenuation, it can be seen that the airborne contribution from the engine is far from insignificant. If an appropriate multi-path analysis approach were used, the difference between the typical interior noise level and the predicted engine contribution would be appreciably less. However, further investigations are necessary before a more accurate figure for the radiated noise contribution from the engine is estimated.

4. CONCLUSIONS

4.1 Current methods for measuring 2nd order radiated engine noise are often grossly inaccurate especially at the lower engine speeds. It is speculated that this is a combination of the poor acoustic performance of many engine test chambers at low frequency and inadequate piping for both the intake and exhaust gases.

4.2 Using theoretical radiation efficiencies based upon an oscillating sphere allows a fair estimation of 2nd order radiated noise to be predicted from a few (6 - 9) engine vibration measurements.

4.3 2nd order engine radiated noise is unlikely to be a significant contributor to exterior noise in the foreseeable future.

4.4 By predicting noise from vibration, an improved estimate may be made of the 2nd order airborne noise contribution to vehicle interior noise.

4.5 2nd order engine radiated noise could be a significant contributor to interior noise in the near future.

5. REFERENCES

1. Priede, T. Noise and vibration problems in commercial vehicles. *Journal of Sound and Vibration* (1967)5 (1) 129 - 154.

2. Chan, C.M.P. and Anderton, D. Correlation between engine block surface vibration and radiated noise of in-line diesel engines. *Noise Control Engineering*, 16 - 24 Winter 1974.

3. Anderton, D., Baker, J.M and Clark, C. Determination of noise radiation from surface vibration as a practical tool. *Proceedings of the 26th International Symposium on Automotive Technology and Automation, Mechatronics, Aachen,* 1993, 681 - 92.

4. Eds Richards, E.J. and Mead, D.J. Noise and Acoustic Fatigue in Aeronautics, (Pub, John Wiley & Sons Ltd).

5. Rhodes, D.M., Phillips, A.V. and Abbot B. Prediction of engine vibration from combustion pressure measurement and examination of the effect of combustion variations in the frequency domain. *Proceedings of the IMechE "Advances in the Control and Refinement of Vehicle Noise" Paper C28/88.*

6. Phillips, A.V., Dixon, D., Rhodes, D. and Hughes, H. Recent developments in powertrain noise simulation. *Proceedings of Autotech '93, Recent Advances in NVH Technology,* Birmingham. Institution of Mechanical Engineers, 1993, paper C462/36/198.

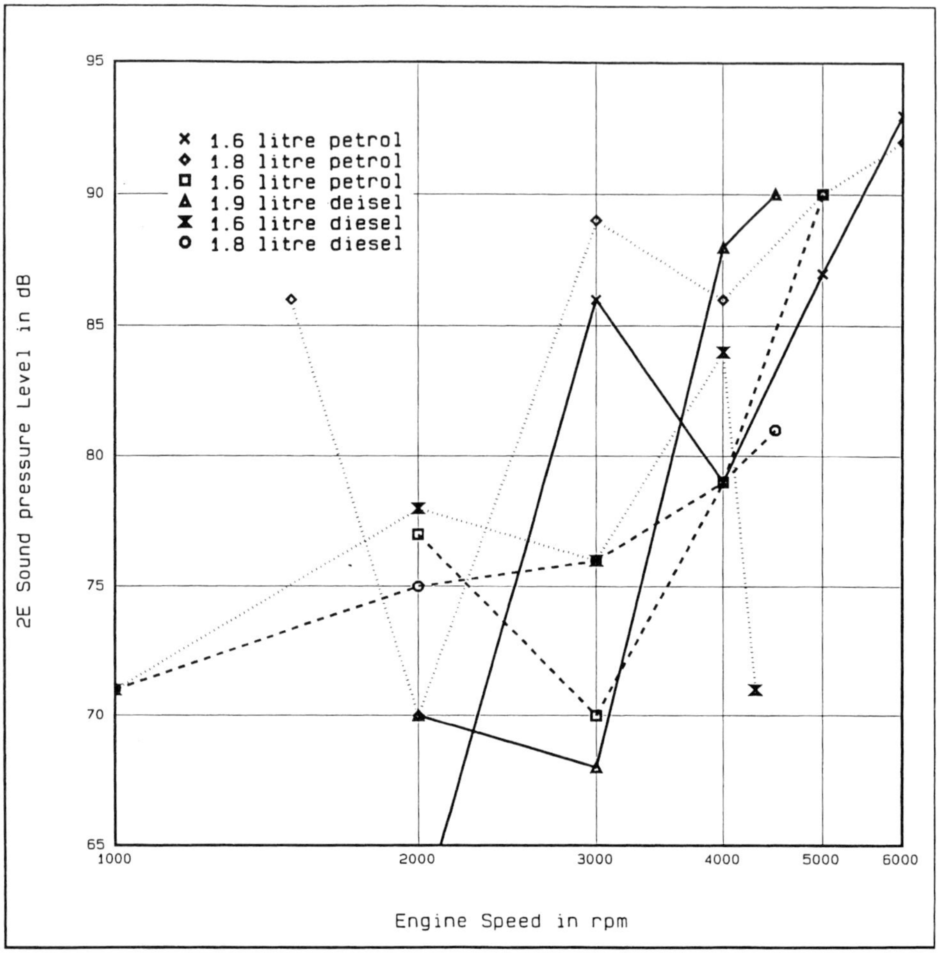

FIGURE 1

Automotive Design Advisory Unit

COMPARISON OF SECOND ORDER NOISE OF VARIOUS 4 CYLINDER ENGINES

FULL LOAD NOISE AT 1m
LEFT HAND SIDE

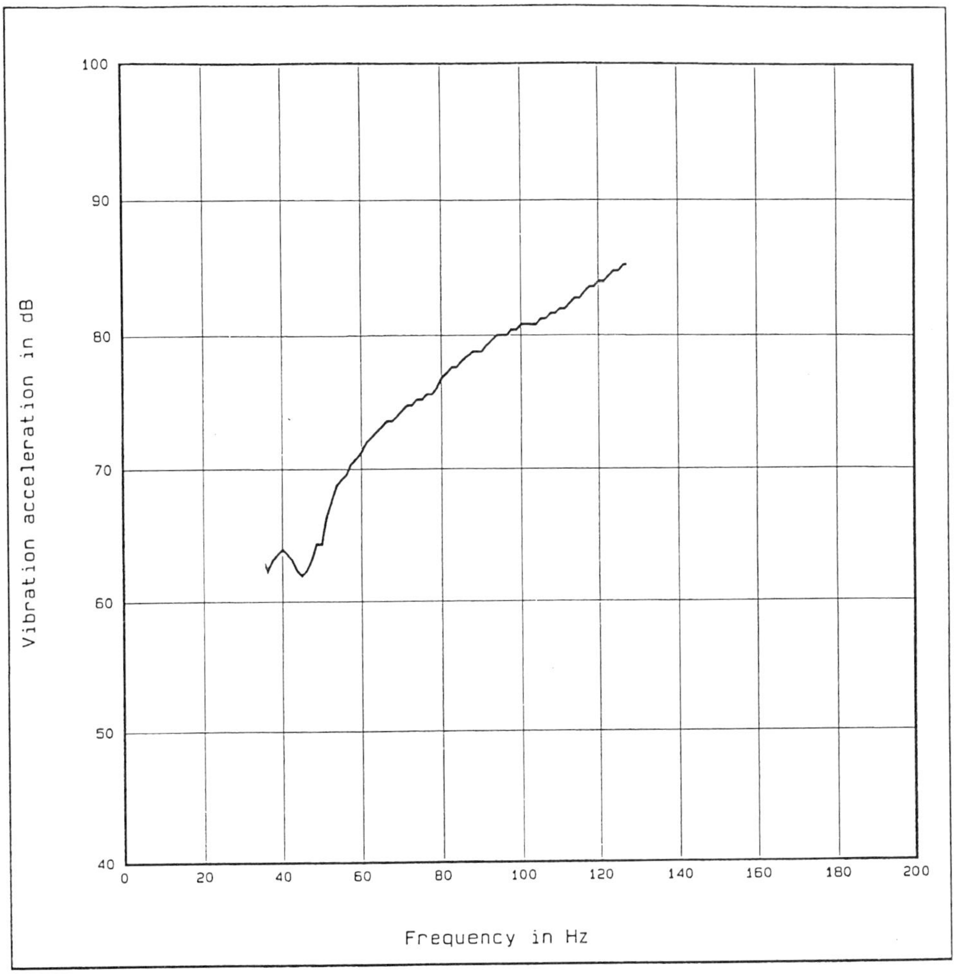

FIGURE 2

AUTOMOTIVE DESIGN ADVISORY UNIT

2nd ORDER VIBRATION OF THE MOTORED ENGINE

VERTICAL VIBRATION AT THE FRONT OF THE VALVE COVER

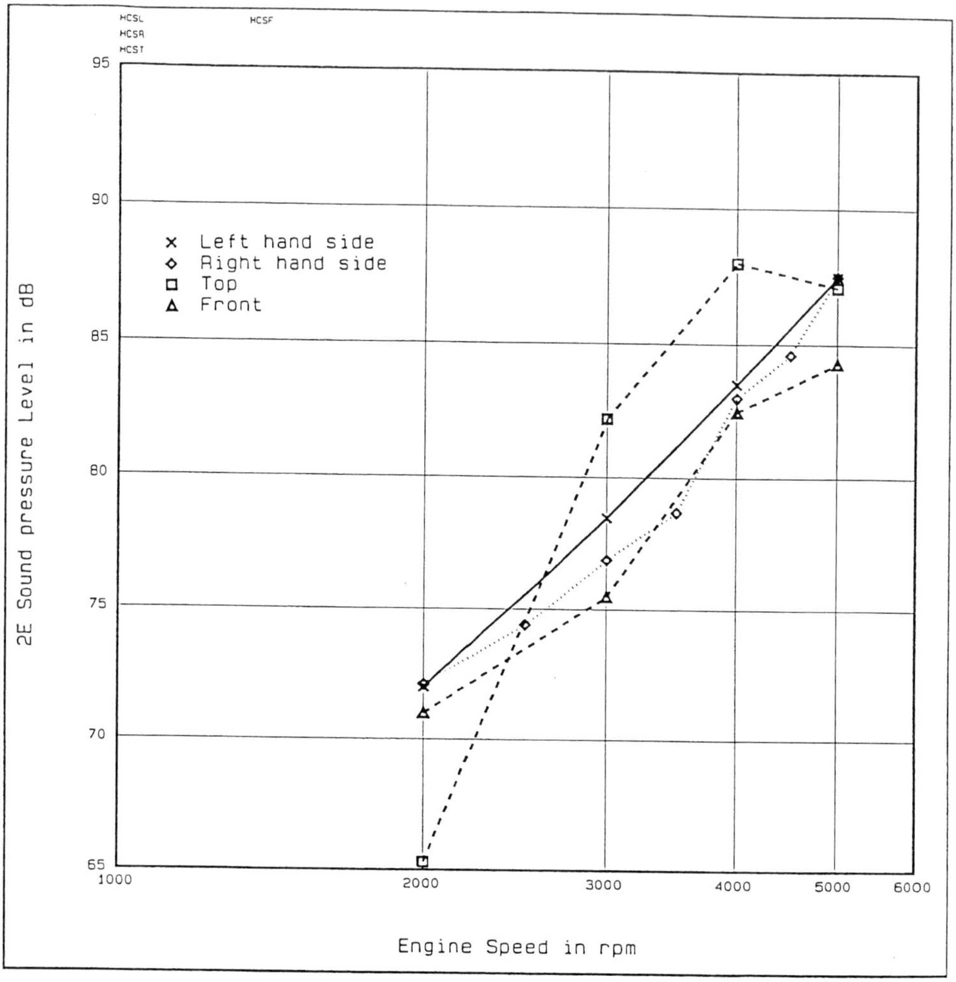

FIGURE 3

AUTOMOTIVE DESIGN ADVISORY UNIT

EFFECT OF MICROPHONE POSITION ON SECOND ORDER (2E) NOISE
MOTORED ENGINE
NOISE AT 1m

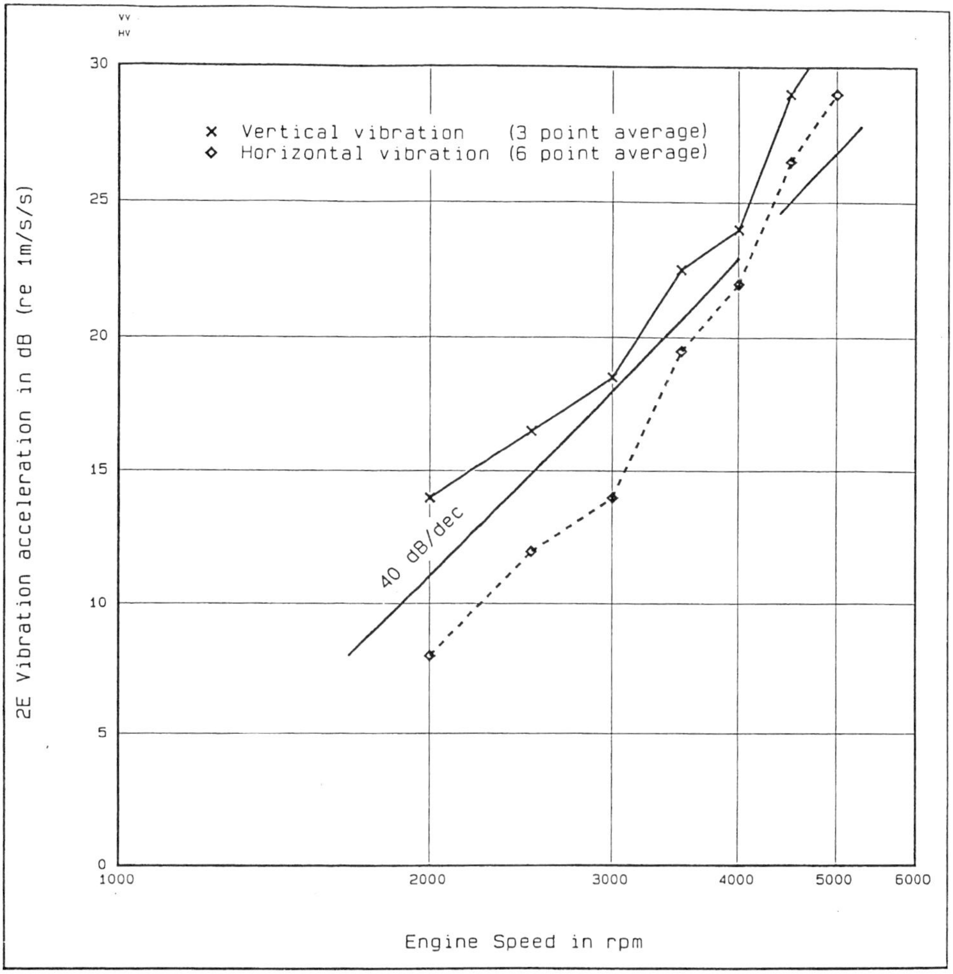

FIGURE 4

AUTOMOTIVE DESIGN ADVISORY UNIT

COMPARISON BETWEEN VERTICAL AND HORIZONTAL SECOND ORDER VIBRATION
MOTORED ENGINE
AVERAGE VIBRATION ACCELERATION

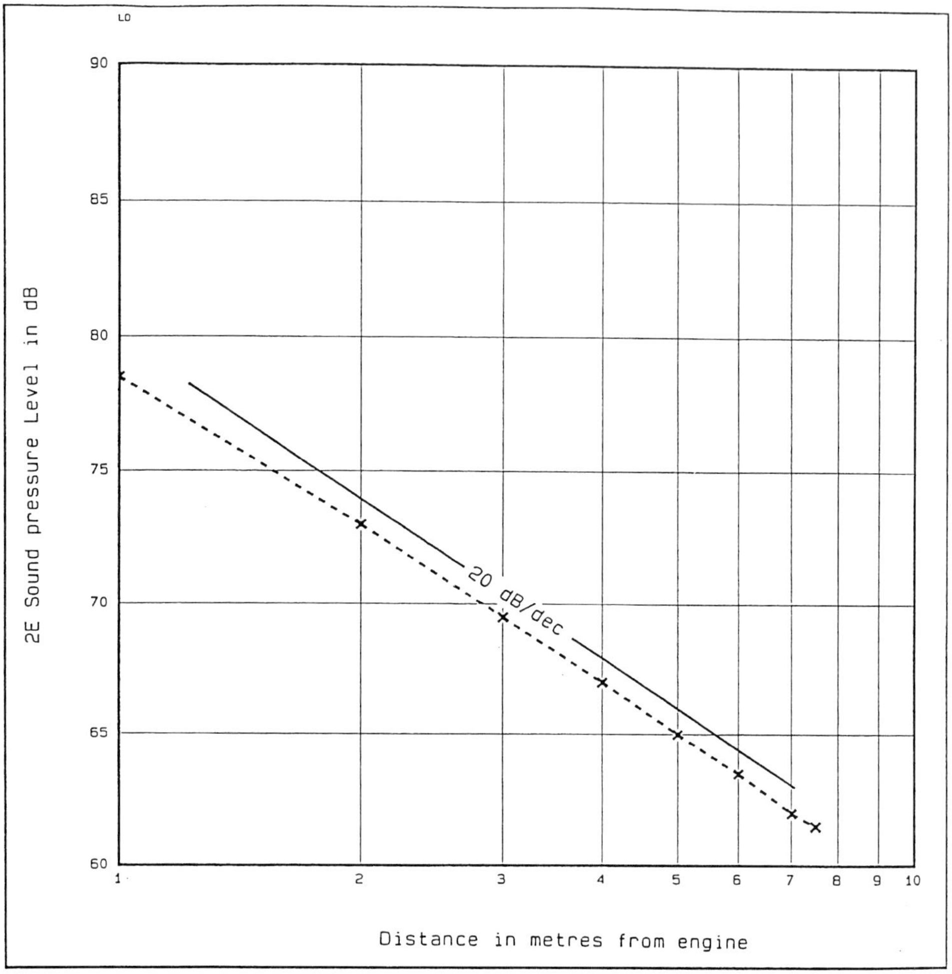

FIGURE 5

Automotive Design Advisory Unit

EFFECT OF MICROPHONE DISTANCE ON MEASURED SECOND ORDER NOISE
MOTORED ENGINE
LEFT HAND SIDE
3000 rev/min (100 Hz)

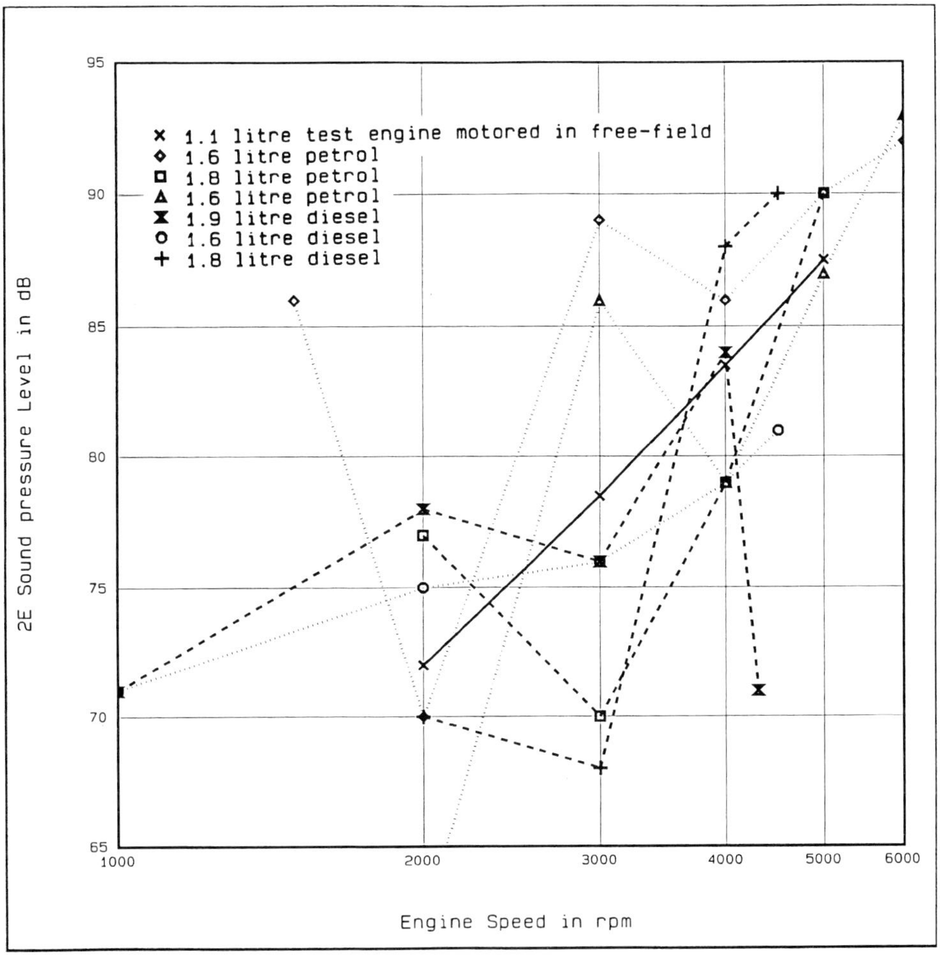

FIGURE 6

AUTOMOTIVE DESIGN ADVISORY UNIT

COMPARISON BETWEEN MOTORED TEST ENGINE AND TEST CELL NOISE DATA BASE
LEFT HAND SIDE NOISE
NOISE AT 1m

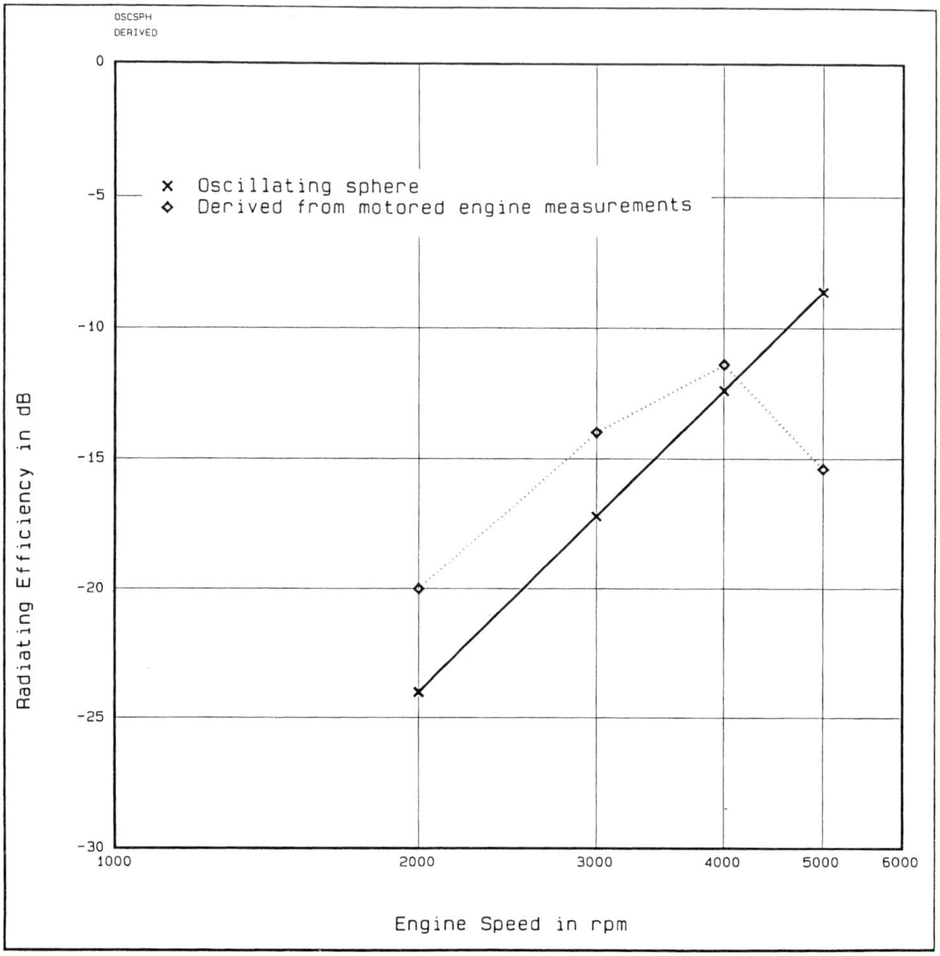

FIGURE 7

AUTOMOTIVE DESIGN ADVISORY UNIT

COMPARISON BETWEEN THEORETICAL AND DERIVED RADIATING EFFICIENCY.

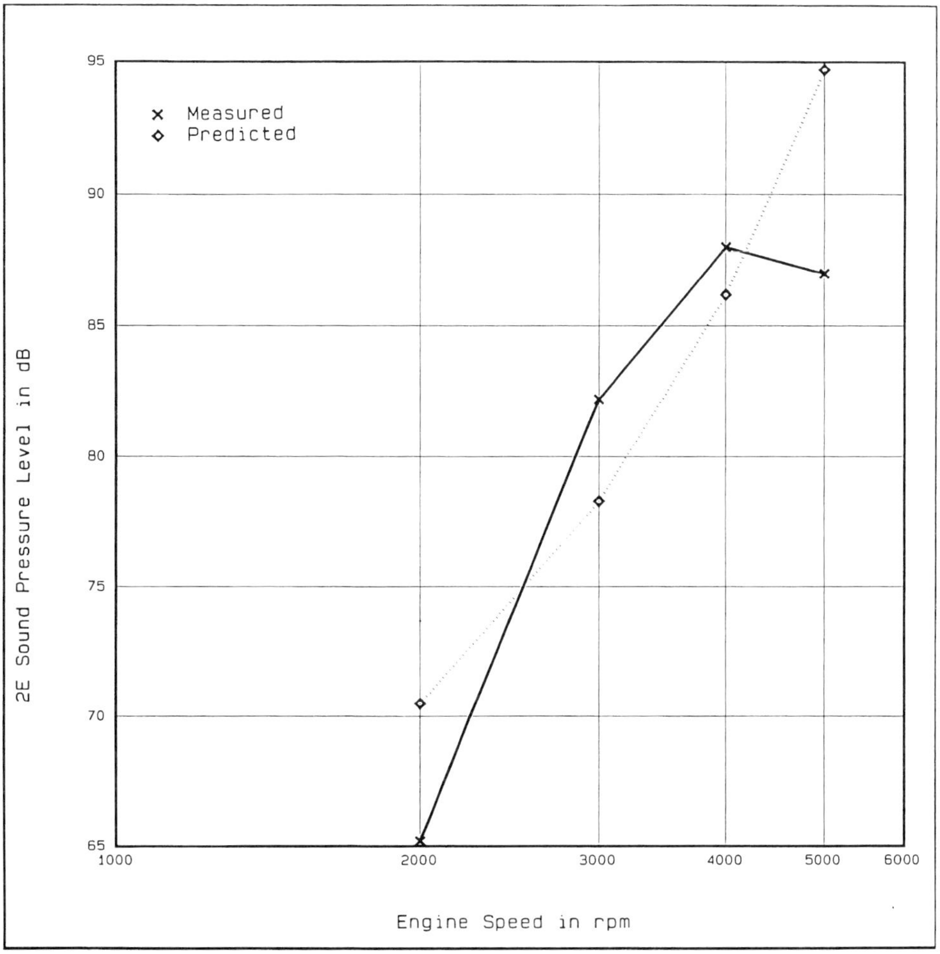

FIGURE 8

Automotive Design Advisory Unit

COMPARISON BETWEEN MEASURED AND PREDICTED NOISE.
TOP MICROPHONE POSITION NOISE AT 1m.

C521/034/98

Optimization of SI combustion noise – development process and basic analyses

M WILHELM, N ALT, and **N WIEHAGEN**
Institute for Combustion Engines (VKA), Aachen Technical University, Germany
and FEV Motorentechnik, Aachen, Germany

Abstract

Combustion is a significant noise excitation of modern thermodynamically optimised S.I. (Gasoline) engines. It has been analysed in detail with regard to global noise emission as well as to sound quality aspects. Interdependencies of thermodynamics and acoustics are special points of emphasis of this study impelled by the necessity for an improved control of the traditional trade-offs. A special test bench has been installed for this reason which enables simultaneous acoustic and thermodynamic measurements, thus e.g. cycle related analysis of both types of parameters.

The procedure for accurate determination of direct and indirect combustion noise is explained and results of cyclic fluctuation analyses are discussed. The correlation between acoustic and thermodynamic parameters is demonstrated with examples and an outlook for future approaches for NVH related development of the combustion process is given.

The paper represents the latest results of a public research work ordered by the German Research Association for Internal Combustion Engines (FVV), conducted at the Institute for Combustion Engines (VKA) and at FEV Motorentechnik (FEV), Aachen, Germany. The project will be in progress until May 1998.

1. Introduction

In spite of significant progress in the field of engine acoustics within the last 10 to 15 years, modern engines are still assessed as being too loud and - with regard to their sound character - uncomfortable and subjectively annoying. The combustion contributes a significant share to the noise excitation of an engine. The crankshaft, the crankcase and the cylinder head for instance are excited to vibrate due to the cylinder pressure excitation which leads - together with other noise sources - to the total engine noise

A distinction has to be made between direct and indirect combustion noise (CN): The direct combustion noise is due to gas forces (combustion) directly acting on the structure assuming that no clearances are involved. The indirect combustion noise is due to gas forces causing components to move through their clearances resulting in impacts (examples: piston slap , crankshaft rumble). The direct combustion noise is transmitted within the engine via the outer path (cylinder head and upper part of the engine block/ liner) and even more intensively via the inner path (crank train components, main bearings and crankcase) (Figure 1). The recent trend to light weight design negatively influences the transmission of vibrations.

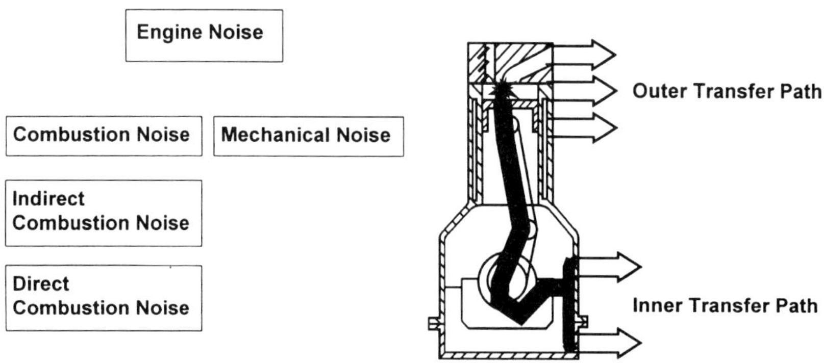

Figure 1: Transfer Paths of Combustion Noise

Excellent work has been done on diesel engine combustion noise since long. Keystones date from the 50ies and 60ies (1), (2) and even at present, diesel engine combustion noise is subject of investigations (3). Just recently, engineers are also interested in S.I. engines concerning this item.

The relevance of the combustion noise in case of S.I. engines increased for instance due to measures for combustion optimisation with regard to fuel consumption, performance and exhaust gas emissions which result in higher flame velocity and higher cylinder pressure gradients (4). The four-valve technique, variable valve timing and intake systems as well as measures aiming at the increase of charge motion (tumble, swirl) should be mentioned in this context (5), (6).

Each single engine make has to be treated individually since differences even of the basic noise behaviour exist. Fact is for instance, that the ratio of sound pressure level of no load and full load of various engines differs between 0 to 10 dB(A) (7). Figure 2 shows two examples of modern 4 valve 4 cylinder engines (2 litre class) with different load influence. It should be taken into account that this behaviour may partly be due to both, differences in the combustion noise and in the mechanical noise.

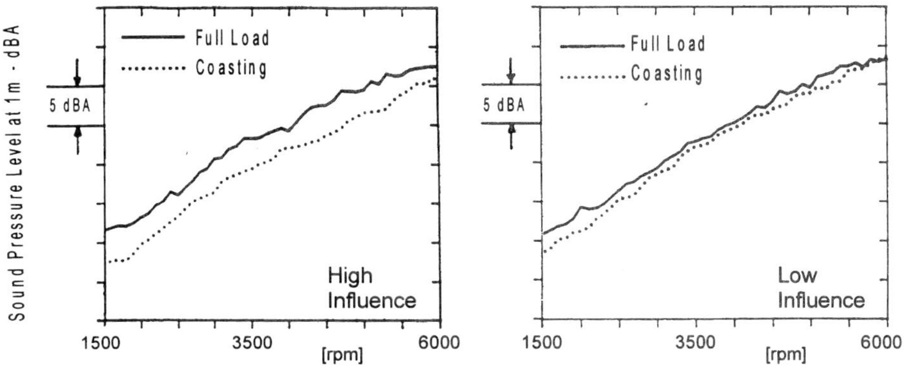

Figure 2: Influence of Load on S.I. Engine Noise

The current knowledge of the generation of both the direct and indirect combustion noise seems not to be sufficient. Following main questions are still to be answered:

- How important are combustion fluctuations (from cycle to cycle and from cylinder to cylinder) with regard to comfort and sound quality (also at idle) ?
- What are the relevant evaluation parameters for combustion excitation ?
- What is the noise improvement potential of measures on the combustion itself in comparison to measures on the engine structure ?

These are the main topics of a public research project started in May 1996 at the Institute for Combustion Engines (VKA), Aachen Technical University, Germany, in co-operation with the Institute for Engine Technology (FEV) in Aachen . The program is initiated and supported by the German Research Association for Internal Combustion Engines (FVV - Forschungsvereinigung Verbrennungskraftmaschinen e. V.). The project will be in progress until May 1998. This paper includes latest results concerning the development procedure and basic analyses. The investigation program includes simultaneous acoustic and thermodynamic measurements, thus represents an innovative approach which allows for example a cycle related analysis of acoustic and thermodynamic parameters as well as their interactions. The experimental investigations are additionally supported by FE-calculations of the dynamic behaviour of the engine's structure aiming at the evaluation of the combustion noise improvement potential by means of structural modifications.

2. Quantification of Combustion Noise

Efficient optimisation of the combustion process with regard to the noise behaviour requires the knowledge of the different engine noise shares. In the acoustic/ thermodynamic test cell the engine noise at motoring and under full load condition is measured. The total combustion noise is approximately determined by subtracting the mechanical noise (motoring) from the WOT noise. An exact noise separation is not possible, influences of clearances due to thermal expansion for instance of the piston can scarcely be taken into account (8), (9).

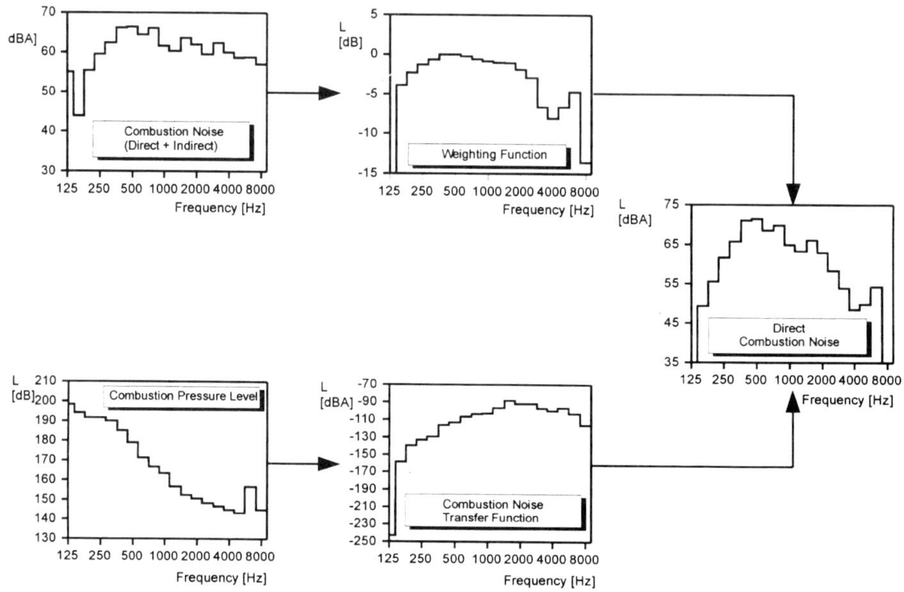

<u>Figure 3:</u> Determination of the Direct Combustion Noise

To distinguish between direct and indirect combustion noise a "weighting function" has been developed by correlation analysis of in-cylinder combustion pressure and far field sound pressure at representative engine mapping points. In Figure 3 the determination of direct combustion noise is visualised in form of a flow chart. The direct combustion noise can be calculated by using this weighting function. Finally the combustion noise transfer function (Fig 3, bottom right) is determined by subtracting the spectra of measured combustion pressure and calculated direct combustion noise. The combustion noise transfer function comprises the A-weighting curve and describes the vibration transfer behaviour of the structure ideally independent of engine speed and load. Additionally, the indirect combustion noise can be predicted by subtracting the direct combustion noise from the total combustion noise. With the knowledge of the transfer function, once established in the way as explained, predictions of the combustion noise share can be then conducted easily.

The accuracy of combustion noise transfer function depends on the weighting function. For a high quality, function correlation analyses are useful. So, for each third-octave band the cylinder individual combustion pressure level has been monitored versus the cylinder individual sound pressure level (time window analysis) for 100 consecutive working cycles.

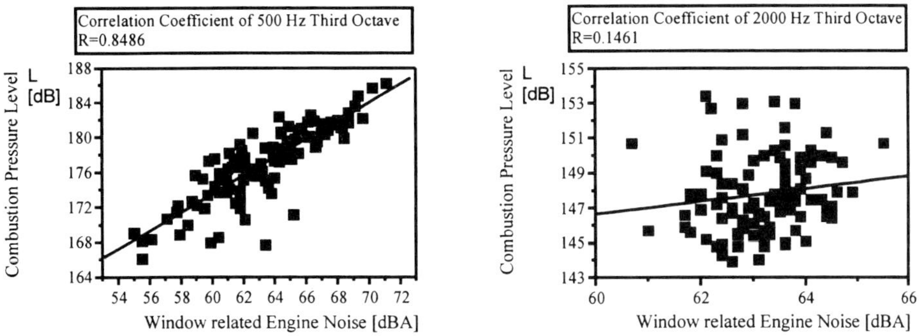

Figure 4: Determination of the Weighting Function for the Direct Combustion Noise

Figure 4 shows the determination of the correlation between combustion pressure level and sound pressure level for two third-octave bands. It can be recognised, that at the 500 Hz third-octave (left diagram) we find a rather good correlation (R = 0.85) and at the 2 kHz third-octave (right diagram) a worse one. A correlation coefficient of 1 means that all points are located on one straight line, that the dependence of sound and cylinder pressure level is linear and extraneous noise is absent. Based on these analyses for all 1/3 octave bands and FEV experience the weighting function is established. The idea is, that the engine block structure is a linear transmission system and that the „extraneous noise" is equivalent to the indirect combustion noise (the mechanical noise being constant). Thus, total combustion noise intensity multiplied by R -band by band- results in direct combustion noise (the rest is indirect combustion noise). The 10x logarithmic notation yields the weighting function.

Figure 5 shows on the left hand side the noise share percentages at WOT for ignition timing at knock borderline and 8° CA retarded. At knock borderline (left hand side, top) beyond 2000 rpm direct combustion noise is the highest noise share. Below 2000 rpm the noise radiation is dominated by the indirect combustion noise. The mechanical noise becomes dominant above 4500 rpm.

In Figure 5 at right hand side the noise shares are shown in detail as third-octave spectra. Between 500 and 2000 Hz the direct combustion noise dominates the noise behaviour, differently the mechanical noise shows high third-octave levels around 2000 Hz. High shares of indirect combustion noise are found in the high frequency area from 3000 to 4000 Hz.

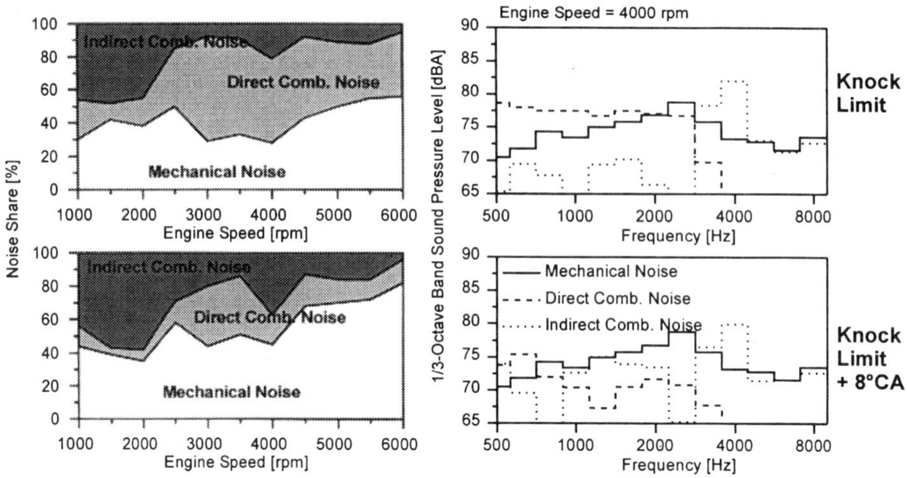

Figure 5: Influence of Engine Speed and Ignition Timing on CN-Shares

The influence of retarded ignition timing can be clearly identified. The direct combustion noise is reduced in the whole engine speed range. Mechanical noise becomes dominant at even lower engine speeds. Ignition timing influences the engine noise mainly in the frequency range 500 Hz - 2 kHz inclusive, as shown by the reduction in direct combustion noise.

3 Cyclic Fluctuations

At operating conditions with load influence on engine sound pressure level combustion noise becomes significant. Two cylinder pressure related parameters may be acoustically relevant, the mean spectrum and its cyclic fluctuations. According to the structure attenuation curve as weighting function for the cylinder pressure spectrum the share of the combustion noise in the total noise may differ in respect to magnitude at the frequency bands.

Engine noise comprises random components due to alterations of the boundary conditions of the various noise generation mechanisms. One major effect in this context is the combustion process itself which is very sensitive to ignition and early flame propagation. Thus, the cylinder pressure signal varies from cycle to cycle. The indirect combustion noise as well as the direct combustion noise are influenced (10), (11).

High fluctuations of the combustion noise could result in a poor sound quality and could even be misinterpreted as technical defects of the engine. Thus, the product reliability of the vehicle could be questioned. So, these two different acoustical aspects have to be balanced during the thermodynamic optimisation work of the combustion process.

Figure 6 shows relevant data of the combustion induced noise behaviour. The cyclic fluctuations of combustion become clear by the cycle related maximum cylinder pressure gradient. These values are represented in the diagram at the top on left hand side. For the rating point 4000 rpm/ full load the maximum cylinder pressure gradients of 50 consecutive working cycles of cylinder 2 lie within a broad scatter band of ca. 3 bar/ degree CA width: lowest value ca. 0.5 bar/ degree CA, highest value ca. 4.5 bar/ degree CA.

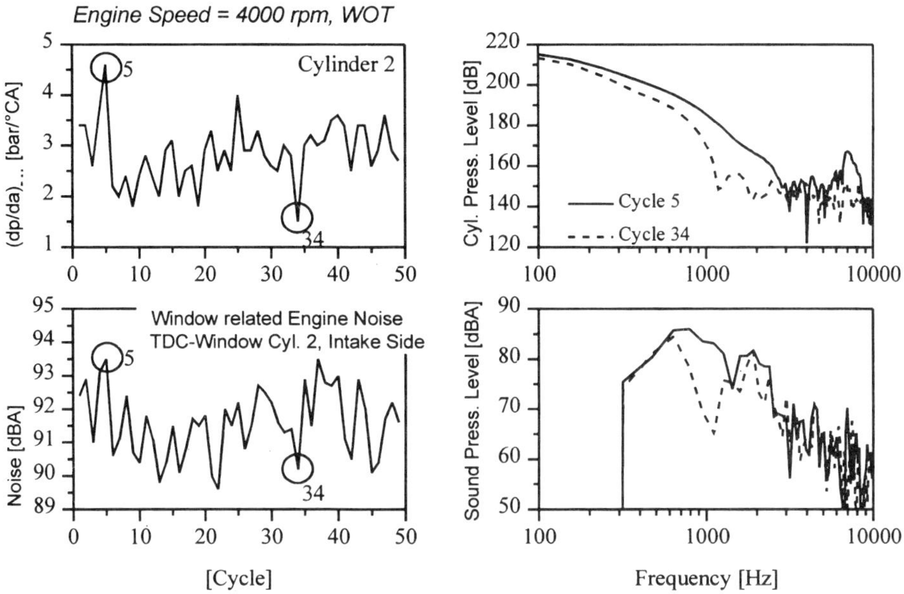

Figure 6: Cyclic Variation of Max. Pressure Gradient and Noise Radiation

The noise emission due to the cylinder pressure fluctuations of the identical working cycles is subject of the diagram below. To focus on cylinder 2 again the engine noise is analysed in a time window of 100 degree CA around firing TDC of the very cylinder. The fluctuations of combustion excitation result in sound pressure level fluctuations in a span of more than 4 dB(A). The correlation between maximum cylinder gradient and window related engine noise is reasonably good. The noise level related to the lowest gradient - cycle No. 34 - is one of the quietest, the working cycle with the highest gradient - cycle No. 5 - coincides exactly with the loudest one.

The diagram on top of the right hand side illustrates the differences between the cylinder pressure excitation spectra of the loud and quiet cycle. The highest level differences are in the frequency band between 0.5 and 2 kHz which is known to be an important frequency range for direct combustion noise.

The diagram below represents the frequency spectra of the engine noise measured at one meter distance on the intake side of the engine. The frequency range of the cylinder pressure spectra influenced by cyclic fluctuations corresponds well with the sensitive range of the window related engine noise spectra. An even higher relationship may not be expected since indirect combustion noise is also influenced by the cylinder pressure. Thus, the indirect combustion noise fluctuates from cycle to cycle, too.

Observation of the random behaviour of the engine thermodynamic process on basis of cycle related indicated mean effective pressure, imep, is in common use. The most important operation parameters influencing the repeatability of the combustion process are air-fuel ratio and ignition timing. An important design parameter in this context especially of 4 valve engines is the tumble level of the in-cylinder mixture motion (12).

The influence of these parameters on direct combustion noise which predominate the total engine noise and on imep standard deviation which is synonymous for engine run stability, and thus, vibration excitation, is subject of figure 7. The shaded band at the abscissa indicates the parameters corresponding to the base engine adjustment. Shifting to the left hand side - richer mixture, advanced ignition timing and intensive tumble level - results in a worse direct combustion noise level on the one hand and better imep standard deviation on the other. A shift of parameters to the right hand side causes a response vice versa.

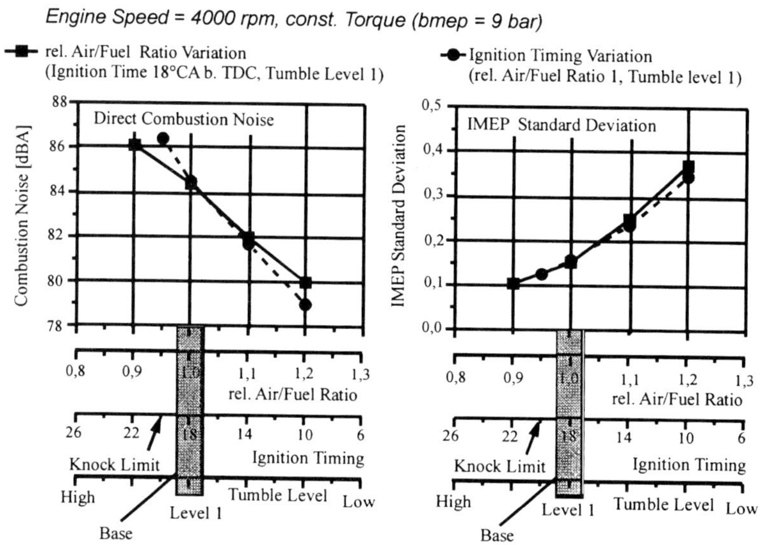

Figure 7: Direct Combustion Noise and IMEP Standard Deviation

Thus, these relationships make clear that a trade-off exists concerning combustion noise and imep standard deviation. More analysis into the interdependencies of the various factors which have to be balanced during combustion process calibration will be reported in the next chapter.

4. Interdependencies of Acoustics and Thermodynamics

The development work of the combustion process becomes even more complex since besides thermodynamic propensities, the acoustical properties like engine noise level and cyclic variations have to be taken into account. For the rating point 4000 rpm/ 8.9 bar bmep figure 8 comprises detailed information about the various aspects. The diagrams at the left hand side represent acoustical parameters, the other ones thermodynamical. The single curves show the influence of increased rel. Air/Fuel ratio (from rich to lean) on the very items. The solid lines stand for fixed throttle position and the dotted lines for constant bmep. In addition the values at rel. Air/Fuel ratio 1.2 and ignition timing for best fuel consumption are indicated in the diagrams.

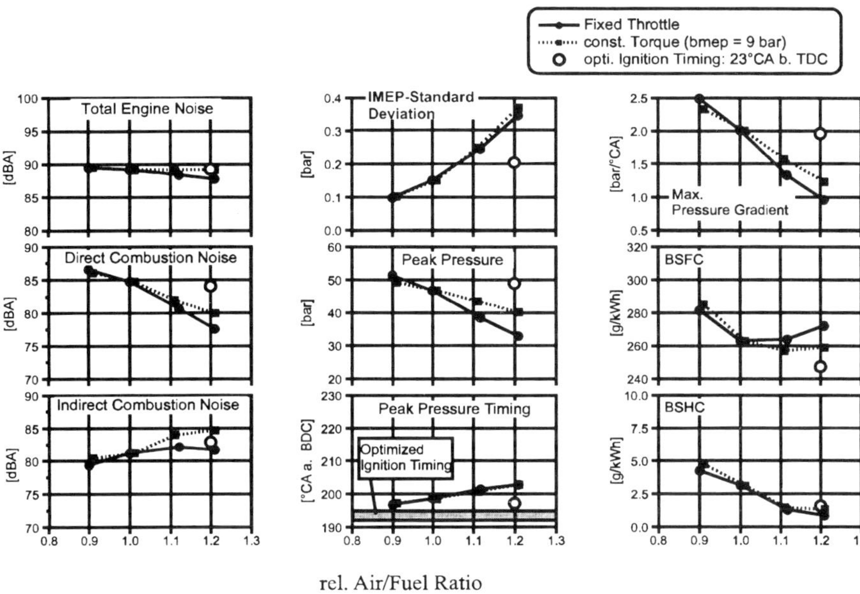

Figure 8: Variation of rel. Air/Fuel Ratio

In case of "fixed throttle", the total engine noise decreases by ca. 1 dB(A) in the interesting rel. Air/Fuel ratio range from 0.9 to 1.2. The influence on the direct combustion noise is strikingly higher: ca. - 9 dB(A) (fixed throttle) and - 6 dB(A) (constant bmep). Whereas the direct combustion noise decreases with leaner ratio the indirect combustion noise increases by ca. 2 dB(A) and ca. 4 dB(A), respectively. The direct combustion noise predominates at rich mixtures but at lean mixtures the indirect combustion noise is more significant. As expected the maximum cylinder pressure gradient shows a similar trend as the direct combustion noise: decrease by ca. 1.5 bar/ degree CA.

Concerning the influence of air fuel ratio on the thermodynamic related parameters different trends are encountered, too. The standard deviation of the indicated mean effective pressure increases drastically by a factor of 3 at lean ratios. This may result in a less smooth engine operation with consequences concerning body vibration excitation. The impact on cylinder peak pressure also is clear: ca. 20 bar and ca. 10 bar, respectively. The peak pressure timing is retarded by ca. 5 degree CA. Due to this deviation from the optimum timing the fuel consumption increases. The HC emissions are reduced as expected.

The diagrams contain the information necessary for optimisation of combustion process. The goal of work depends on a certain extend on the individual weighting of the parameters.

5. Outlook

After completion of this project, in April 1998, important questions concerning the combustion noise of S.I. engines with regard to noise excitation and interdependencies between acoustics and thermodynamics will be answered. This knowledge will serve for future control of the traditional trade-offs. Further action then has to be put on the elaboration of effective and efficient approaches for combustion noise reduction. These approaches and measures finally, have to be developed from the beginning of a new engine design under consideration of both acoustic and thermodynamic requirements.

From the current state of view different approaches for combustion noise reduction exist which nevertheless meet low acceptance with regard to thermodynamics. But this is not the only conflict: Unfortunately some approaches cause a deterioration of combustion stability which has to be taken into account with respect to driving comfort. However, other approaches possess high potential to become acceptable compromises for the future (Table 1) by an accurate balancing of all aspects. But to this end additional research and development work is necessary. Conventional combustion systems as well as the direct injection for gasoline engines are taken into account.

Approach	Measure	Potential
Ignition Timing	Retardation Twin Spark (13)	no realistic
Charge Movement	Decrease of: Tumble Swirl	realistic
Valve Timing	Valve Overlap Increase of EGR	realistic
Compression Ratio	Decrease	no
Combustion Velocity	Decrease Delay at Commencement of Comb.	realistic
Air/ Fuel Ratio	Increase	realistic
Intensive Spark	Plasma Ignition	realistic

Table : Approaches for Combustion Noise Reduction

6. Summary and Conclusion

The combustion processes have been shown to be a significant noise excitation of modern thermodynamically optimised S.I. (Gasoline) engines and have been analysed in detail with regard to global noise emission as well as to sound quality aspects. Interdependencies of thermodynamics and acoustics are special points of emphasis of this study impelled by the necessity for improved control of the traditional trade-offs.

To distinguish between direct and indirect combustion noise a combustion noise transfer function has been developed by correlation analysis. Thus, the direct combustion noise can be predicted simply by using combustion pressure measurements and evaluating their spectra with the function. Additionally, the indirect combustion noise can be separated from the total combustion noise.

At knock borderline beyond 2000 rpm the direct combustion noise contributes most to the engine noise. Below 2000 rpm the noise radiation is dominated by the indirect combustion noise. The mechanical noise becomes dominant above 4500 rpm. Further results show that there is a trade-off between combustion noise level and imep standard deviation, which is synonymous for engine run stability and thus cyclic sound pressure level variations.

The development work of the combustion process becomes even more complex, since besides the thermodynamic propensities the acoustical properties have to be taken into account. For rel. Air/Fuel ratio variations between 0.9 and 1.2 e.g. it is shown that the direct combustion noise predominates at rich mixtures but at lean mixtures the indirect combustion noise is more significant.

From the current state of view different approaches for combustion noise reduction exist which may be unacceptable with regard to thermodynamics. But this is not the only conflict: Some approaches unfortunately cause a deterioration of combustion stability which has to be taken into account with respect to driving comfort. Nevertheless measures like optimisation of charge movement, valve timing and air-fuel ratio are promising approaches.

7. Literature

(1) A. E. W. Austen, T. Priede,
Origins of Diesel Noise,
Symposium on Engine Noise and Noise Suppression. Instn.Mech.Engnrs. London 1958

(2) T. Priede,
Some Studies into Origins of Automotive Diesel Engine Noise and its Control,
Internationaler Automobiltechnischer Kongreß, 12.-16. Juni 1966, Fisita, München, Paper C 12

(3) J. Wolschendorf,
Zyklische Schwankungen im Verbrennungsgeräusch von Dieselmotoren und ihre Ursachen,
Dissertation 1990, TH Aachen

(4) M. Bargende,
Verbrennungs- und Ladungsanalyse,
Stuttgarter Symposium Kraftfahrwesen und Verbrennungsmotoren 1995

(5) H. Endres, R. Wurms, J. Neußer,
Verbrauchs- und Emissionsminderung bei Mehrventil-Ottomotoren durch Anwendung von Drall und Tumble,
3. Aachener Kolloquium Fahrzeug- und Motorentechnik, 1991

(6) H. Endres,
Combustion System Development Trends for Multi-Valve Gasoline Engines,
SAE 900652

(7) R. Künzel, U. Essers,
Die Lastabhängigkeit des mechanisch verursachten Motorgeräusches,
MTZ 54 (1993) 9

(8) J. Borggräfe, J. Berthold,
Verbrennungsgeräusch des Ottomotors, ATZ 86 (1984) 7/8, S. 333 - 338

(9) J. Kodweiß,
Strategien zur akustischen Beherrschbarkeit eines rauhen Motorlaufs, der durch einen schnellen Verbrennungsprozeß verursacht wird,
Vortrag im Rahmen der Tagung "Verbrennungsmotorenakustik" im HdT, Essen, 1993

(10) J. Wolschendorf, M. Dürnholz, K. Schmillen,
The IDI Diesel Engine and Its Combustion Noise Variations,
SAE 910228

(11) K. Schmillen, J. Wolschendorf,
Cycle-To-Cycle Variations of Combustion Noise in Diesel Engines,
SAE 890129

(12) K. Gregotsch, R. Herrmann, J. Kodweiß,
Engine Acoustics in Rapid-Combustion 4-Valve SI-Engines - A Contribution to the Acoustic Tuning of the Combustion Process,
SAE 920411

(13) H. Niefer, H. K. Weining, M. Bargende, A. Waltner
Verbrennung, Ladungswechsel und Abgasreinigung der neuen Mercedes-Benz V-Motoren mit Dreiventiltechnik und Doppelzündung

Acknowledgements

The authors are indebted to the Bundesministerium für Wirtschaft (BMWi, Bonn) and the Arbeitsgemeinschaft industrieller Forschungsvereinigungen (AiF, Köln. AiF-Nr. 8385) for the financial support that made this research project possible. The investigations have been initiated by the Forschungsvereinigung Verbrennungskraftmaschinen e. V. (FVV), Frankfurt and accompanied by a council under the chairmanship of H. Borggräfe, BMW AG München. Their support is also acknowledged with thanks.

Systems

C521/006/98

Optimizing the power steering components to attenuate noise and vibrations

G E SMID, M S QATU PhD, MASME, MASCE, MSAE, MAAM, MPE, and **M L DOUGHERTY**
Technology Center, Echlin Automotive, Michigan, USA
J DREW PhD
Hobourn Automotive Limited, Rochester, UK

Higher demands for less noise from automotive customers coupled with the reduction of noise and vibration levels from major sources like the engine necessitate better performance of other sources of noise and vibrations in a vehicle. One of these sources is the power steering system. Noise can be transmitted directly from its source, like the pump in a hydraulic system. Objectionable fluid borne noise and vibrations exhibit as a pressure ripple or flow pulsation through fluid lines. This will become an excitation force to the structure of a vehicle or the steering gear and is a considerable source of discomfort to the vehicle passengers. Current power steering technology attenuates this ripple along the pressure line connecting the pump to the steering gear. Finding the optimum configuration for the components (hose, tuner, tube ...) has been a matter of experience-based trial and error.
A fluid borne noise simulation software program has been developed. It uses the MATLAB language to determine the optimal configuration for a specific application. This optimization procedure determines, among other things, finding the optimal hose length in a power steering configuration. Optimization is accomplished on the basis of ripple attenuation at any point along the fluid line. The output results have been verified and validated against measurement data and other validated simulation software.

1. INTRODUCTION

Hydraulic systems in a vehicle are sources of objectionable noise. The noise can originate from pressure and flow fluctuations generated by the pump. When the fluctuations are transmitted along the hydraulic lines, they cause forces at pipe bends and various couplings and the steering gear, which result in mechanical vibrations that in turn cause objectionable audible sounds.

Research can be done to the internal design of pumps to minimize the ripples in the output pressure and/or flow. In many practical applications this can be a costly task. Furthermore, most hose assembly manufacturers do not have a choice on the pump selected by the Original Equipment Manufacturer (OEM). It is thus the objective of this study to configure the fluid line between the pump and the steering gear such that the pressure ripple at any specified location is minimized.

In a typical application, the pump type and location, the steering gear type and location, and possible alternative routings for the fluid lines are given. In these alternative routings, the fluid line may be connected to the structure at various locations. The procedure described in this paper allows optimizing the fluid ripple with respect to a specific point along the fluid line. At the present time, many of the design rules are based on the experience and knowledge of the Noise Vibration and Harshness (NVH)-engineers.

The work that is presented in this paper will provide NVH-engineers with a simulation software program that can predict system performance in terms of pressure and flow ripple anywhere between the pump and the steering gear. It can also predict important features like transmission loss for each component in the system.

The next two sections explain how the pump, steering gear and other components are modeled and characterized. In a following section, the optimization procedure is explained. Conclusions of this study are then described at the end of the paper.

2. POWER STEERING SYSTEM MODEL

The governing relations for input pressure and flow and output pressure and flow are needed for each component at any given frequency to assemble the system. We will first study the impedance of each of these components and then concatenate the system.

2.1. Pump

The pump model is obtained by finding the impedance matrix coefficients in the frequency domain for a particular set of environmental conditions like fluid viscosity, mean flow, pressure and temperature. This will necessitate a series of measurements of input and output flow and pressure ripple amplitude and phase. Johnston and Edge [2] demonstrated that recorded measurement data can be incorporated in the system model. The experimental procedure required to obtain the impedance coefficients was discussed by Johnston and Drew [3]. It assumes linearity in the environmental conditions that are further detailed in the references.

A series of pumps have been characterized and can be selected from the simulator library. The library contains the non-parametric data for different pressures, speeds (i.e., flow) and temperatures. An example of the model data is shown in Figure 1.

2.2. Tube

Tube is a major component of the power steering system. It is typically made of steel and relatively inexpensive compared with other power steering components.

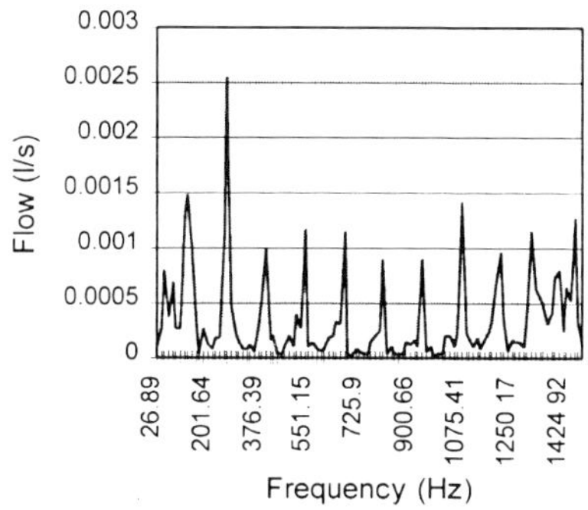

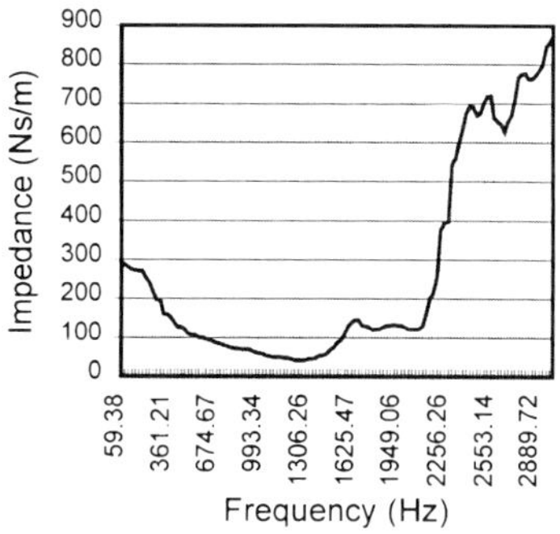

Figure 1 Typical data for the pump flow ripple and the internal impedance. This data represents a typical pump at 800 rpm, 60 bar mean pressure and 70°C.

The complex amplitudes of the pressure and flow ripple at a position x along the pipe can be determined using the following equations:

$$P_x = P_F e^{-\gamma x} + P_R e^{\gamma x}, \qquad Q_x = P_x / Z_o$$

where P_F and P_R are the pressure waves at x=0 traveling in the forward and reverse directions respectively. The damping coefficient γ and the pipe impedance Z_0 are given by

$$\gamma = \frac{j\varpi}{c_0}\xi, \qquad Z_0 = \frac{\rho c_0}{A}\xi \qquad \text{where} \quad c_0 = \sqrt{\frac{B_E}{\rho}}$$

where ρ is the fluid density, A is the cross sectional area of the pipe, B_E is the effective bulk modulus of the fluid and the pipe and ϖ is the frequency. It should be noted that the effective bulk modulus B_E should be used as this considers the bulk modulus of both the fluid and the tubing system [1] . ξ is defined by

$$\xi = \left[1 - \frac{2}{zj^{3/2}} \frac{J_1\left(zj^{3/2}\right)}{J_0\left(zj^{3/2}\right)}\right]^{-\frac{1}{2}}, \qquad where \qquad z = r_p\sqrt{\frac{\rho\varpi}{\mu}}$$

μ is the absolute viscosity, r_p is the radius of the pipe, and J_o and J_1 are Bessel functions. Letting l_h be the pipe length, the transfer matrix for the pipe can be formed as

$$Z = Z_0\begin{bmatrix} \coth(\gamma l_h) & \operatorname{cosech}(\gamma l_h) \\ \operatorname{cosech}(\gamma l_h) & \coth(\gamma l_h) \end{bmatrix}$$

2.3. Hose

Various types of hoses are available in the market place that can be used as a part of the power steering assembly. These include low and high expansion hoses with various types of brading construction and materials. Some hoses may include a tuner. These are proven to attenuate more fluid ripple than others.

The work presented in this paper can be used for various types of hoses. We will however focus solely on optimizing the length of a plain hose. For a plain hose, an impedance model must be derived relating input pressure and flow ripples to the output pressure and flow ripples. Two physical relations define the impedance of a plain hose. They represent the pressure-volume relation and the fluid motion respectively and are given by [1]

$$\frac{A_b}{A_r}\gamma U + \frac{1}{B_E}P = 0, \qquad U + \left(\frac{(1-j)}{r_b \varpi^2 \rho}\sqrt{\frac{2\mu}{\rho\varpi}} - \frac{1}{\varpi^2 \rho}\right)\gamma P = 0$$

A_b is the bore cross sectional area of the hose (i.e., the area containing the fluid), A_r is the cross sectional area contained by the reinforcement. U is the complex fluid motion and P is the complex pressure and

$$\frac{1}{B_E} = \frac{A_b}{A_r B_{\text{fluid}}} + \frac{A_l}{A_r B_{\text{lining}}} + \frac{2r_r\left(1 - \nu_x \nu_y\right)}{E_y}.$$

where A_l is the cross sectional area of the lining, B_{fluid} is the fluid bulk modulus, B_{lining} is the lining bulk modulus, ν_x and ν_y are Poisson's ratios in the x and y directions respectively, E_y is the modulus of elasticity in the y-direction. Assuming a harmonic solution $U = U_o e^{\gamma x}$ and $P = P_o e^{\gamma x}$, a characteristic equation in γ can be written as:

$$\frac{B_E A_b}{A_r}\left(\frac{1-j}{r_b \varpi^2 \rho}\sqrt{\frac{2\mu}{\rho\varpi}} - \frac{1}{\varpi^2 \rho}\right)\gamma^2 + 1 = 0$$

The general solutions allows for U and P to be written at the boundaries as:

$$U_{x=0} = K_1 + K_2 \qquad P_{x=0} = K_1 - K_2$$

$$U_{x=l_h} = K_1 e^{\gamma l_h} + K_2 e^{-\gamma l_h} \qquad P_{x=l_h} = \frac{B_E A_b}{A_r}\gamma\left(K_1 e^{\gamma l_h} - K_2 e^{-\gamma l_h}\right)$$

K_1 and K_2 can be solved for by considering the boundary conditions at the two hose ends. With these formulations, the impedance matrix for the hose can be formulated as

$$Z = \frac{B_E A_b^2}{j\varpi A_r}\gamma \begin{bmatrix} 1 & -1 \\ e^{\gamma l_h} & -e^{-\gamma l_h} \end{bmatrix} \begin{bmatrix} K_1 \\ K_2 \end{bmatrix} \left(\begin{bmatrix} 1 & 1 \\ e^{\gamma l_h} & e^{-\gamma l_h} \end{bmatrix} \begin{bmatrix} K_1 \\ K_2 \end{bmatrix} \right)^{-1}.$$

A more detailed fundamental discussion on wave propagation along a fluid filled hose is given by Longmore [4], with a more detailed account of the theoretical approach given by Tuc [5]. The derivations that also incorporate hose wall motions can be found in [1].

2.4. Steering gear

At this point in time, the steering gear is considered to be the end of the fluid transmission line. This is true as far as high pressure is concerned. The steering gear is not modeled as a transfer or impedance matrix, but as a "point impedance", represented by a single frequency dependent equation for pressure and flow.

The relation will be read from a library of characteristic measurement files for various brands steering gear-assembly. Johnston and Edge [6] outline the experimental method of measuring component impedance.

2.5. System concatenation

A two-by-two matrix relating input and output pressures and flows will represent the components in the fluid line. This matrix can be written either as a transfer matrix or as an impedance matrix:

$$\begin{bmatrix} P_{in} \\ P_{out} \end{bmatrix} = \begin{bmatrix} z_{11} & z_{12} \\ z_{21} & z_{22} \end{bmatrix} \begin{bmatrix} Q_{in} \\ Q_{out} \end{bmatrix} \qquad \begin{bmatrix} P_{out} \\ Q_{out} \end{bmatrix} = \begin{bmatrix} t_{11} & t_{12} \\ t_{21} & t_{22} \end{bmatrix} \begin{bmatrix} P_{in} \\ Q_{in} \end{bmatrix}$$

where $t_{11} \ldots t_{22}$ are the elements of the transfer matrix and $z_{11} \ldots z_{22}$ are the elements of the impedance matrix.

3. SIMULATING PRESSURE AND FLOW

Once the various components of a power steering system have been modeled, it is possible to derive the pressure and flow ripple at any point along the fluid line. Generally, the high-pressure line of a power steering system will consist of a hose, connected to the steering gear and the pump with steel pipe. The model for such a system is shown in Figure 2. To compute the pressure and flow ripple P_1 and Q_1, at the outlet of the pump, first equivalent impedance must be derived for the transmission line and the steering gear. This can be found as

$$Z_{eq} = \frac{t_{22} R - t_{12}}{t_{11} - t_{21} R}$$

where R is the impedance of the steering gear and $t_{11} \ldots t_{22}$ are the elements of the equivalent transfer matrix of the pressure line. This matrix can be found by multiplying the component transfer matrices. The pressure at the outlet of the pump is

$$P_1 = \frac{Z_i Z_{eq}}{Z_i + Z_{eq}} Q_s$$

where Z_i is the internal impedance for the pump and Q_s is the pump flow ripple. The flow ripple at the pump outlet can then be found by dividing the pressure ripple by Z_{eq} as

$$Q_1 = \frac{P_1}{Z_{eq}}.$$

Pressure and flow ripples at different locations can be found by multiplying P_1 and Q_1 with the transfer matrix of the components between the pump and the particular location.

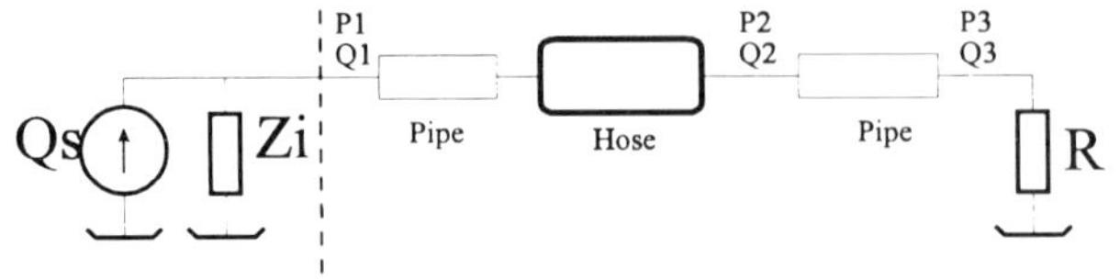

Figure 2 General model representation of a power steering system

4. SIMULATION AND OPTIMIZATION

A simple power steering configuration will be taken in the application of optimizing for noise and vibrations attenuation. The pressure line of the system consists of a pipe of 0.4 m, a hose of 1m and a pipe of 0.4 m. A model for this system is shown in Figure 2. It can be seen from the simulation in Figure 3 that the pressure ripple only decreases slightly due to the hose, for high frequencies, and is much worse for the 10th and 20th order ripple.

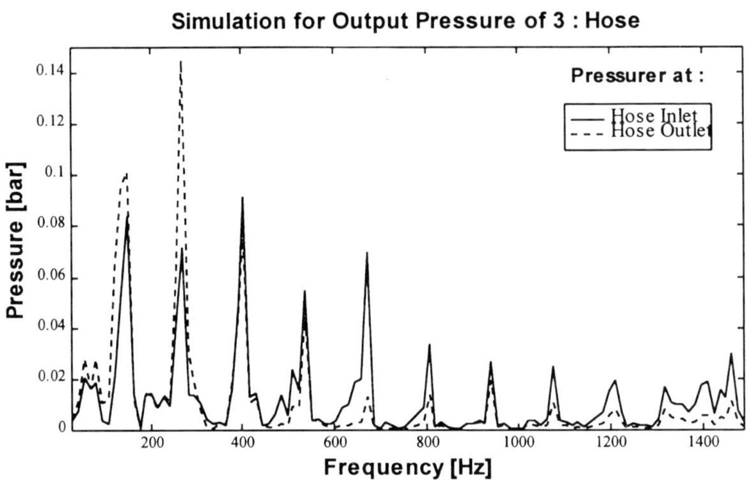

Figure 3 Pressure ripple at the inlet and the outlet of the hose.

It is the purpose of optimization to look for the hose length that minimizes the ripple (i.e. maximizes the ripple attenuation) in the hose. Pressures at the outlet for different lengths of hose show the effect of attenuation. The simulation for 4 different hose lengths is plotted in Figure 4. The graph shows that a hose length of 0.6m performs best for 133 Hz, which is the 10th order of the pump that consists of 10 vanes. However, the same length shows worst result at the 20th order.

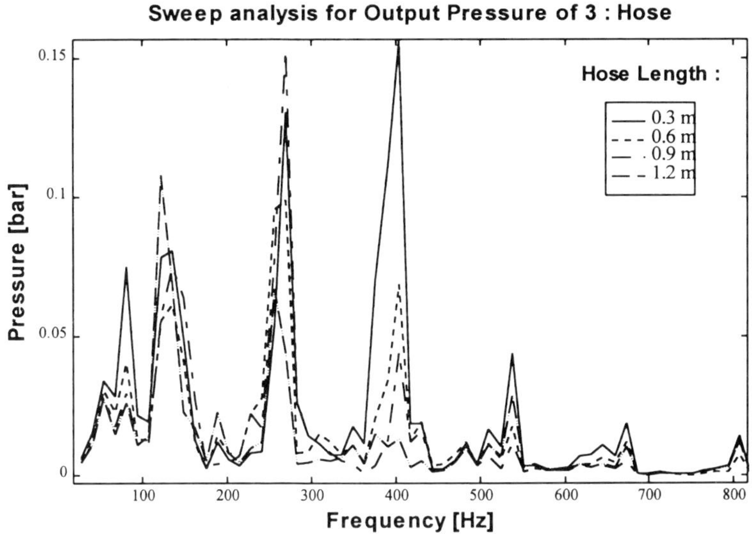

Figure 4 Pressure at the outlet of the hose for different hose lengths.

The speed at which the pump operates depends on the engine's speed. It is thus of interest to study the attenuation at the higher order frequencies of the motor speed. Results of the simulation of the pressure ripple at the outlet of the hose with respect to its length for these particular frequencies are given in Figure 5. The results show indeed that 0.7m has poor performance at the 20th order frequency. It is interesting to note that a slightly shorter hose would perform excellent for the same frequency. The graph in Figure 5 shows that best attenuation can be achieved for a hose length of approximately 0.63m. This could be expected, since it is close to the ½ wavelength of the particular frequency. At this wavelength the pressure ripple forms a standing wave inside the hose, that will absorb the energy from the pressure ripple. Since the effective bulk modulus is 2.5 10^8 N/m^2 and the density of the fluid is 840 kg/m^3[1], the speed of sound inside the hose is

$$c = \sqrt{B_E / \rho} \approx 546\ m/s$$

For the 20th order frequency at 800 rpm, which is (800/60)*20=266 Hz, this means that the ½ wavelength can be found by solving

$$l_{0.5} = \frac{\lambda}{\pi} = \frac{1}{f_n \pi}\sqrt{\frac{B_E}{\rho}}$$

which is just 0.63 m. It can also be seen, that close to, but not exactly at the ½ wavelength, the attenuation is worse than elsewhere. This is due to the fact that the pressure ripple at the hose ends are close to being in phase with the first order reflected image of the pressure ripple inside the hose. This means that a pressure ripple will enter the hose, suffers reflection at the hose end and travel almost one wavelength when it reaches the same point in the hose. Therefore the pressures will be almost in phase and thus they will add. Only very little energy will be absorbed, since there is no standing wave. The effect can be seen most clearly for the hose length analysis at the 20th order in Figure 5.

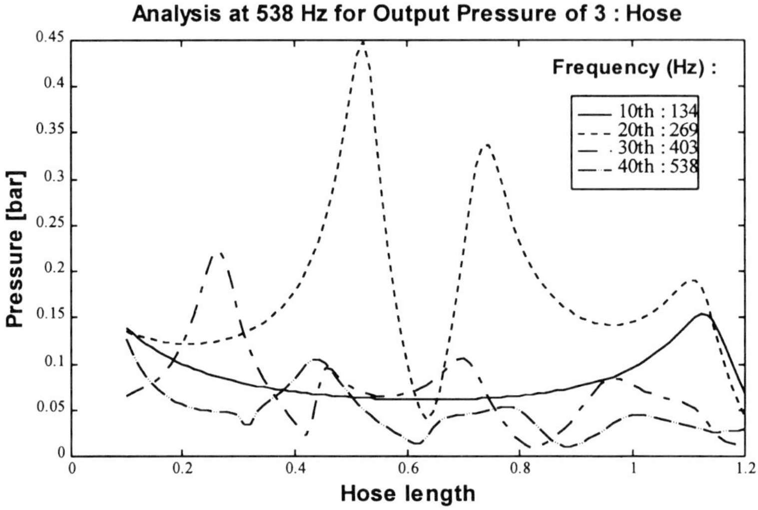

Figure 5 Pressure ripple at the outlet of the hose for varying hose lengths at excitation frequencies.

Selecting the hose length to be 0.65m, the pressure ripple is shown in Figure 6.

In order to automate this optimization procedure, weight functions W_f can be applied to each of the pressure ripple curves for the frequency modes in Figure 5, denoting the relative need to suppress the noise at the particular frequency. These weight functions are based on the noise levels, or vibration levels, produced by each of these frequencies in actual testing conditions. They can also be based on the experience of the engineer interested in minimizing noise in the power steering, or other hydraulic, systems. The pressure function is

$$P(l_h) = \sum_f W_f P_f(l_h).$$

A gradient search algorithm could then check for the best hose length. This will be the

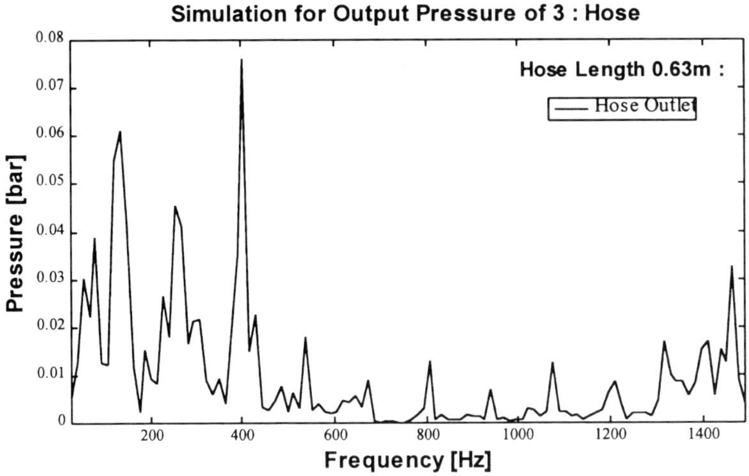

Figure 6 Simulation of the pressure ripple at the hose outlet for l_h=0.65m.

hose length where the pressure ripple is the smallest of the local minimum found for P_{lh}:

$$\hat{l}_h = \min_{P(l_h)} \left\{ l_h \left| \frac{\partial P(l_h)}{\partial l_h} = 0 \cap \frac{\partial^2 P(l_h)}{\partial {l_h}^2} \geq 0 \right. \right\}$$

Obviously, the curves in the graphs above will show different results for different speeds of the pump. One needs to know at what frequency excessive objectionable noise occurs, so that the model can be set up accurately. After finding the best attenuation from the graph in Figure 5, a frequency sweep can be performed to understand where the system generates most noise or vibration. These frequencies should not coincide with mechanical resonance in the vehicle. The results of a simulated spectrum are given in Figure 7. It can be seen there that hose length of 0.63m shows very good performance for 266 rev/min. However it always needs to be considered that performance might degrade at other frequencies. Performance degradation did not occur in this particular case.

5. CONCLUSIONS

A model of fluid axial flow in the high pressure line of a power steering system has been written in MATLAB. This allows the travel of hydraulic pressure pulses to be calculated. Simulation for different hose lengths show that plain hose attenuates best when the hose length coincides with the pressure ripple ¼ wavelength, and least efficient when it co-incides with the ½ wavelength.

The use of optimization algorithms will contribute heavily in the automation process of the simulator. Engineers could then import the vehicle characteristics, together with the pump and steering gear data, whereas the simulator would simply return the optimal

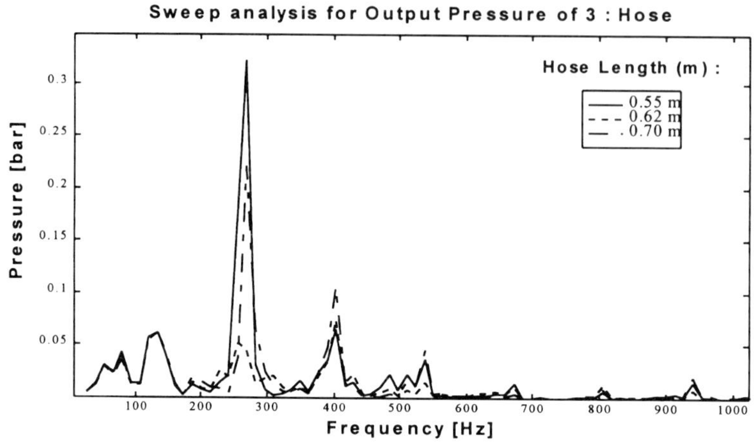

Figure 7 Frequency sweep simulation for varying hose length.

configuration for a particular vehicle. Considering various other components like a tuner, a restrictor, or a multi-chamber hose is the focus of ongoing research.

It should always be considered that the optimal hose length should attenuate all frequencies of pressure ripple that coincides with the harmonics of the mechanical structure. During the presentation, simulations will be conducted live, to show the powerfull capabilities that can be at hands in seconds for the engineers.

REFERENCES

[1] **Drew, J.E.** The use of flexible hose to reduce pressure ripple in Power Steering. *Ph.D. Thesis, University of Bath, 1997.*

[2] **Johnston, D.N., and Edge, K.A.,** Simulation of the pressure ripple characteristics of hydraulic circuits, *Proc. Inst. Mech. Eng., 1996, Vol 203, 275-282*

[3] **Johnston, D.N. and Drew, J.E.** Measurement of positive displacement pump flow ripple and impedance. *Proc. Instn Mechanical Engineers, 1996, Vol. 210, 65-74.*

[4] **Longmore, D.K.** The transmission and attenuation of fluid borne noise in hydraulic hose. *Inst. Mech.. Eng. Seminar on Quiet Oil Hydraulic Systems, 1977, C267/77, 127-138.*

[5] **Tuc, B.** The use of flexible hose for reducing pressure ripple in hydraulic systems. *Ph.D. Thesis, University of Bath, 1981.*

[6] **Johnston, D.N., and Edge, K.A.** The impedance characteristics of fluid power components: transfer matrix characteristics of two-port hydraulic components. *ASME Winter Annual Meeting, Chicago, Nov. 1994.*

[7] **Kojima, E. and Edge, K.A.** Experimental determination of hydraulic silencer matrices and assessment of the method for use as a standard test procedure. *7th Bath International Fluid Power Workshop, Bath, Sept 1994.*

C521/007/98

Vibration diagnosis and optimization of a test bench to characterize power steering pump noise using vibration measurements

G SCIORTINO and **D BAMDAD-SOUFI**
DELPHI Automotive Systems, Paris, France

1- ABSTRACT

An audit test bench was used to perform power steering pump noise evaluations, with vibration measurements setting the process specifications. The test bench used was originally dedicated to evaluating pump functional performance, based on effective pressure cycles and measurements.

1.1- Objective of the work

This steering pump noise evaluation work had several key objectives. They included the following:

- Evaluation of the existing test bench adequacy for vibration measurements on power steering pumps, using an **Order Tracking Procedure**.
- Identification of the disturbance sources from the test bench, which can influence steering pump vibration measurements and result in unacceptable production reject rates of the pump.
- Optimisation of the test bench characteristics for more reliable measurements with the following constrains :
  - No change in the customer measurement procedure, processing methods and performance specifications,
  - Cost savings, because of the re-design of the existing test bench.
  - Reasonable improvement in the confidence levels concerning production variations.
- Validation of the up-graded test bench operation.

The test bench, as shown in Figure 1, is constits of :

1- A main supply pump driven by an asynchronous motor with two pole pairs, an internal gear pump with nine drive teeth, and an idler-gear with seven teeth;
2- A hydraulic-fluid reservoir;
3- A test bench pump driven by an asynchronous motor with two pole pairs and a straight gear pump with ten drive teeth;
4- A complete valve assembly for the hydraulic circuit allow separate testing of the power steering pump in a closed hydraulic circuit;
5- An automatic quick-clamping mechanism for making hydraulic circuit connections to the steering pumps.
6- A drive motor for a speed run-up from 900 rpm to 4000 rpm, as specified by the procedure;
7- A flexible coupler between the motor and the pump;
8- The steering pump being tested; and,
9- A pneumatic/mechanical clamping system to aid back-support mounting of the pump.

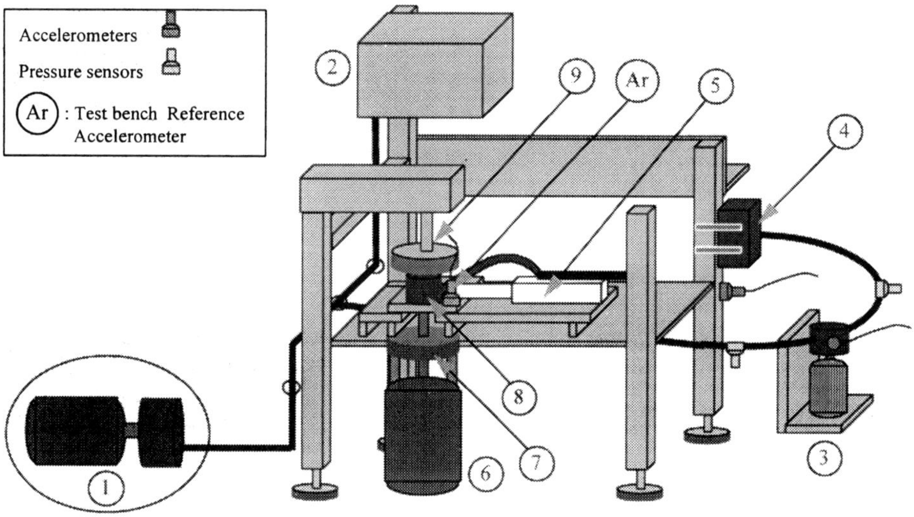

Figure 1: Overview of the test bench description and measurement points.

Figure 1 provides an overview of this test bench assembly with main accelerometer measurement points and locations identified. In addition to these accelerometers, three pressure sensors were placed at the outlet of the main supply pump and at the inlet and outlet of the test bench pump to monitor static and dynamic pressures.

A tachometer was also installed at the shaft of the pump being tested to monitor speed for the order-tracking process. Measurements were recorded using a multi-channel DAT recorder and processed with a multi-channel analyser.

1.2- Approach to the problem resolution

Four successive experimental campaigns have been conducted. These stages of investigation included:

1. Preliminary investigation of the system behaviour;
2. Analysis of the test bench adequacy for meeting the specifications by proper system diagnosis;
3. Improvement and validation of the optimal test bench configuration for achieving the given objectives; and,
4. Implementation of the methods in production by reception testing of the bench using customer test procedures and performance specifications.

Preliminary investigations and their results have enabled a global evaluation of the sources contributing to the steering pump vibration disturbances identified on the test bench.

This work has been performed by analysing and studying different test bench physical parameters. These involved all functional components - such as the hydraulic elements and circuits, the pneumatic units, and the rotating machines and components - as well as the structural response of the bench.

Then, by using a thorough diagnostic approach, the test bench behaviour characteristics and the testing improvement potential were identified for providing appropriate adequacy of meeting the steering pump testing specifications.

Reasonable vibration improvement was achieved by implementing the following:

- Corrective actions on the hydraulic elements of the machine and reduction of the dynamic pressure ripple generated by the test bench components;
- Vibration insulation of the pump hydraulic lines from the test bench structure; and,
- Improvements in the design of the test bench itself.

The validation of the re-designed test bench in the pump production process resulted in an acceptable adjustment of the component reject rate compared to customer specifications.

2- INTRODUCTION

"End Of Line" performance testing of components in automotive engineering has become, for the characterisation of noises, a major concern, because of its impact on the quality of the end product.

Choosing a proper physical measurement parameter is relatively difficult, because of the severity and variability of the requirements in the manufacturing environment. These include ever-increasing automation and productivity goals.

In addition, the different steering pump noise generating mechanisms and their corresponding transmission paths also contribute to the complexity of identifying appropriate ways to evaluate concurrent test parameters.

Defining the specific measurement parameters should be representative of the major source contribution of the noise.

Further, the influence of the disturbances in the measurement results should be kept to a minimum and not vary with time to insure the proper evaluation of manufacturing quality.

Four test benches were dedicated to the **"End Of Line Performance Testing"**, in one selected manufacturing process line for the power steering pumps. Flow and pressure measurement methods were investigated.

One of these four test benches was assigned to perform vibration testing to characterise the steering pump noise performance per a customer request.

The evaluation parameter selected was the acceleration level measured, in the axial direction of the driver shaft, at the hard mounted steering pump bracket position.

The requested method for the characterising noises in the vane-type pumps was the use of **Order Tracking,** as the pump speed increased from 900 rpm to 4000 rpm.

The fundamental vibration frequency and its harmonics of the order 10 that corresponded to the vane passing frequency were order-cut to evaluate a metric criteria for acceptable vibration level with respect to the perceived in-vehicle noise.

3- TEST BENCH DESCRIPTION AND EXPERIMENTAL SET-UP

Supplying adequate hydraulic fluid to these test benches is insured by two main supply pumps. Both of these were an internal-gear type, having both a rotor drive shaft with nine teeth and an idler-gear with seven teeth.

The rotational speed of the main hydraulic-fluid supply pumps was approximately 1450 RPM, mainly because of the use of asynchronous motors, with two pairs of poles.

The gap between the outside diameter of the idler and the inside diameter of the rotor was sealed by the **Crescent** for delivering each a flow rate of 190 litters per minute. This would have generated an average outlet pressure of 3.5 bar at the total operational load.

The test bench being investigated was approximately 25 meters from the main supply pumps. This would have resulted in a fluid-supply pressure, at a full load condition, below 3 bar. This was caused by a pressure drop over the hydraulic circuit and by the distance between test bench and main supply pumps.

4- PRELIMINARY INVESTIGATION AND ANALYSIS

4.1- Evaluation of Test Bench Adequacy

Using vibration measurements on an initial configuration of the test bench, to comply to the customer specifications, a poor correlation of results to the subjective in-vehicle evaluations performed by the customer has been observed.

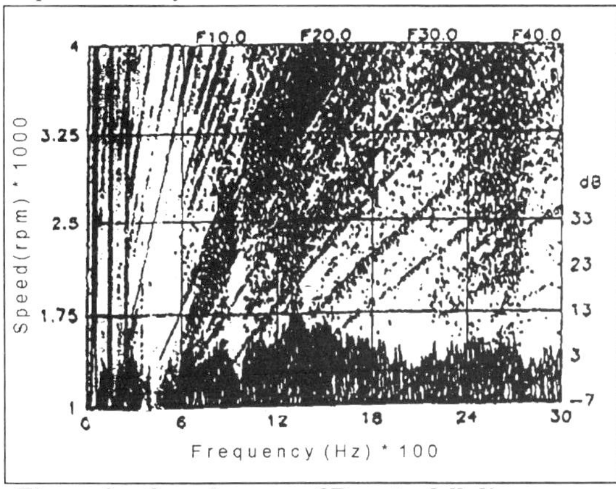

Figure 2a: Speed run-up 3D waterfall diagram of the acceleration measurement for an in-vehicle subjectively rated "quiet" steering pump.

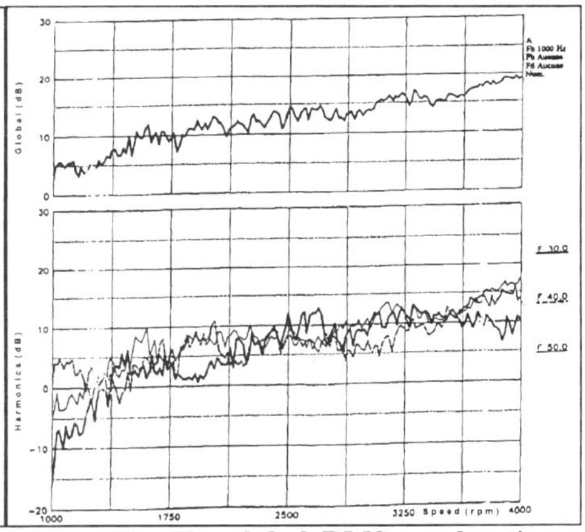

Figure 2b: The global RMS acceleration level in dB (upper plot) and the order-cut of the 3rd, the 4th and 5th vane passing frequencies (lower plots) from Figure 2a.

Acceleration measurement results for two reference pumps, subjectively rated on an application platform as a "quiet" pump and a borderline "less quiet" pump, are shown in Figure 2a/2b and 3a/3b, respectively.

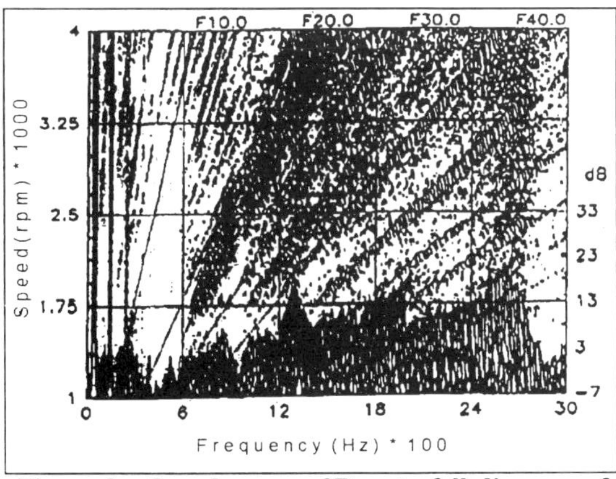

Figure 3a: Speed run-up 3D waterfall diagram of the acceleration measurement for an in-vehicle subjectively rated "less quiet" steering pump.

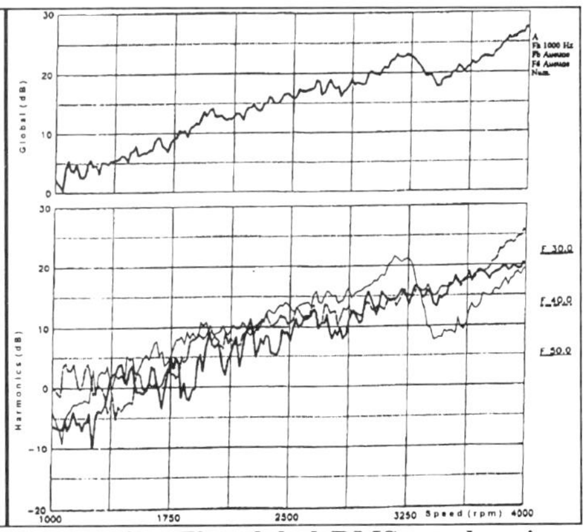

Figure 3b: The global RMS acceleration level in dB (upper plot) and the order-cut of the 3rd, the 4th and 5th vane passing frequencies (lower plots) from Figure 3a.

The results show a very small difference in the vibration levels ("noise") between a "quiet" and a "less quiet" pump. The maximum difference is less than 5 dB, as shown in Figure 2b and 3b, exception made for a behaviour observed around 3250 RPM for 50th Order.

On the 3D diagrams, for quiet as well as less-quiet pumps, a response to broad band excitation is observed around 2708 Hz as background level. This response to background level excitations is amplified by the discrete 50th order components around 3250 RPM for higher excitation level generated by less-quiet pump.

The influence of the background level can also be observed on the 3D Waterfall diagrams, shown in Figure 2a and 3a. Even though the harmonic orders of 30, 40, and 50 were higher for the less quiet pump, they hardly emerged from the background level.

The importance of this background level does not allow reliable evaluation of the vane passing harmonics order-tracking.

Furthermore, because of this inconsistency, previous reports claimed that not only were the production reject rates too high, but the reject rate also increased over time.

4.2- Identification of the sources of disturbances

With the above conclusions, a matrix of experiments were defined to identify and evaluate the main contributors to the vibration disturbances, measured at the test pump.

Some selected results of these experiments have helped to identify that the test bench hydraulic pump was a major source of disturbances during power steering pump testing.

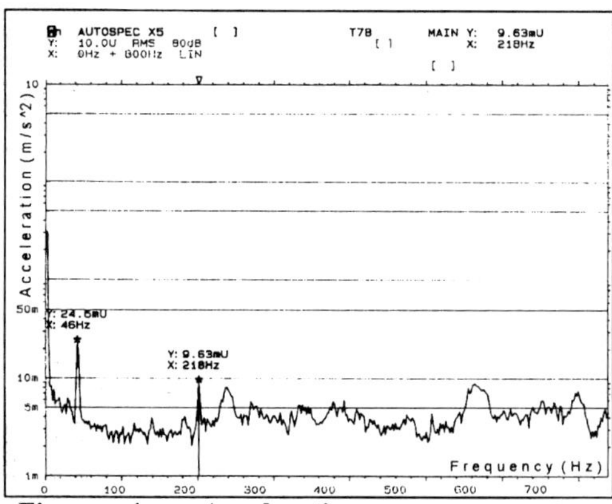

Figure 4 : Acceleration measurements when the test bench is in stand-by and the bench hydraulic pump is turned off.

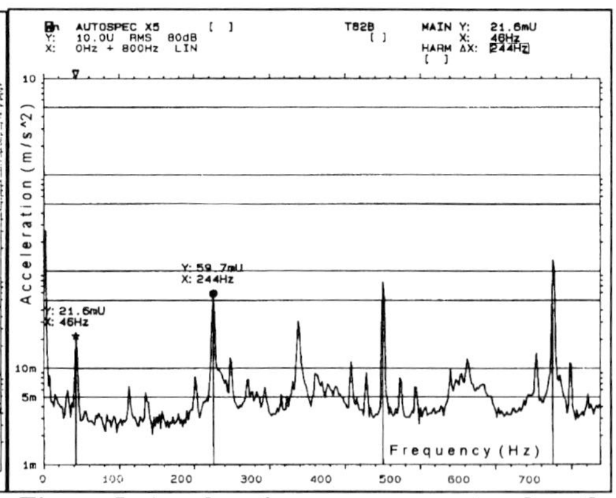

Figure 5: Acceleration measurements when the test bench is in stand-by with the bench hydraulic pump functioning.

Figure 4, for instance, shows the reference measurement, when the test bench hydraulic pump is not functioning and the pump being tested is in a "stand-by" condition at 50 rpm.

Two vibration components can be identified; with one at 46 Hz, and another at 218 Hz, with relatively low level of vibration (approximately less than 3 mg and 1 mg respectively) hardly emerged from the background wide-band level. These components have been identified as related to the driven motor and the main supply pump of the test bench.

Comparing data in Figure 5 to that in Figure 4, shows the predominance of a fundamental components at 244 Hz and all the corresponding harmonics modulated by 24.4 Hz.

Measurement results for the bench hydraulic pump as well as for the hydraulic valve assembly, which was rigidly mounted to the test bench chassis in the same configuration used for Figure 5 data, are shown in Figure 6a and 6b.

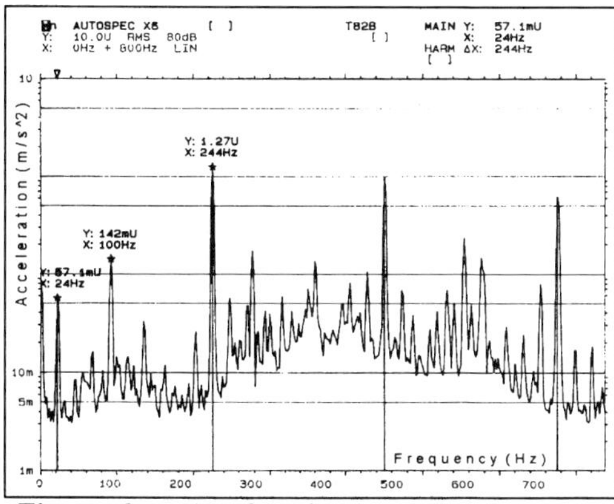

Figure 6a : Acceleration measurement at the test bench hydraulic pump in the same configuration as in Figure 5.

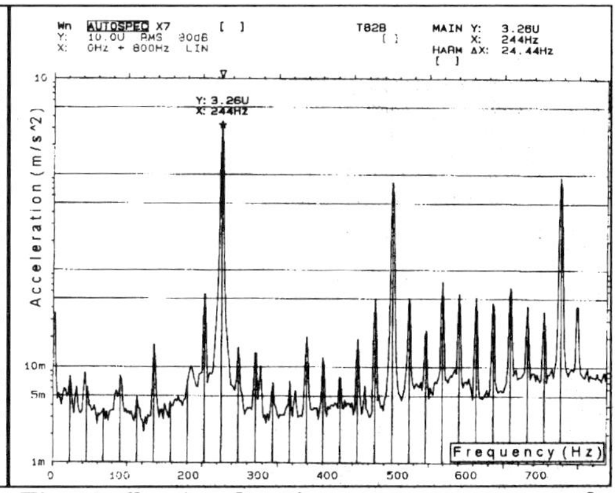

Figure 6b: Acceleration measurement at the test bench hydraulic valve assembly using same test configuration as used to generate data for Figure 5 and 6a.

From Figure 6b, the impulsive vibration characteristics of the test bench hydraulic pump can be observed by the comb of frequency components at the 24.4 Hz relative to the test bench hydraulic pump speed.

A component can also be observed at 100 Hz, corresponding to twice European line frequency.

In addition, the amplitude of the gear-meshing of the test bench hydraulic pump at 244 Hz increases at the valve assembly position due to the response of the system.

The frequency response function of the test bench frame column has shown a structural response at approximately 210 Hz.

5- INTERPRETATIONS

The above investigations and analysis have shown that the test bench, in its initial configuration, is not adequate for the characterising power steering pump noises using vibration measurements and the corresponding order-tracking signal processing associated with them. This is due to the following:

- The excessive level of vibration spectral components generated by the test bench hydraulic pump;
- The vibration disturbances generated by the hydraulic circuit and the valve assemblies; and,
- The vibration spectral components of the steering pump match those generated by the test bench. This is caused by the application of the test specifications.

6- FAULT DIAGNOSTICS

The previous mentioned vibration sources partially explain the non-appropriate measurement conditions for assessment of a "quiet" pump versus a "less quiet" pump.

However, this identification of the source of disturbances does not adequately explain either the excessive level of vibrations generated or the increase in the reject rate caused by an increased level of these disturbances.

Using the vibration diagnostics approach to the system behaviour for the test bench hydraulic pump, the following fault generating mechanisms have been identified.

The existence of the 100 Hz component (twice the European line frequency) is related to the stationary electro-magnetic fault of the asynchronous pump motor.

The impulsive vibration characteristics of the pump speed generating a comb of components at 24.4 Hz and a side bands around the gear meshing component at 244 Hz, including its harmonics, are related to "localised" faults on the gear teeth.

Further investigations were conducted on measuring the dynamic inlet and outlet pressure on both the original test bench pump and a replacement pump. The results of these are shown in Figure 7a/7b (outlet pressure comparisons) and in Figure 8a/8b (inlet pressure comparisons).

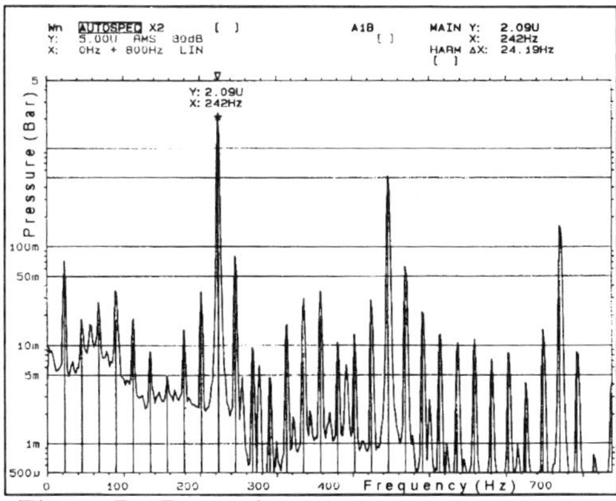

Figure 7a: Dynamic pressure measurement at the outlet of the original gear pump.

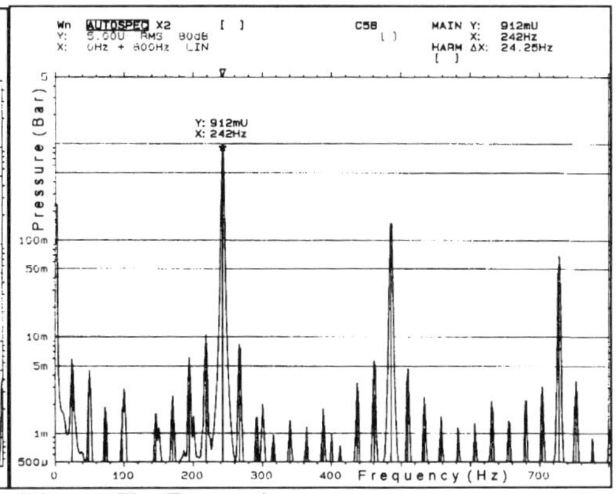

Figure 7b: Dynamic pressure measurement at the outlet of the replacement gear pump.

In comparing the data provided in Figures 7a and 7b, a reduced level of the gear-meshing spectral frequency components (Approximately 50 % less), of the pressure ripple measurement at the pump outlet, is seen when the original test bench gear pump is replaced.

Further, the comparison of data shown in Figure 8a and 8b, corresponding to the pressure ripple measurements at the pump inlet, shows an increased level of gear meshing spectral components, when replacing the test bench gear pump.

It is also shown that the harmonics and side-bands around the meshing frequency, caused by the rotational components at 24.4, were substantially reduced when the original gear pump was replaced.

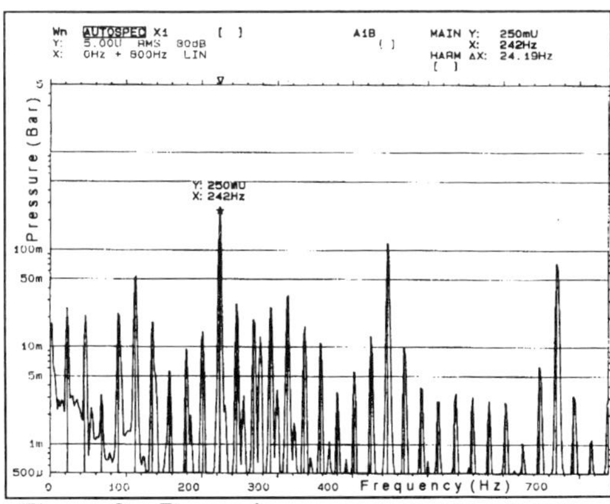

Figure 8a: Dynamic pressure measurement at the inlet of the original gear pump.

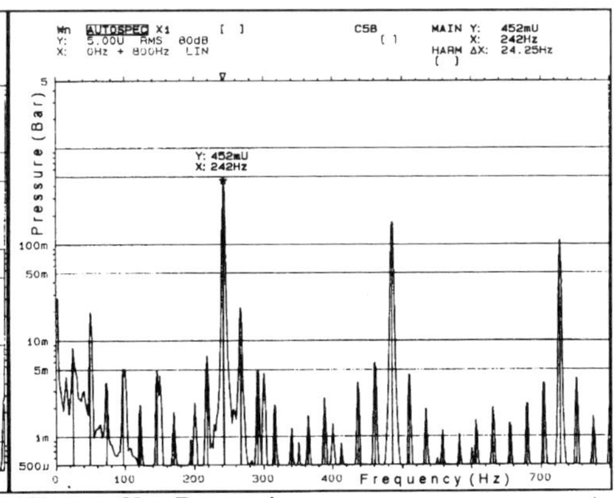

Figure 8b: Dynamic pressure measurement at the inlet of the replacement gear pump.

The excessive level of the gear meshing components as well as the inter harmonics, has indicated the presence of an excessive load at the inlet of the pump. This load exceeded the supplier's specification requirements.

We also observed that the gear meshing pressure component of the pressure ripple increases at the inlet of the pump and decreases at the outlet of the pump. Consequently, the modulation index, shown by the extent of the side bands, decreases as well assuming no change in the form of the modulation signal.

In the visual inspection of the gear assembly itself, some "localised" worn teeth as depicted in Figure 9 were found.

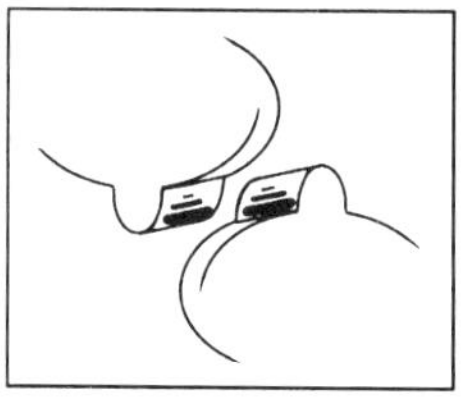

Figure 9 : Illustration of the gear mesh localised fault.

Based on the preceding results and evidence of damaged teeth on the original (used) pump, it was evident that the gear pump was subjected to an excessive inlet pressure load, which would have readily caused non-stable pump performance.

This condition generated an excessive static air-gap in the pump motor, which was indicated by the 100 Hz components.

The valve assemblies are subjected to an excessive pressure pulsation generated by the test bench gear pump anomalies (worn-out teeth and/or excessive inlet pressure loads). In addition, their hard-mount connection to the test bench produces a high level of vibrations through a transmission path to the power steering pump being tested.

As a result, further experiments were performed using acceleration measurements at both the hydraulic pump position and the test bench frame column position. These studies made measurements with both hard-mounted valve assemblies and with valve assemblies insulated from the vibration sources using rubber bushings. The results are shown in Figure 10a and 10b (hard-mounted assemblies) and in Figures 11a and 11b (bushing-insulated assemblies).

Comparing the various data in two views of Figure 10 and Figure 11, the level of the noise transmissibility ratio at the gear meshing frequency was reduced (improved) by a factor of four, when using insulator bushings for the valve assembly configuration on the test bench hydraulic pump.

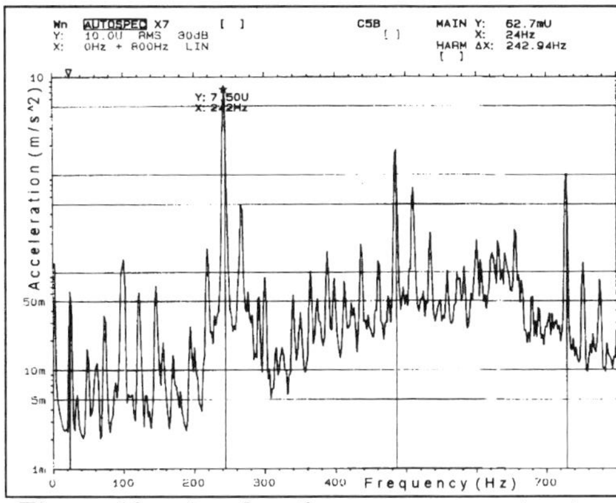

Figure 10a: Acceleration measurements at the hydraulic pump position with the hard-mounted valve assembly.

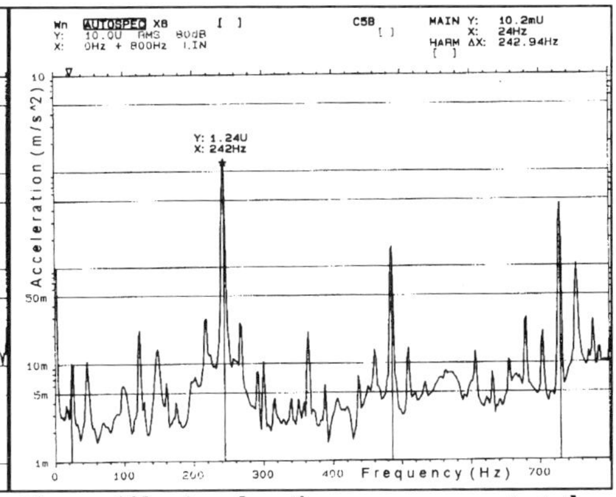

Figure 10b: Acceleration measurement at the chassis column position by the hard-mounted valve assembly.

This vibration transmissibility ratio represents the response measured at the test bench, close to the valve assembly (Figure 10b and 11b), versus the excitation level measured at the hydraulic pump position (Figure 10a and 11a).

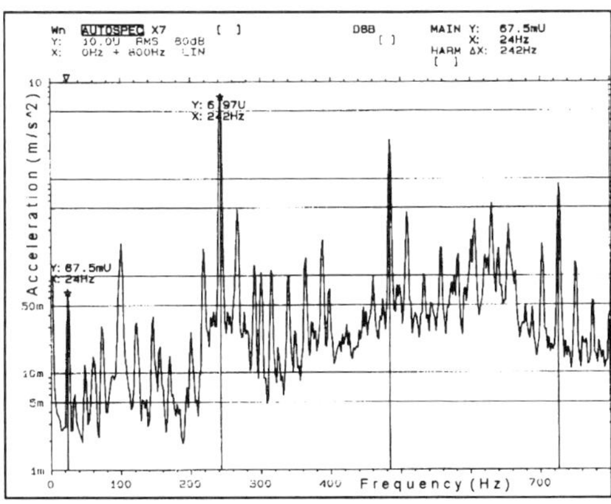

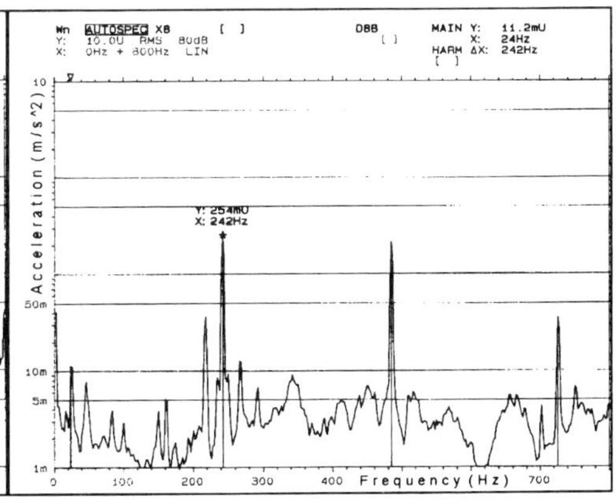

Figure 11a: Acceleration measurement at the hydraulic pump position with the valve assembly configuration insulated using rubber bushings.

Figure 11b: Acceleration measurement at the chassis column position with the valve assembly configuration insulated using rubber bushings.

7- REMEDIAL SOLUTIONS & CORRECTIVE ACTIONS

Based on the findings of this work, the following corrective actions have been implemented:

1- Disconnection of the hydraulic systems (valves, lines, and so on) from the test bench:
   - A new design sub-frame, dedicated to all the hydraulic systems as well as to the test bench pump;
   - The new sub-frame was insulated from the machine foundation with the vibration insulation mountings;
   - Flexible hoses were used to connect the sub-frame to the main test bench. As such, the hydraulic circuits coming from and going to the main test bench were structurally de-coupled.

2- Replacement of a the test bench hydraulic pump.

3- Introduction of a pressure reducing valve at the inlet of the test bench pump for controlling the static inlet pressure to 2 bar.

4- Installation of a hydraulic de-coupling system (i.e.: tuner) at the inlet and outlet of the test bench pump.

By implementing these corrective actions, a reasonable improvement was achieved in the vibration performance and sensitivity/reliability of the test bench. The on-going vibration measurements then showed a higher degree of accuracy in the evaluation and validation of the power steering pumps.

8- RECEPTION TESTING AND VALIDATION

The final validation of the test bench, with all the corrective actions implemented, was performed on the two referenced pumps and the related results are shown in the two views each for Figure 12 and 13.

By comparing the data in Figure 12a with that shown previously in Figure 2a, the following were noted:

- The acceleration levels shown in Figure 12a are low, compared to those achieved with the original test bench configuration (refer to Figure 2a); and,
- Consequently, in the Waterfall diagram, the related principal harmonic frequency components emerge from the background levels.

Such results will enable definite, accurate detection of the amplitude level of vibration measured on the steering pumps.

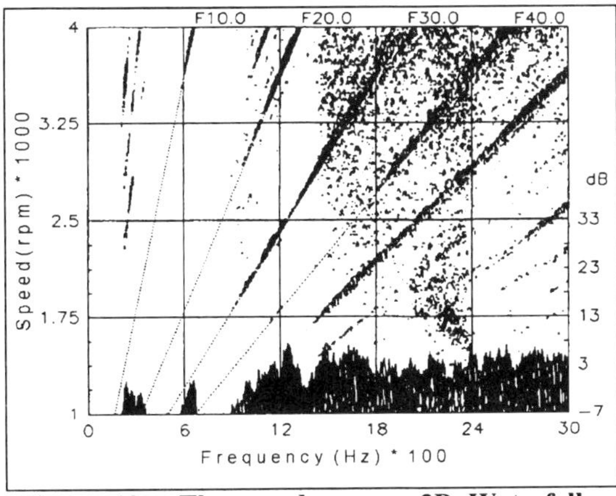

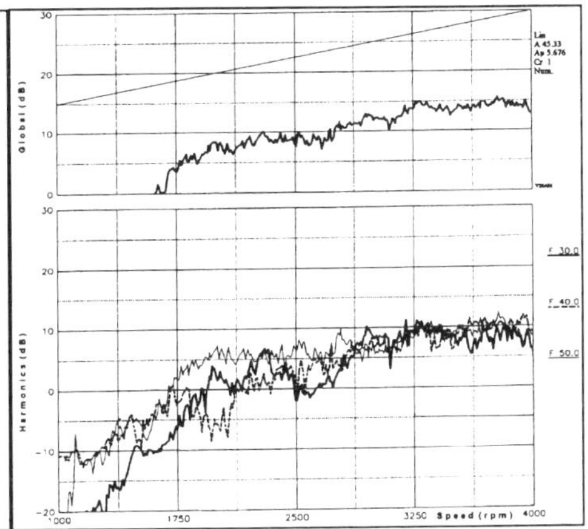

Figure 12a: The speed run-up 3D Waterfall diagram of the same "quiet" steering pump used in Figure 2a after implementation of various test bench corrective actions.

Figure 12b: The global RMS acceleration level, as well as the order-cuts obtained from Figure 12a and to be compared to Figure 2b.

The global acceleration levels and the order-cuts, shown in Figure 12b and previously in Figure 2b, indicate that the measurement levels with the original test bench configuration were not representative for the tested pump over a 5 dB level.

Comparing the speed run-up 3d Waterfall diagrams for the "less quiet" pump configuration - shown for the new test bench in Figure 13a, and for the original configuration in Figure 3a - also indicates a high improvement of the background levels.

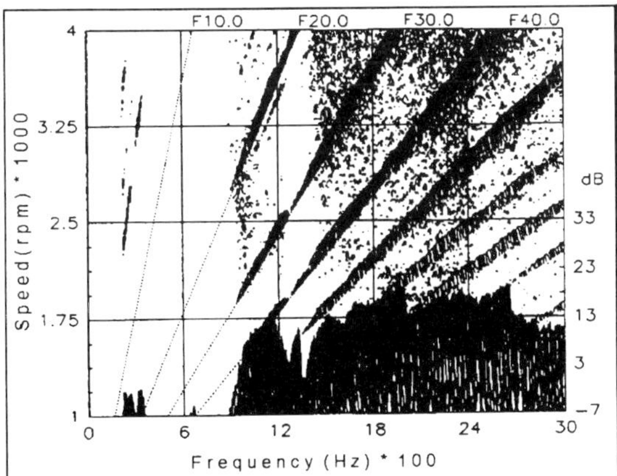

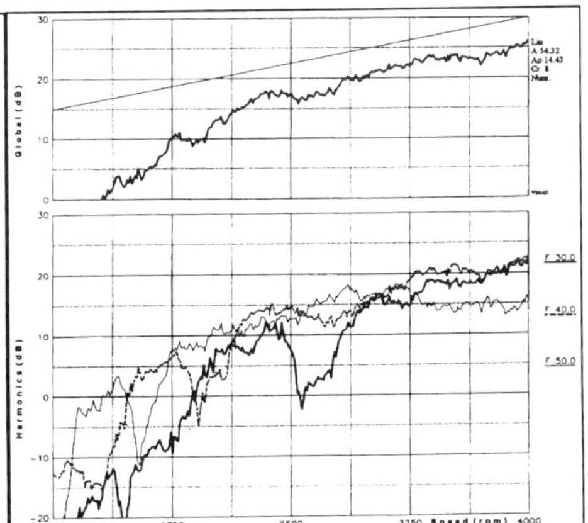

Figure 13a: The speed run-up 3D Waterfall diagram of the same "less quiet" steering-pump used in Figure 3a after implementation of various test bench corrective actions.

Figure 13b: The global RMS acceleration level, as well as the order-cuts obtained from figure 13a and to be compared to Figure 3b.

However, the evaluation of the order-cut results, relative to those in Figure 13b versus those in Figure 3b, does not show a significant difference in the level of vibration amplitudes, especially in higher speed conditions such as above 2000 rpm. This is due to the relatively high level of vibration, intrinsic to the less-quiet steering pump.

9- CONCLUSIONS

9.1- Results and Achievements

The gear pump used on the test bench was subjected to an excessive high inlet pressure load. This caused a high level of vibration spectral components and a deterioration of the gear teeth of the pump over the long term range. A pressure reduced valve was implemented as a result for controlling the static inlet pressure to 2 bar as a result.

Insulating the inlet pipe hydraulic lines from the vibration was implemented as well.

The vibration levels measured next to the valve assembly at its test bench mounting was reduced to an optimum level of insulation (>= 75 %). This was mainly achieved by insulating the valve assembly from the main test bench.

The approach to the problem solution was based on the following action plan:

- Identification of the test bench inadequacy;
- Localisation of the sources of vibration disturbances;
- Diagnosis of the fault generating mechanisms and then implementation of various corrective actions;
- Validation of the procedure with the customer and then re-adjustment of the acceptable reject rate for the production manufacturing lines producing power steering pumps.

The results of this action plan had the following impacts on our continuing business:

- Maintain business levels with major customers;
- Well-assured image of our technological capabilities;
- Proved technical competence to the customer and the world-wide organisation,
- Standardisation of world-wide testing procedures and published reliable, alternative diagnostic procedure.

9.2- Lessons

Even though the repeatability and reproducibility of a specific measurement system and configuration can be very high, the measurement system should consider the capability of the method to distinguish between a "good" and a "less good" performance of the product. This process is well known as ISOPLOT™.

In order to localise the sources of perturbations a matrix of experiments should be defined by using various test configurations, such as the Design Of Experiments (DOE^{TM}).

The diagnostic approach requires a good knowledge of the system behaviour under investigations and expertise in various engineering disciplines (i.e.: structural mechanics, fluid dynamics, pneumatics, electro-mechanics, advanced signal processing, etc.)

In the characterising vibration of the rotating equipment, the manufacturing engineering should take into consideration the specific requirements for vibration measurement in order to obtain consistent results.

9.3- Perspectives

In addition to the noise and vibration requirements for test bench adequacy, the application of an alternative signal processing method using vibration measurement would improve the standardisation of the results obtained on different testing platforms as well as provide higher levels of confidence.

This specific signal processing method - based on the Cepstrum analysis - is in process of development, validation and correlation to in-vehicle measurements.

10- REFERENCES

The work presented in this paper has been extracted from a set of internal technical reports by authors, who performed these development activities for DELPHI Automotive Systems, at Technical Centre in Paris, France.

11- DEFINITIONS, ACRONYMS, ABBREVIATIONS

Order tracking : The signal processing method by which the measurement parameter is re-scaled with respect to the rotational component in order to extract a given integer harmonics of the signal.

Waterfall : Three dimensional representation of spectral processing where the three axis represent amplitude level, frequency (Hz), and time (in this application the third axis represents the speed increase).

Modulation index : The percentage of amplitude modulation or phase modulation of the carrier signal by the modulating signal. The amplitude and the extent of the side band components around a carrier frequency depends not only to the form of the modulating signal but also the modulation index.

DOE^{TM} : Systematic statistical approach in the definition of experimental campaign in order to identify the major contributors in the total variation levels.

Isoplot™ : A statistical method which quantifies the ratio of measurement variation with respect to the product dispersion.

Crescent: Adjuster part in the internal gear pump to generate sealing between rotor and idler.

Stationary electro-magnetic fault: Induced magneto motive flux, generated by twice the line frequency. This category of fault is originated by the stator, in particularly in the case of the induction motor, which could be related to the inadequate winding spiral of the stator, or electrical insulation of the spiral adjacent turns or also excessive static air-gap

Static Air-gap in electric motor: the static air-gap is defined as the eccentricity of the rotor versus the stator of the motor.

C521/008/98

Development of hydro-elastic strut mountings for vehicle secondary ride enhancement: from concept to prototype evaluation

M A SHAW* BEng, AMIMechE and **J DARLING** BSc, PhD, MIMechE
University of Bath, UK

Synopsis

Hydroelastic mounts are becoming commonplace in powertrain mounting systems, but their characteristics make them suitable for solving other vibration problems in vehicles. This paper describes the development of a hydroelastic mount that improves secondary ride when used as a strut mount on suspensions with telescopic dampers. The performance limitations of conventional bushes are discussed, leading to an explanation of the rationale behind the hydroelastic strut mount. A ride model is described, including hydroelastic mounts, with ride predictions for suspensions using conventional and hydroelastic bushes. Finally, the results of these trials are presented, demonstrating an improvement in the ride quality of the vehicle.

1 Introduction

1.1 The conventional suspension system

A conventional suspension system consists of a spring, usually a steel coil or leaf, and a damper unit in which a fluid is forced through small orifices to provide a damping action. The design and construction of dampers vary depending upon the application, and their characteristics can be tuned by the manufacturer, however all dampers exhibit an increasing dynamic stiffness with frequency. In order to produce acceptable noise and harshness performance, it is common practice to fit elastomer bushes in series with the damper unit, thus limiting the maximum stiffness attained by the suspension. These bushes progressively reduce the effectiveness of the damper with increasing frequency of excitation.

Figure 1 shows the effect that varying the stiffness of the damper bushes has on the ride performance of a vehicle. Stiff bushes which force the damper unit to deflect are advantageous for primary ride performance, maximising the effect of the damper and controlling the body resonance, but increase the transmission of vibration over the rest of the

*Also at Avon VMS, Chippenham, UK.

frequency range including secondary ride and noise regions, producing higher levels of vibration and noise. Soft bushes tend to deflect themselves, in preference to the damper unit and especially at high frequency, providing improved isolation at the expense of primary ride control. This compromise is an inherent limitation of the conventional suspension.

In addition to the problem of finding suitable damper bush rates, the ride engineer must also consider the damper characteristic and its effect (ref. 1). The critical damping required for the two major modes of vibration, body bounce and wheel hop are, for a typical vehicle, 2300Ns/m and 4200Ns/m respectively. As a first approximation, a road profile may be considered to be a constant velocity spectrum, thus for a particular road, the damper operates at a similar point on its force - velocity characteristic for all frequencies of excitation. Under these conditions it is impossible to damp both modes equally; either the wheelhop mode will be poorly damped, or the body mode excessively damped, or both. The effect of the damper bushes, by reducing the effect of the damper with increasing frequency, makes this shortcoming worse by further reducing the damping for the wheelhop mode.

1.2 The requirements for an enhanced system

What becomes apparent from figure 1 is that the ideal suspension would be relatively soft, with heavy damping around the resonant frequencies only, thus providing control of the body and wheel modes, but isolating the higher frequencies and the important secondary ride region between the main modes, in the frequency range 6 - 12 Hz. With the conventional system the first of these requirements, control of the body mode, can be achieved satisfactorily, but increasing the damping to suppress the wheel hop mode introduces a trade - off between isolation and control of the wheel.

2 The Hydramount

2.1 A Typical Configuration

The hydramount is a hydroelastic device which is currently used predominantly for powertrain mounting applications. A mount, shown in figure 2, normally consists of a main spring that contributes the static stiffness and contains the working fluid, a valve plate, which together with the main spring forms a working chamber, and a bellows which contains fluid within a compensation chamber. In addition, the valve plate contains a diaphragm, used to modify the volumetric compliance of the working chamber, and a channel connecting the working chamber to the compensation chamber.

2.2 Operation of the hydramount

In operation, deflection of the main spring causes a change in the volume of the working chamber. At low frequency, fluid passes through the channel to compensate for the difference, and the stiffness of the mount remains low. As the frequency of the excitation is increased, the mass of fluid in the channel and the volumetric compliance of the working chamber form a resonant system, generating large pressure fluctuations in the working chamber as the resonance frequency is approached. The pressure generated acts upon the inside of the main spring, producing a damping force. At frequencies above resonance fluid motion through the channel subsides, the change in the working chamber volume is accommodated by the volumetric compliance and an increased dynamic stiffness is observed.

2.3 Hydramount characteristics

Figures 3a and 3b show a typical hydramount characteristic. The static stiffness in this case is 200 N/mm, this value rising to a maximum, then settling to a high frequency stiffness around 600 N/mm as the excitation frequency is increased. Note that the damping, measured as a loss angle, can be more than 90 degrees and is confined to a relatively small frequency band, in this case centred on 10 Hz. This combination of features is exactly that which was identified as desirable when considering figure 1; a controlled level of damping in a narrow, tuneable frequency range, and an upper limit on the stiffness achieved at high frequency. The characteristics are also amplitude rather than velocity dependent, offering further scope for tuning the system.

3 Vehicle modelling

A vehicle model was produced to investigate the replacement of the conventional damper bushes with a hydroelastic strut mounting. A quarter vehicle model was used, incorporating wheel and body only. The effect of powertrain motion was omitted to improve the clarity of the prediction, but it was also felt that the strut mount would be equally applicable to front and rear suspensions, and so the powertrain would not necessarily be present in all installations.

The wheel and body were modelled as point masses, the tyre as a lightly damped spring, and the main suspension as a constant rate spring. The damper was modelled as an idealised viscous damper with a nonlinear force - velocity profile, incorporating a bump / rebound bias and knee points. The damper bush was modelled as a lightly damped spring, the strut mount as a hydramount unit. The hydramount model was a simplified version of Avon's hydramount prediction software, incorporating the main non linear geometrical and fluid flow effects. Further details of hydramount modelling techniques are given in refs. 2 and 3. A time stepping routine was used with a Newton - Cotes integration scheme to model the response in the time domain, with a Fourier analysis to generate the frequency response plots.

A range of strut mount characteristics were investigated and the performance compared to determine an optimum mount condition. The intention was to preserve the primary ride and noise performance, and to maximise the improvement in the control of the wheel hop, thus reducing secondary ride vibration. Figure 4 compares the model prediction for the original condition with elastomer bushes with that of the hydroelastic strut mount equipped vehicle. The model predicts a 25% reduction in the wheel hop vibration transmitted to the body.

4 Vehicle trials

4.1 Prototype manufacture

The next stage of development was to manufacture prototype mounts for vehicle trials. The mounts were designed to fit an existing vehicle with minimal modifications, and initially to be fitted to the front suspension only. A range of mounts were produced with varying stiffness and damping characteristics, nominally the optimum condition predicted by the model with positive and negative variations.

4.2 Benchmarking and Initial Tests

Benchmark measurements were made with the standard vehicle for sinusoidal sweep rig tests and objective and subjective road tests. The front hub, damper rod, strut tower, floor and seat rail were instrumented for the initial trials, with the rear hub and rear corner of the sill being added during the investigation.

The first tests with hydroelastic mounts were performed exclusively on a four poster rig using constant velocity sinusoidal sweep inputs to gain an insight into the response of the vehicle with the strut mounts. These tests confirmed the model predictions to be reasonably accurate.

4.3 Optimisation - front suspension

The rig tests continued in parallel with objective and subjective road tests to optimise the system. For the road tests, a specific section of road was used, and tests were carried out at a range of speeds to assess the effect of wheel base filtering, due to relative phasing of the inputs to the front and rear wheels. Care was taken to ensure minimal change in the fuel load of the vehicle. The results of these investigations were promising, but also a little disappointing because the impact of the hydroelastic mounting was reduced. The effect of the changes to the strut mount diminished as the measurement location moved away from the front of the vehicle. This is not surprising, since there is a progressively increasing contribution from the standard rear suspension and a reduction in the contribution of the front suspension. As a consequence, the benefit as measured at the strut tower was considerable, but at the seat the effect was much smaller and difficult to detect subjectively.

4.4 Optimisation - front and rear suspension

In order to demonstrate the full potential for ride improvement, more prototypes were produced and fitted to the rear suspension in addition to the front. This configuration was also optimised through rig and road testing, and using the knowledge gained from the work on the front of the vehicle. The same testing procedures were followed, with additional accelerometers fitted to the rear of the vehicle.

5 Test results

5.1 Rig tests

Figure 5 shows a summary of the seat rail responses of the vehicle, measured on the 4 poster rig. Responses are shown for the standard benchmark condition, for a vehicle fitted with the optimised front hydroelastic strut mounts only and for a vehicle with optimised mounts front and rear. The front only condition shows some improvement across the whole secondary ride region, from 3 to 30 Hz, the greatest effect occurring around 17 Hz. This is consistent with the tuning of the mounting used and the model prediction. It was found that none of the hydroelastic mountings tested made the secondary ride performance worse than that of the standard vehicle, however the static stiffness of the bush had an effect on the primary ride, and the dynamic stiffness influenced the high frequency response, as predicted in figure 1.

The result for mounts fitted front and rear shows further gains in secondary ride performance, since the improvement is now present front and rear; the standard ride performance of the rear suspension is not diluting the improvements made at the front. A reduction in the vibration of one third is achieved across a broad range of frequencies from 15 to 25 Hz. The effect on primary ride is not so good, however, exhibiting a worsening of the body bounce mode. In fact this is not due to the damper bush, but can be explained by the modifications made to install hydroelastic mounts in the rear suspension. It was necessary to remove the rear bump stops to fit the mountings, and their absence allowed greater travel of the rear suspension.

5.2 Objective road tests

Figure 6 shows the road test results for the same optimised mounting conditions as figure 5. Once again seat rail response is used as a measure of the suspension performance. The benefit of the mounts is less marked in these tests, but an improvement over the benchmark condition can still be seen for both hydroelastic mounting configurations. The effects arc limited to the secondary ride region, above 15 Hz.

The main differences between the rig and road tests are the tyre condition, longitudinal input and input content. For the rig tests, the tyre is not rolling; the effect of sidewall stiffness and filtering due to the contact area are changed relative to the normal operating conditions on the road. The 4 poster rig applies vertical excitation only, there is no facility for replicating the longitudinal inputs experienced on the road. Finally, the rig test utilises a sinusoidal input at a single frequency and the response is allowed to settle, whilst the road excitation is almost exclusively random, producing predominantly transient response from the vehicle. With these considerations, the ability to identify similar trends in both rig and road tests is encouraging.

5.3 Subjective road tests

Subjective assessment of vehicle ride is considered to be an important technique for quantifying vehicle ride performance. The acid test for any proposed ride - improving product is that its effect be noticed by passengers in the vehicle. To this end a group of ride engineers made subjective assessments of each set of strut mounts fitted to the vehicle. Figure 7 shows a summary of the average ratings awarded to the same optimised configurations as presented in figures 5 and 6. The assessment panel were asked to rate the vehicle performance against a standard 1 to 10 scale for noise, primary and secondary ride and general vibration. The noise assessment was intended to ensure that the effects of high frequency stiffness were considered. This revealed a similar small increase in the noise experienced by passengers for both front only and front and rear installations. The primary and secondary ride characteristics were judged to have been progressively improved by the addition of front and rear hydroelastic mountings, and the general vibration, used to assess rattles, trim vibration etc. was also considered to be better.

The overall effect, then, was perceived to be improvement in the primary and secondary ride of the vehicle and a small reduction in the general vibration.

5.4 Model prediction comparison

Figure 8 shows a comparison of the model prediction and actual performance improvements measured in the vehicle tests. The improvement is presented as the vibration measured with hydroelastic mounts as a percentage of the vibration in the standard condition. The model prediction shows similar performance below 12 Hz, with a reduction in vibration between 12 and 20 Hz. This trend tallies with the results for the rig tests, the performance of the front only condition is very similar and the front and rear installation shows considerable improvement. This suggests that the mounts have been more effective than expected, though this may be attributable in part to the damping produced by the prototype being greater than predicted by the model. There is also some discrepancy around 10 Hz, since the model does not include the powertrain effects.

6 Conclusions

It is known that the conventional suspension system exhibits design conflicts which result in a compromise in the primary and secondary ride performance of passenger cars. However it is shown that it is possible to partially resolve these conflicts, increasing the wheel hop damping, by using a hydroelastic strut mounting in place of the conventional elastomer bush.

This development is capable of improving the vehicle ride when fitted to the front of a vehicle only, but maximum benefit is obtained with mountings fitted to all the wheel stations. The front and rear of the vehicle interact at the frequencies associated with secondary ride, hence the mountings for front and rear suspensions must be tuned as a system for maximum benefit.

The improvement is measurable both objectively and subjectively, the improvement achieved in this study amounting to approximately half a point for primary ride and one point in secondary ride, measured on a subjective scale of 1 to 10. This is a significant gain in performance.

Acknowledgements

The authors wish to thank the directors of Avon Vibration Management Systems for permission to publish this study.

References

1, Best, A. "Vehicle Ride - Stages in Comprehension", Physical Technology, vol. 15, 1984

2, Shaw, M. & Darling, J. "Hydroelastic mountings for Vehicle Vibration Control: Modelling the Nonlinear Characteristics", Land Transport Hydraulics Seminar, S499, University of Bath, 15th September 1997

3, Bosworth, R. & Fursdon, P.M.T. "A Frequency Domain Representation of Hydramounts For Ride Prediction", 5th International EAEC congress, 1995

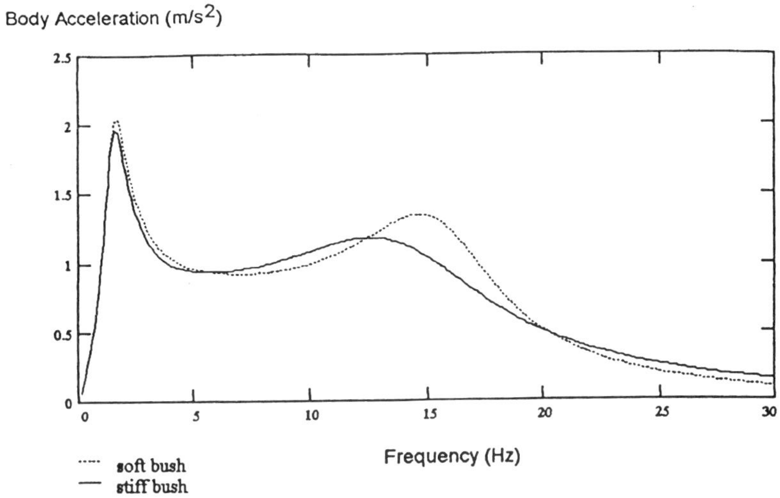

Fig. 1 - Effect of damper bush rate on transmissibility

Comparison of soft and stiff damping bush for 0.1 m/s input

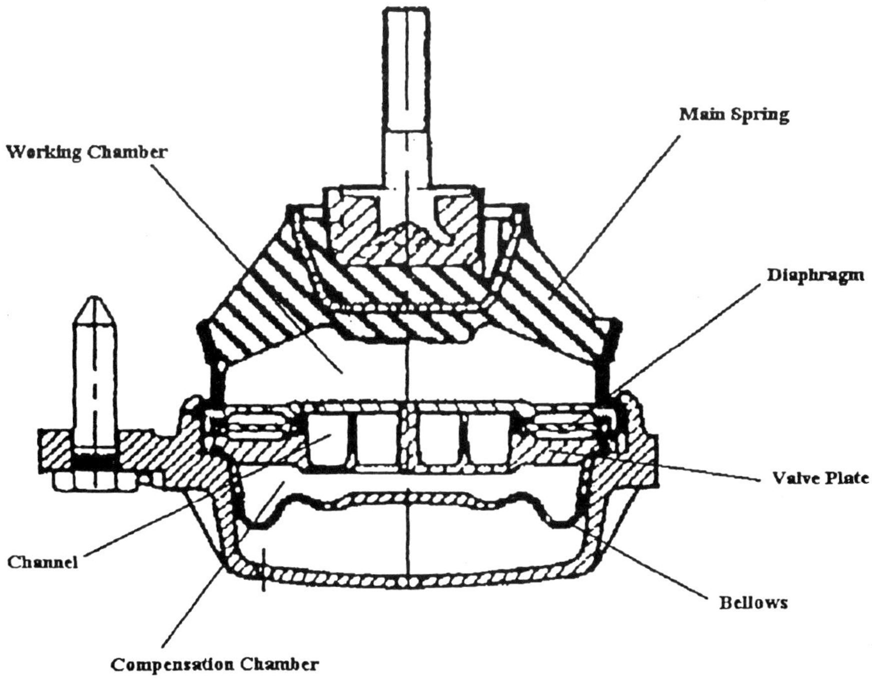

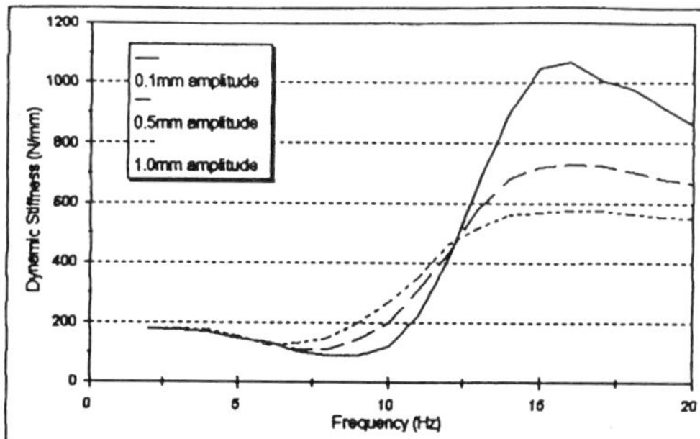

Fig. 3a Hydramount characteristic
Typical dynamic stiffness

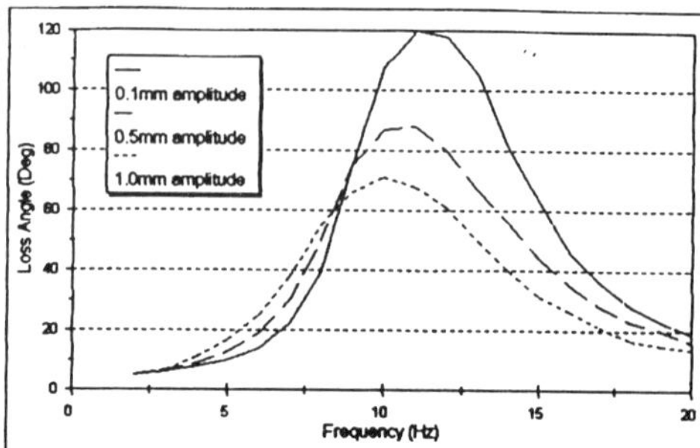

Fig. 3b Hydramount characteristic
Typical loss angle

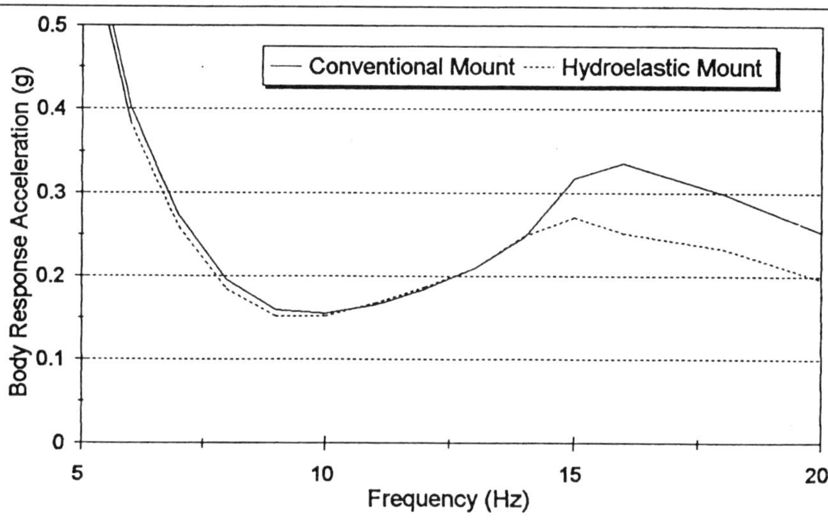

Fig. 4 Hydroelastic damper mount model
Comparison of secondary ride effects

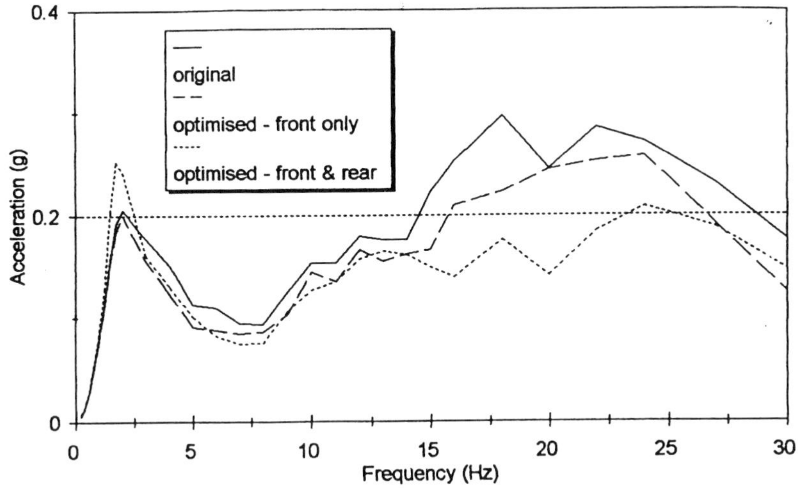

Fig. 5 Comparison of rig results
Optimised conditions - seat rail

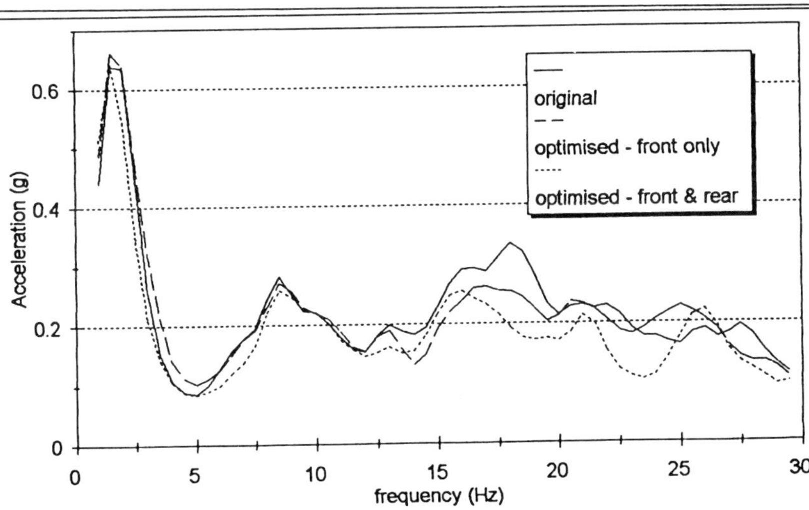

Fig. 6 Comparison of road results
Optimised conditions - seat rail

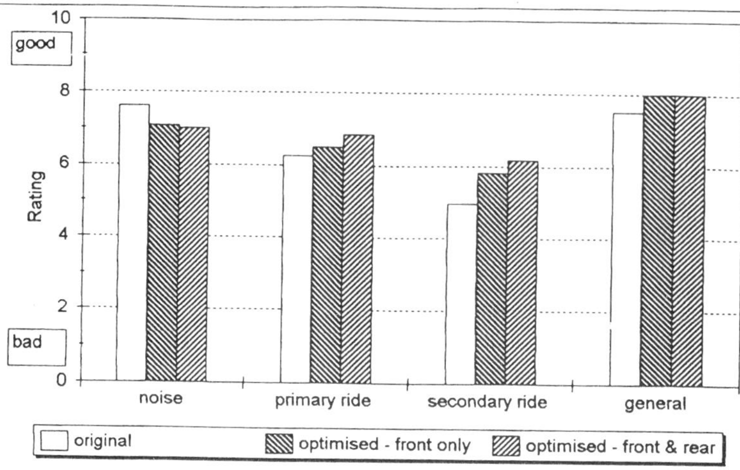

Fig. 7 Comparison of road results
Optimised conditions subjective rating

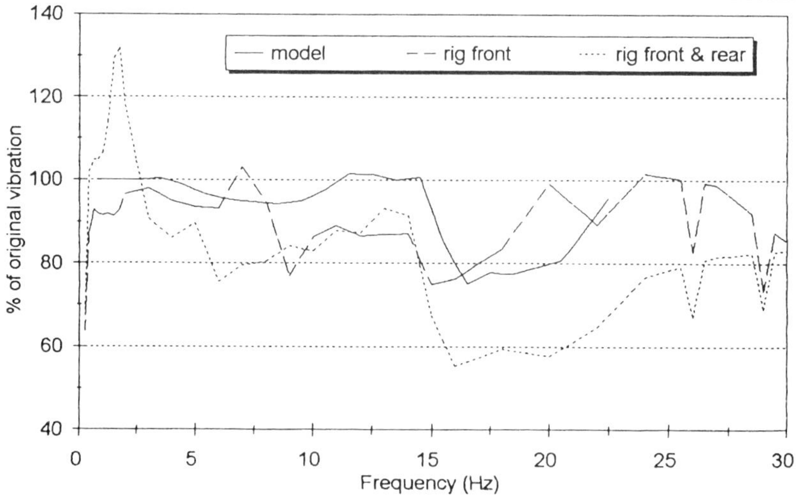

Fig. 8 Model prediction comparison
Vibration reduction relative to std.

Body and Intake/Exhaust

C521/003/98

Acoustic double baffle utilizing programmed expandable foam as body cavity seal

R J CHANG PhD, MSAE, ACS, MIEEE, **K DAWES** PhD, MSAE, ACS, SPE, **L SCHWARTZ**, and **R FICK**
Raychem Corporation, California, USA

SYNOPSIS

Single baffles based on heat activated foams have found increasing use as body cavity filler to inhibit noise propagation through hollow body channels of passenger cars. While the benefit of double decoupled layers in noise reduction is well known, no double baffle design was economically feasible previously, due to the limitations of conventional heat activated foams. A double baffle design based on preprogrammed heat activated foam technology is introduced. The preprogramming imparts directional control that enables more effective filling and makes double baffle design possible. The differences between preprogrammed and conventional foams are reviewed, the advantages of preprogrammed foams are presented, and the benefits of the double baffle over the single baffle design are demonstrated by acoustic testing.

1. INTRODUCTION

In modern unibody car construction, the body is built by hollow steel framework. Unless the hollow channels are effectively blocked, they provide a natural path for water or fume transport and noise propagation. With the increasing market demand for quieter cars[1], most car manufacturers have adopted various channel sealers at key locations such as engine rails, pillars, rockers, and body hinges, to block the noise propagation. Fig. 1 depicts such applications.

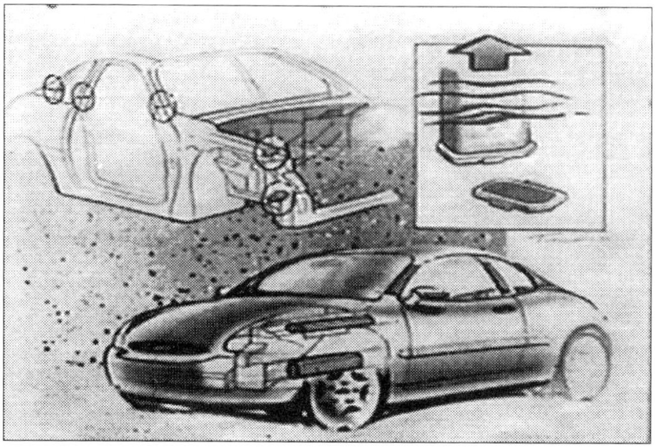

Fig. 1 Body Cavity Seals for Passenger Cars

The main channel sealers used today include stuffers, two-part polyurethane foams, and baffles[2,3,4,5]. The stuffers are bags of glass fiber or foam that are inserted into the body cavity after the painting processes and before the general trim operations. Likewise, the two-part polyurethane foam is metered and injected into the body cavity just before the general trim operation. The baffles, on the other hand, are heat activated foams that are attached to the body channels before the painting processes. As the car body goes through the E-coat bake oven, heat activates the foam to expand and fill the çavity.

The baffles provide many advantages over two-part polyurethane since they can be installed easily and require no capital investment of ventilation hoods and dispensing equipment to handle toxic materials. However, if not sandwiched by rigid plates such as nylon or metal, the current commercially available heat activated foams tend to sag and deform by heat before foaming and expand at random during foaming. The effectiveness of channel seal is therefore compromised. The sandwich designs increase the sealing reliability to a certain extent by constraining foam deformation and directing the foam expansion toward the channel walls. Still, some sharp pinch-welded corners of body channels are hard to access using conventional foams even with the sandwich designs. The sandwich designs also unnecessarily add to the part weight and costs since some of the nylon back plates have to be injection molded to closely match the channel cross-section.

A new preprogrammed heat activated foam has been introduced recently by Chang et al[6]. The same study[6] also demonstrated the sealing effectiveness of the preprogrammed foam by observing the dynamics of foam expansion in a model channel. The preprogramming provides its own directional control during expansion without complex back plates. The directional control of foaming drastically improves the sealing reliability by driving the foam toward sharp channel corners that are otherwise difficult to access.

In this paper, the key characteristics of preprogrammed foam, including isotropic expansion, uniform expansion over wider temperature range and directional control are first reviewed. Next, the use of preprogrammed foam in the design of double baffle is presented.

The benefit of the double decoupled layers in noise reduction is well known in the industry[7]. For example, double layer grommets have been widely used and demonstrated superior acoustic performance over the single layer counterparts. However, a double baffle design for body cavity seal has not been available commercially due to the limitation of the conventional foams. The conventional foams expand randomly, and it is difficult to keep two layers totally decoupled without using constraint plates on both layers. The complete decoupling of the two layers is essential for the superior acoustic performance. Although it is technically possible to sandwich both layers of conventional foams during foaming to keep them separated by adding constraint rigid plates on each layer, the solution, however, incurs significant cost disadvantages and makes it unattractive for the automotive application. The preprogrammed foam, due to its expansion characteristics, offers an ideal solution to design and makes cost effective double baffles. The design of double baffle based on programmed foams is presented and the acoustic performance is compared to the single baffle on the equal weight basis.

2. CHARACTERISTICS OF PROGRAMMED HEAT ACTIVATED FOAM

2.1 Isotropic Expansion

The difference between the programmed foam and the conventional foam is illustrated in Fig. 2. Starting from a 2.45 cm x 2.45 cm x 0.64 cm pre-expanded dimension, the conventional foam rises like a loaf of bread during foaming. The lateral expansion is minimum. In sealing a hollow body channel without constraint, the conventional foam will grow preferentially along the channel's axis without much lateral expansion to drive the foam toward the channel walls and corners. On the other hand, the programmed foam expands isotropically; it imparts the self-directional control to expand laterally, which will drive the foam more consistently toward the channel walls and corners.

Fig. 2 Comparison of Programmed vs. Conventional Foam Expansion at 175° C for 30 minutes.

2.2 Higher Lateral Expansion

The programmed foam also exhibited higher lateral expansion at all temperatures as shown in Fig 3. The data suggest that the programmed foam will drive material more effectively toward the channel walls and corners, will be more tolerant to the temperature variation of E-coat ovens, and will expand more at the low temperature of 145°C.

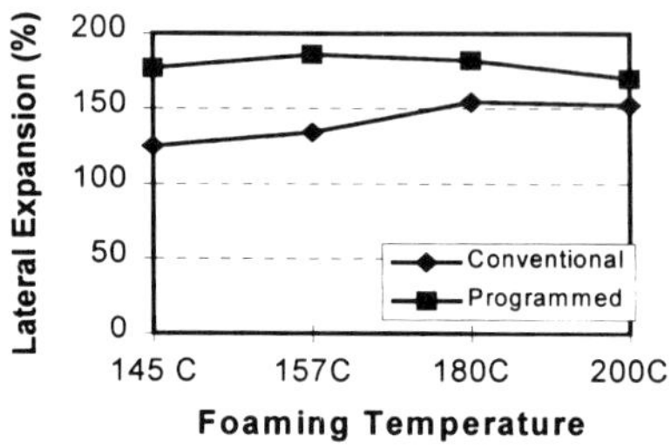

Fig. 3 Change of Lateral Expansion with Foaming Temperature

2.3 Uniform Expansion over Wider Oven Temperature

It is also highly desirable to have a foam that expands to the same degree over a wide temperature range. Since baffles are placed at different locations within the car body, the temperature that each baffle experiences in the E-coat bake oven varies widely, typically in the range of 145°C to 200°C. Most conventional foams expand to a different degree at different temperature. Also, some foams do not expand sufficiently at lower oven temperatures of 145° C. Therefore, the variation of oven temperature causes different degrees of foam expansion, which in turn, affects the consistency of channel seal. The programmed foams expand more uniformly over a wider temperature range to account for the variation of oven temperature. The comparison of expansion ratio of the conventional and programmed foams from 145°C to 200°C is shown in Fig. 4.

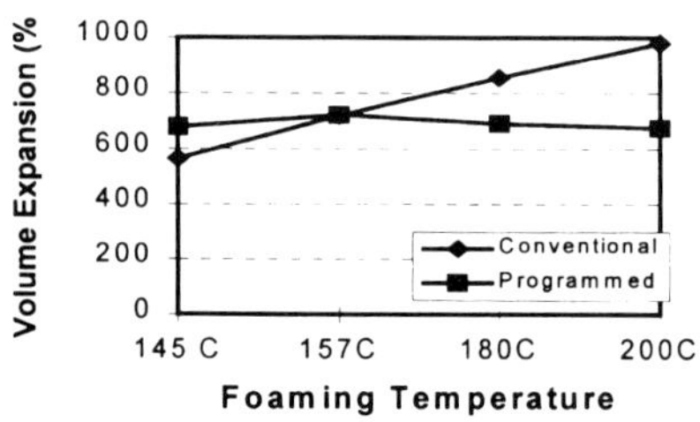

Fig. 4. Change of Volume Expansion with Foaming Temperature

2.4 Preprogrammed Expansion

Fig. 5 shows that the preprogramming can be used to control the degree and direction of foam expansion as well as to change the final foam shape as desired. This feature is particularly useful for the design of the baffle to seal hard-to reach corners.

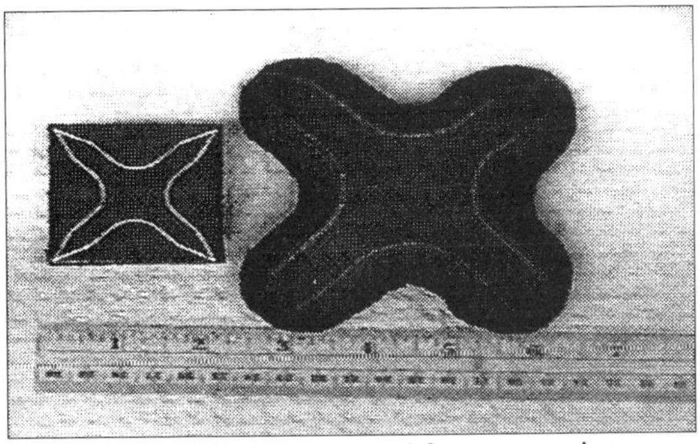

Fig. 5 Preprogrammed foam expansion

3. DOUBLE BAFFLE DESIGNS

A typical double baffle after foam expansion is shown in Fig. 6. It is noted that the preprogrammed foams can be conveniently fastened by mechanical means through the material. Also, two layers can be anchored by a single attachment. Two layers remained well decoupled after foaming.

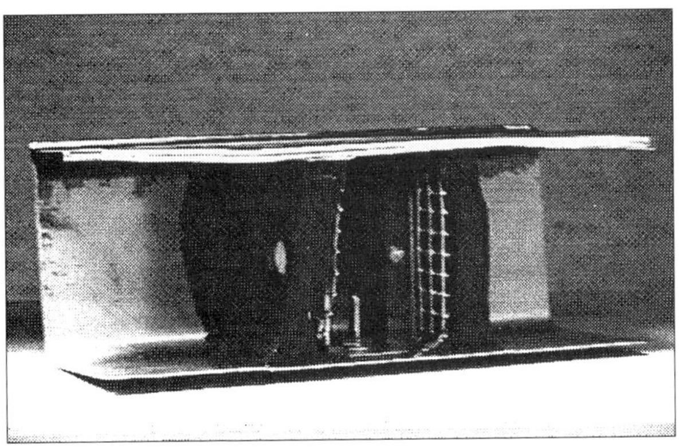

Fig. 6. Cross-section of a double baffle after foam expansion

4. ACOUSTIC MEASUREMENTS

Both the single and double baffle samples were prepared by using preprogrammed foam to seal a proposed SAE test channel as shown in Fig. 7. For single baffles, three samples with different pre-expanded thickness of 6.3 mm, 9.5 mm, and 12.7 mm were prepared and tested. The sample weights were 30, 45, and 60 grams, respectively. For double baffle, each layer had 6.3 mm pre-expanded thickness, and the two layers were separated by 25 mm. Thus, the total thickness and weight of the heat activated foam in the double baffle are comparable to the single baffle with the thickness of 12.7 mm.

The acoustic testing was conducted according to a test channel being reviewed for a SAE test method. The details of the test methods have been described[8,9]. The channel had a 75 mm x 75 mm cross-section and 250 mm length. The baffle was placed near the end of mounting bracket and expanded at 180°C for 30 minutes. The sample was secured, by the mounting brackets, on a wall that separates a reverberate room and a anechoic room. A noise source with frequency from 25 Hz to 10,000 Hz was generated in the reverberate room. A microphone is placed 100 mm behind the sample or the mounting bracket in the anechoic room to measure sound pressure level. The insertion loss(IL) is reported. IL is defined as the difference in sound pressure level between a test channel containing expanded preprogrammed foam with mounting bracket and a control channel containing mounting bracket only.

Fig. 7 Proposed SAE Test Channel

The insertion loss of the three single baffles is shown in Fig. 8. The results showed that the preprogrammed foams provided effective acoustic channel seal over the frequency of 300 to 10,000 Hz. The effectiveness was further increased by increasing foam thickness or weight.

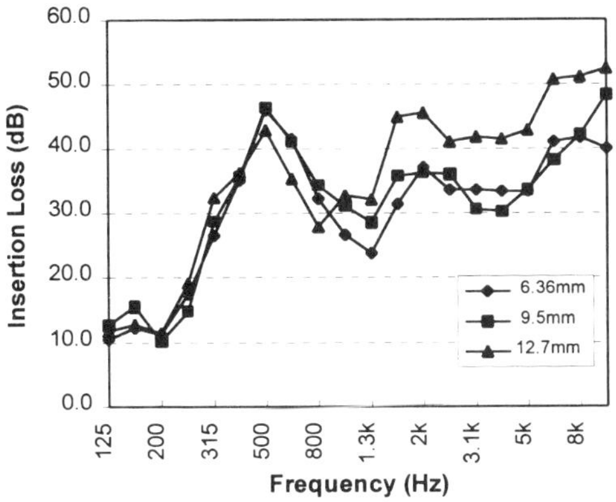

Fig. 8 Insertion Loss of Programmed Foams with Various Thickness

The comparison of single and double baffle under the same total foam thickness(and same part weight) is shown in Fig. 9. The double baffle design further improved the acoustic performance over the single baffle. Particularly, the frequency dependence from 500 Hz to 10,000 Hz was greatly reduced.

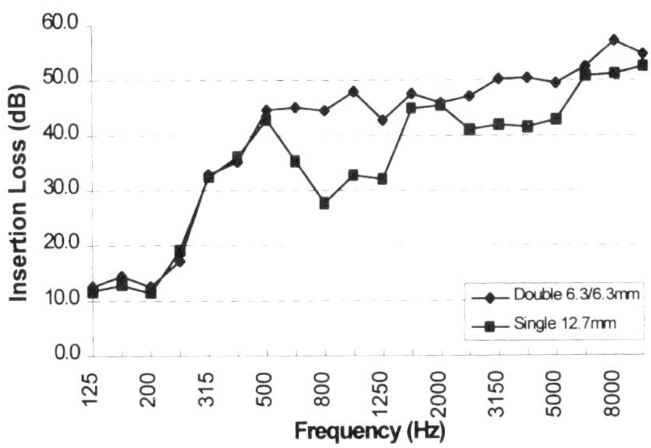

Fig. 9 Acoustic comparison of double and single baffle with the same total thickness

5. CONCLUSIONS

1. The preprogrammed heat activated foams exhibited key characteristics of isotropic expansion, higher lateral expansion, and more uniform expansion over a wider temperature range than the conventional foams.
2. The key characteristics impart self directional control to drive foam expansion toward desired locations.
3. The single baffles prepared by the preprogrammed heat activated foams exhibited good noise reduction over a wide frequency range.
4. By using the preprogrammed heat activated foam, the double baffles can be economically made and attached in the body cavity.
5. The double baffle showed superior acoustic performance to the single baffle, when compared at an equal weight(or thickness) basis.

6. ACKNOWLEDGMENTS

The authors would like to thank Raychem Corporation for permission to publish this paper. The acoustic measurements presented in this paper were conducted by Kolano and Saha Engineers, Inc., Waterford, Michigan.

7. REFERENCES

1. Saha, P., and Meyers, R. D., "Importance of Sealants for Interior Noise Control of Automobiles," Paper 920412, presented at International Congress & Exposition, Detroit, Michigan, Feb. 24-28, 1992
2. Saha, P. And Chahine, J., "Sound Package Materials in Automobiles"- Proceedings of the ASME Noise Control and Acoustics Division, NCA-Vol. 22, 1996.
3. Gettys, G. A., "New Concepts in Acoustical Baffles"- Automotive Interior and Exterior Systems, IBEC(International Body Engineering Conference), 1993.
4. Goldman, A., and Kurt, L., "NVH Improvement with Two-component Polyurethane Foam"- Automotive Interior and Exterior Systems, IBEC, 1994.
5. Otto, D., and Hanley, J. L., "From Art to Science: A New Acoustic Baffle Material"- Paper 951245, Proceedings of the 1995 SAE Noise and Vibration Conference, Traverse City, MI, pp. 95-101, May 15-18, 1995.
6. R. J. Chang, Keith Dawes and Lawrence Schwartz, "Expandable Foams with Programmed Expansion as Body Cavity Filler", paper 980590, accepted for presentation at Society of Automotive Engineers International Congress and Exposition, to be held in Detroit, Michigan, USA, February 23-27, 1998.
7. T. Hirabayashi (SAE Sound & Hear Insulation Committee), "Application of Noise Control and Heat Insulation Materials and Devices in the Automotive Industry", SAE Noise & Vibration Conference & Exposition, Traverse City, Michigan, May 10-13, 1993.
8. Saha, P., "The Need of a Standardized Acoustical Evaluation Method for Body Cavity Fillers"- Automotive Interior and Exterior Systems, IBEC, 1995.
9. Visintainer, A. P. And Saha, P., "Acoustical Study of Cavity Fillers for Vehicle Applications", Paper 971931, Proceedings of the 1997 SAE Noise and Vibration Conference, Traverse City, MI, pp. 503-509, May 20-22, 1997.

C521/029/98

An investigation of machine excitation analysis as a possible tool in the control of interior noise quality in light commercial vehicles

R H CORNISH BSc, CPhys, MInstP
Faculty of Engineering and Computer Technology, University of Central England in Birmingham, UK

SYNOPSIS

As part of an initiative to improve interior noise in light commercial vehicles an experimental and analytical program has been underway to link interior noise quality with vehicle build. Interior noise was recorded at steady-state conditions in two identical vehicles. This was followed by frequency and order analysis. Comparison of the two vehicles revealed differences in 2nd and 4th order noise. These differences were shown to be significant with tests to quantify the measurement errors. Machine analysis was then used to decide whether path or source variability causes interior noise variability. In general, path effects are shown to be responsible for the variability of interior noise.

1. INTRODUCTION

Efforts to reduce interior noise in light commercial vehicles are hampered by the variability in noise levels observed between nominally identical vehicles. Small differences in manufacture and assembly can have a large effect on the noise in these vehicles due to the relatively small amount of damping material per unit area of the body structure compared to passenger cars. However, this does not mean that variation in body construction is necessarily the cause of the variability. The amount of excitation from the drive-line may also vary between identical vehicles, and it is the separation of these two effects that this work attempts to achieve.

By analyzing interior noise alone it may be possible to separate the effects of path and source variabilities. A successful procedure have practical application to vehicle quality control. Any build faults could be rectified and corrective action taken on production lines to ensure that high quality is maintained. The work described here assesses one possible technique.

2. METHOD

Two identical vehicles were selected and instrumented with a noise and vibration measurement and analysis system (1,2,3,4). The system digitally sampled the noise at 12.5 kHz and stored it on hard disc in a notebook computer. The test procedure consisted of constant speed runs at 2250 to 4000 rpm in 250 rpm intervals along a straight and level section of public highway. The vehicles were both preconditioned by a warm-up period of normal driving.

Reproducibility of the data was examined by repeat tests on two consecutive days. Scatter due to uncertainties in microphone position were assessed using a three dimensional array of four microphones. The results were signal processed within the Matlab programming environment on a personal computer. Programs were written to manipulate, frequency and order analyze the data.

3. SIGNAL POST-PROCESSING

The noise data was converted to its frequency components using Fourier Analysis. Fig. 1 shows the noise components versus frequency with A-weighting at a defined speed of engine rotation. These components were then replotted against frequency expressed as multiples of the engine rotation frequency, which had a fixed value for each particular test. The engine rotation frequency is given by:

$$f_{rot} = rpm / 60$$

where rpm is the number of revs per minute for the test. The result of this rescaling applied to data in Fig. 1 is shown in Fig. 2 and shows that the noise components lie exactly on multiples of the engine rotation frequency, and importantly at the half positions. These multiples are called the Engine Orders and the half multiples are called the Engine Half-Orders. Each test result was processed to give the amount of each Engine Order and Half-Order present. Levels in each order were plotted against engine speed for each vehicle (Fig 3).

A further step in the analysis borrowed an idea from work with industrial machines. The engine order and half order components were plotted as a function of frequency. The result of this is shown in Fig. 4. Here, results for all the engine speed tests are used to show the acoustic response of the vehicle to each engine order and half-order separately. This is a method of finding the acoustic response of the vehicle using the vehicles' own engine harmonic vibration as the 'shaker' device.

4. REPRODUCIBILITY

Scatter of measurement results was found to depend critically on the degree of speed control achieved in the tests. Care was also needed in the selection of the part of the highway used for the tests. A small effect of plus or minus 2 dB(A) was produced by microphone position uncertainty. A larger error of plus or minus 4 dB(A) was produced by repeat tests on consecutive days. However, further examination of data showed that in these repeat tests, speed control had been poor. The figures for test repeatability are therefore the worst case.

5. DISCUSSION

The interior noise is caused by engine vibration being transmitted through the structure and to the driver via the air inside vehicle. Order noise below second were found to be insignificantly small. If the order magnitude is plotted as a function of engine speed in each test, order sections such as in Fig. 3 are produced. Figure 3 shows that Second Order noise was very significant above 3500 rpm in both vehicles. The vehicles differed in the amount of Second and Fourth Order noise. Second order noise variation has been reported at 10 dB(A) for a small family vehicle in track tests (5). The causes of each Engine Order and Half-Order noise for light vans is shown in Table 1.

Engine Half-Order noise above second order became significant at engine speeds higher than 3500 rpm. The cause of Half-Orders is cylinder to cylinder combustion differences, which is related to fueling consistency, and which would be expected to deteriorate at the higher speeds. The most likely cause of fourth order excitation is exhaust and intake noise, primarily from their open apertures, but also from their boundary walls as 'shell noise'. Shell noise has been reported to vary by up to 20 dB at certain frequencies (6). Intake noise with strong components at 2nd, 4th and 8th Engine Orders has been reported for a medium-sized 4-cylinder petrol engine (7).

Figure 3 is equivalent to the order sections produced in a run-up test with an infinitely long run-up time because each test was at a steady speed. Plotting the magnitude of each engine order and half order as a function of frequency in Fig. 4 provides an additional insight analogous to the method of machine excitation in gear noise analysis. An artificial offset in noise level is used to separate the curves to improve clarity.

The method examines system response to integral vibrating machinery when harmonics of the excitation are fairly constant across the range of rotation speed. This would normally be the case for instance with a gear-tooth source of vibration at constant driven load. It will be appreciated that the vehicle tests here were all at light load but air resistance, for instance, will increase with vehicle speed. The vehicle body should be unable to distinguish between excitations at the same frequency, so a similarity of response shape should appear where adjacent orders from the same vehicle overlap. Table 2 gives the orders which have a similar overlap for a single vehicle. The lack of response shape similarity at overlap elsewhere suggests that the excitation of the orders may not be simply related to engine speed indicating resonances in the drive-line system.

The differences and similarities between the two vehicles on the other hand point to differences and similarities in the acoustic response of the bodies. Both vehicles have similar response shapes to 5th order. Overlapping orders from one vehicle may have a similar shape yet be quite different from those of another identical vehicle.

An example of the insight provided by machine analysis is provided by examining Fig. 3 more closely. A peak in the 4th order noise occurs in vehicle 1 at 3250 rpm that is not present in vehicle 2. The hypothesis that the vehicle bodies cause this difference can in principle be tested by examining Fig. 4. The peak in fourth order at 3250 rpm translates to a frequency of 210 Hz, indicated by the vertical chain dotted line in Fig. 4.

If body response caused the peak, then any overlapping orders should also peak at 210 Hz. The overlapping orders at 210 Hz are 3.5, 4, 4.5, and 5. There is a shared general peak in these orders for vehicle 1 (solid line) indicating a common body response which is absent from vehicle 2 (dotted line). This feature of the plots suggests that the body response is different in the two vehicles at 4th order frequencies. As fourth order noise is strong in these vehicles, the analysis supports the hypothesis that the vehicle body is the cause of interior noise variability.

6. CONCLUSIONS

The work reported shows that 2nd and 4th order noise was the most important, both for the variability and for the overall high level of interior noise. When the reproducibility of the tests are considered, the variability of vehicles is shown to be a real effect. The comparison of noise in the two vehicles strongly suggests that body acoustic response to vibration is significantly variable. Similar results could be caused by systematic variability in adjacent integer and half integer order excitation which is unlikely.

Reproducibility of the results depends on careful control of noise tests. To make this method more reliable, more detail in the responses are needed by testing at smaller engine speed increments. Also, steps need to be taken to control engine load which can seriously effect engine vibration levels.

7. ACKNOWLEDGMENTS

The Author would like to thank the directors of LDV Ltd for supporting the program of research work described.

8. REFERENCES

1. Cornish, R.H. and Daniels, M. Improving Interior Noise in Light Commercial Vehicles. 5th International EAEC Congress, Conference II: Chassis-Body Engineering, Strasbourg, June 22nd 1995.
2. Cornish, R.H. and Atkinson, G. A. An Investigation of Production Variability of Light Commercial Vehicle Interior Noise. ROVA '95, Bolton, 11-13 September 1995.
3. Cornish, R.H., Daniels, M. and Atkinson, G.A. Reducing Interior Noise in Light Commercial Vehicles. Autotech '95, NEC Birmingham, 7-9 November 1995.
4. Cornish, R.H., Daniels, M. and Atkinson, G.A. An investigation of Interior Noise Variability due to Production Variation Effects Using Signal Processing Techniques. 19th Annual Energy Sources Technology Conference and Exhibition, Houston, Texas, January 1996.
5. Roberts, S. N. and Anderson C. G. Practical Implementation: Noise Path Analysis. IMechE C487/045/94
6. Lonas, T. L., Garcia, P. H. E. and Kuntz, C. Design Parameter on Exhaust System Silencers and Influence on Shell Noise. IMechE C487/012/94
7. Harrison, M. F. and Davies P. O. A. L. Rapid Predictions of Vehicle Intake/Exhaust Radiated Noise. IMechE C487/019/94

Table 1: Summary of knowledge on noise generation and variability in vans

Order:	Importance: 1 to noise 2 to variation	Rev. Range (rpm)	Cause:
2nd	1. High 2. High	3000 - 4000	Normal 4 cylinder unbalanced forces Intake and exhaust noise
2.5	1. Medium 2. Low	3500 - 4000	Cylinder to cylinder combustion differences
3rd	1. High 2. Medium	3000 - 4000	Distortion of the engine block under the influence of half-orders
3.5	1.Medium 2.Medium	2250 - 4000	Cylinder to cylinder combustion differences
4th	1.High 2.High	2250 - 4000	Air-intake and Exhaust Noise *Known to excite panel boom at 3500 rpm*
4.5	1.Medium 2.Medium	2500 - 3500	Cylinder to cylinder combustion differences
5th	1.Low 2.High	2250 - 4000	Distortion of the engine block under the influence of half-orders
5.5	1.Low 2.Low	2250 - 4000	Cylinder to cylinder combustion differences
6th	1.Medium 2.Medium	2250 - 4000	Air-intake and Exhaust Noise
6.5	1.Low 2.Low	2250 - 2750 & 3500 - 4000	Cylinder to cylinder combustion differences
7th	1.Medium 2.Low	3000 - 3500	Distortion of the engine block under the influence of half-orders
8th	1.Low 2.Medium	2500 - 3500	Air-intake and Exhaust Noise

Table 2: Similar response shape of orders from a single vehicle for 2 identical vehicles

Vehicle 1:	Order 2.5 overlaps 3 Order 4.5 overlaps 5
Vehicle 2:	Order 3 overlaps 3.5

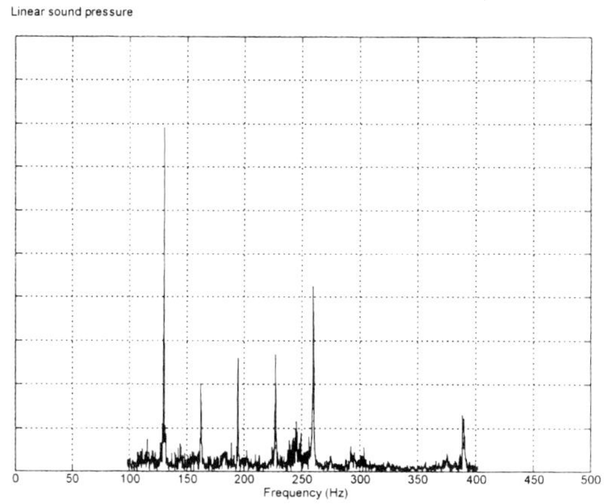

Fig. 1 Frequency analysis of noise in at driver's ear at 3750 rpm

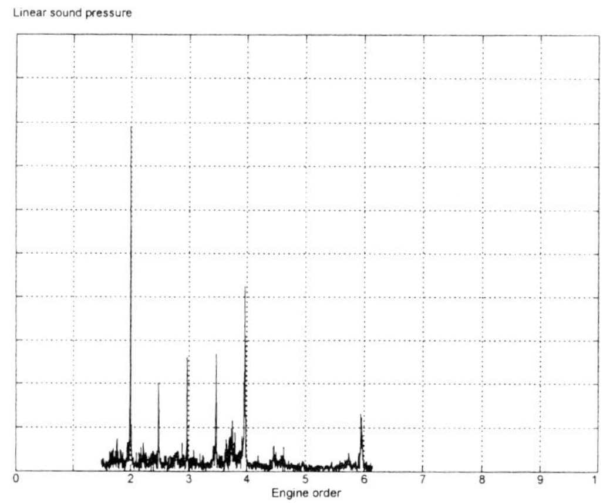

Fig. 2 Order analysis of noise at driver's ear at 3750 rpm

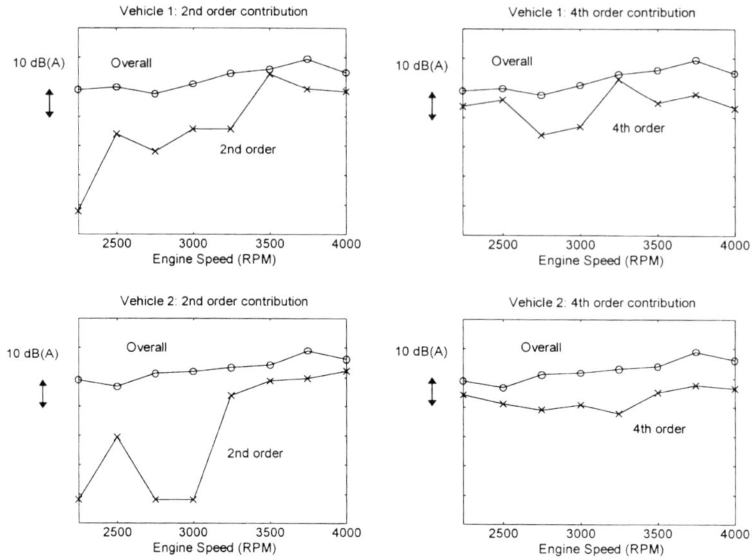

Fig. 3 Comparison of 2nd and 4th engine order noise differences for two vehicles

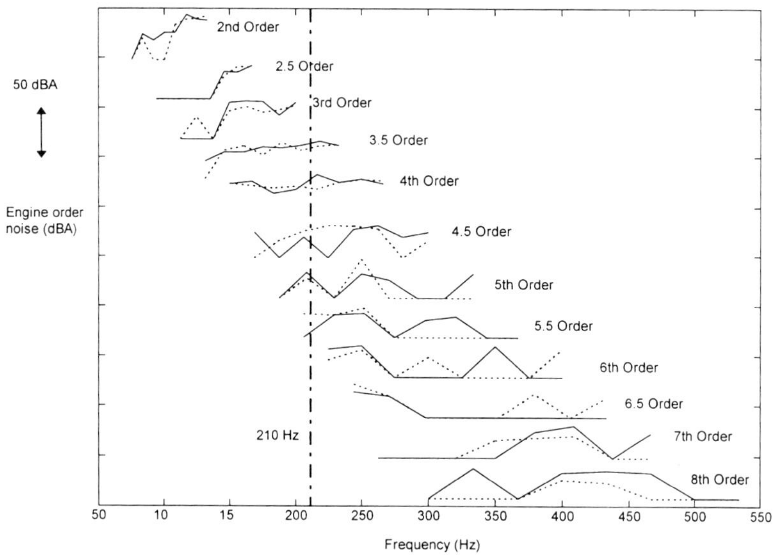

Fig. 4 Engine order noise vs. frequency for vehicle 1 (solid) and vehicle 2 (dotted) with offsets between each order for clarity

C521/031/98

Comfortable seat suspension design for commercial vehicles

M A EL-NASHER, S M EL-DEMERDASH, and **I M IBRAHIM**
Automotive and Tractors Engineering Department, University of Helwan, Cairo, Egypt

Abstract
In this paper, a limited bandwidth active suspension system has been employed for the development of a commercial vehicle driver seat suspension system. An eight degrees of freedom tractor semi-trailer model representing driver seat, cab suspension system and tractor main suspension is introduced. The driver seat is supported by an actuator mounted on a linear spring while a shock absorber is connected between the driver seat and cab base. The data have been considered for widely used tractor semi-trailer vehicles. The road surface irregularities are represented as stationary Gaussian random excitation. Linear optimal control theory has been applied to obtain the feedback gains of the driver seat suspension controller with different frequency' bandwidths. An optimization for the dynamic controller behaviour is achieved considering the driver vertical acceleration and driver seat suspension working space in the system performance index. The generated results from the proposed system are compared with those obtained from the conventional passive one.

Keywords: Commercial vehicles, driver seat suspension, limited bandwidth, active suspension, optimal control.

1. Introduction
In recent years, significant new constructions and developments of motor roads have been apparent. Furthermore, maintaining lower cost of transporting goods using commercial vehicles when it is compared with other transporting facilities, is increasingly challenging. For these reasons, the designers of commercial vehicles pay great attention to the driver ride comfortability. This objective could easily be reached through the improving of driver seat suspension rather than cab or main vehicle suspension.

Many investigations have established that the driver's health, safety and productivity are strongly affected by the vehicle ride behavior. Although commercial vehicle ride has been investigated since the early days of vehicle development, particular attention on vehicle ride was seen in the last three decades. In an effort to solve the problems of vibration isolation in vehicle main suspension, a covariance analysis technique was used to design and optimize the truck main suspension (1). A sensitivity analysis was performed to help choose design variables for suspension optimized parameters (2). Recently active controllers were incorporated with the passive main suspension elements of trucks (3, 4, 5).

For particular reasons there are limits to how far one can go in improving the driver ride quality through the vehicle main suspension, due to the hedge cargo and vehicle structure weights. Therefore the importance of the driver ride quality through the development of active main suspension is not practical and costly. A computer approach for improving the cab suspension of a truck vehicle was introduced based on a mathematical simulation technique for optimizing the values of springs and shock absorbers (6). The optimal active suspension for the cab-in-tractor, semi-trailer vehicle was obtained using linear optimal control theory (7).

In improving of driver seat isolation, special designs have been considered for both passive and active seat suspensions. A recent active seat suspension equipped with electrohydraulic active vibration control system was introduced for off-rood vehicles (8). Another design of optimal driver seat suspension for heavy truck vehicles including the simulation of the human body as malt-segment mass composed of six parts (9). To introduce the optimization of the driver seat parameters the driver vertical acceleration , seat suspension working space and control law have been minimized in the objective function.

In this work, a tractor semi-trailer model is introduced with special limited bandwidth active suspension system for the driver seat. The model design parameters are achieved using linear stochastic optimal control theory based on a limited bandwidth control strategy. The driver accelerations as well as the generated driver suspension working space of the actively suspended driver/seat assemblies are compared with those of the passive system with different vehicle speeds and road conditions.

2. Vehicle model

The vehicle model is presented in Figure 1. The model is eight degrees of freedom two-dimensional structure. The degrees of freedom are :

Driver vertical motion, Z_d

Cab vertical motion, Z_c

Cab pitch motion, Θ_c

Tractor frame vertical motion, Z_t

Tractor frame pitch motion, Θ_t

Front unsprung mass (front wheel and axle assembly) vertical motion, Z_{w1}

Rear unsprung mass (rear wheel and axle assembly) vertical motion, Z_{w2}

Vertical motion of the driver seat actuator connecting point with passive spring, Z_a

Both tractor tyres are represented by linear springs and fitted between the wheel and axle assemblies and the road surface. The cab mass, m_c, is mounted on the tractor frame through two passive suspension systems each of which consists of a shock absorber in parallel with a linear spring. The tractor frame mass, m_t, is assumed to be rigid and mounted on the masses of the wheel and axle assemblies m_{wf} and m_{wr}, through two passive suspension systems each of which contains a shock absorber in parallel with a linear spring.

In order to simplify the mathematical model and reduce the system degrees of freedoms, it has been assumed that the front and rear parts of the semi-trailer are moving in a decoupled manner. Therefore, the mass of the semi-trailer front, m_{sf}, and its mass moment of inertia are only considered in the vehicle model. In the case of the passive system, the driver/driver seat mass, m_d, is suspended on the cab base through a linear spring in parallel with a viscous damper.

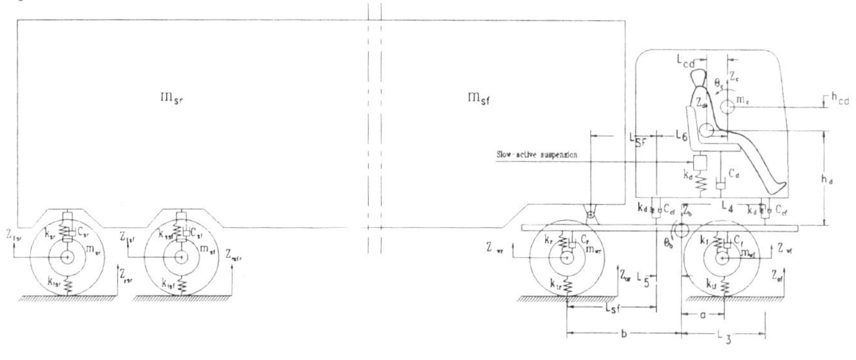

Fig. 1 Schematic drawing for the vehicle model with limited bandwidth active seat suspension system.

3. Limited bandwidth active seat suspension system

The limited bandwidth active suspension system of the driver seat is shown in Figure 2. The driver/driver seat mass is mounted on the controller actuator which is suspended on the cab base through a linear spring while a viscous damper is connected between the driver seat and the cab base. The limited bandwidth active system actuator is used herein as a displacement producer rather than a force producer in the fully active suspension system. The feedback control signals are filtered demand displacements to be used in a low frequency range up to 1-5 Hz according to the required cut off frequency. The filter equation of the limited bandwidth active suspension controller can be expressed as

$$\frac{-Z_a + Z_d}{u} = \frac{\omega^2}{s^2 + 2\zeta\omega s + \omega^2} \qquad (1)$$

where $\omega = 2\pi f$, $f = 3$ Hz

ζ is the filter damping ratio (0.707)

Equation (1) can be rewritten as

$$-\ddot{Z}_a - 2\zeta\omega\dot{Z}_a - \omega^2 Z_a + \ddot{Z}_d + 2\zeta\omega\dot{Z}_d + \omega^2 Z_d = \omega^2 u \qquad (2)$$

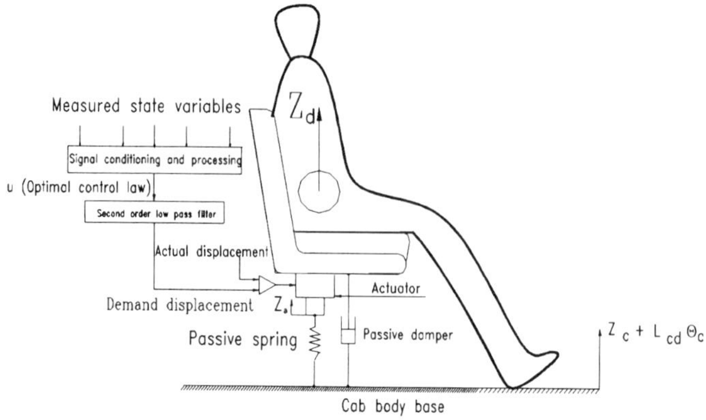

Figure 2 Driver seat equipped with limited bandwidth active suspension.

4. Equations of motion and output variables

The equations of motion of the vehicle system equipped with passive or limited bandwidth active suspension systems can be written in a matrix form as

$$M_{XDD}\ddot{Z} + M_{XD}\dot{Z} + M_X Z = M_U x_o + M_F u \tag{3}$$

where

M_{XDD}, M_{XD}, M_X are vehicle masses, damping and stiffness matrices respectively.
Z is the generalized coordinates vector.
M_U is the distribution matrix of excitation road signals.
M_F is the matrix of active control input.
u is the vector control law.

The equations of motion of the vehicle model are generated using Lagrange's energy equation which is expressed as;

$$\frac{d}{dt}\left(\frac{\partial KE}{\partial \dot{Z}_i}\right) + \frac{\partial DE}{\partial Z_i} + \frac{\partial PE}{\partial Z_i} = F_i \tag{4}$$

where

KE, DE and PE are kinetic, dissipative and potential energies respectively.
Z_i is generalized coordinate number i,
F_i is generalized force number i.

For passive systems, there are no generalized forces acting on the generalized coordinates. The vehicle system energies in this case are evaluated as follows:
Kinetic Energy:

$$KE = \frac{1}{2}m_b\dot{Z}_b^2 + \frac{1}{2}m_c\dot{Z}_c^2 + \frac{1}{2}I_b\dot{\Theta}_t^2 + \frac{1}{2}I_c\dot{\Theta}_c^2 + \frac{1}{2}m_d\dot{Z}_d^2 + \frac{1}{2}m_{wf}\dot{Z}_{wf}^2 + \frac{1}{2}m_{wr}\dot{Z}_{wr}^2 \tag{5}$$

Potential Energy:

$$PE = \frac{1}{2}k_f\left[Z_{wf} - (Z_b + a\Theta_b)\right]^2 + \frac{1}{2}k_r\left[Z_{wr} - (Z_b - b\Theta_b)\right]^2 + \frac{1}{2}k_{cf}\left[(Z_b + L_3\Theta_b) - (Z_c + L_4\Theta_c)\right]^2$$

$$\frac{1}{2}k_{cr}\left[(Z_b + L_5\Theta_b) - (Z_c - L_6\Theta_c)\right]^2 + \frac{1}{2}k_d\left[(Z_c + L_{cd}\Theta_c) - Z_d\right]^2 + \frac{1}{2}k_{tf}\left[Z_{of} - Z_{wf}\right]^2 +$$

$$\frac{1}{2}k_{tr}\left[Z_{or} - Z_{wr}\right]^2 \qquad (6)$$

Dissipative Energy:

$$DE = \frac{1}{2}C_f\left[\dot{Z}_{wf} - (\dot{Z}_b + a\dot{\Theta}_b)\right]^2 + \frac{1}{2}C_r\left[\dot{Z}_{wr} - (\dot{Z}_b - b\dot{\Theta}_b)\right]^2 + \frac{1}{2}C_{cf}\left[(\dot{Z}_b + L_3\dot{\Theta}_b) - (\dot{Z}_c + L_4\dot{\Theta}_c)\right]^2$$

$$+\frac{1}{2}C_{cr}\left[(\dot{Z}_b + L_5\dot{\Theta}_b) - (\dot{Z}_c - L_6\dot{\Theta}_c)\right]^2 + \frac{1}{2}C_d\left[(\dot{Z}_c + L_{cd}\dot{\Theta}_c) - \dot{Z}_d\right]^2 \qquad (7)$$

For the active system, the term for the potential energy generated in the driver seat suspension will be modified to become $\frac{1}{2}k_d\left[(Z_c + L_{cd}\Theta_c) - Z_a\right]^2$ and the external force acting on the driver seat is expressed as $k_d\,[Z_c + L_{cd}\Theta_c - Z_a]$.
Substituting the partial derivatives of the above energy equations in the Lagrange's equation, the equations of motion can be generated.

The following variables have been chosen to examine the ride quality of the driver seat:
1. Driver Vertical Acceleration = $\ddot{Z}_d$,
2. Driver seat suspension working space = $Z_c + L_{cd}\Theta_c - Z_d$.

5. System optimization

The limited bandwidth active suspension system is designed based on a 3 Hz bandwidth analogue controller. The optimal control theory has been applied to generate the system feedback gains based on full state information. For this purpose the equations of motion are rearranged in a state space formulation. More details can be found, for example, in (10). In order to improve the ride quality of the driver, the following performance index is chosen

$$J = \int_0^\infty [q_1\ddot{Z}_d + q_2(Z_d - Z_c - L_{cd}\Theta_c) + \rho u^2]dt \qquad (8)$$

Where q_1, q_2 and ρ are weighting constants.

6. Results and discussion

The model results are derived for driver vertical acceleration and seat suspension working space for both limited bandwidth active suspension and conventional passive system. The vehicle model parameters are adapted from widely used commercial tractors/semi-trailer vehicles (11).
The road input is represented as a power spectral density in the form of $psd = GV^{n-1} / f^n$ where G is the road surface roughness coefficient, V is the vehicle speed, f is the frequency and the exponent n is equal to 2.5.

Figure 3 shows the power spectral density of the driver vertical acceleration for the active seat suspension with different actuator bandwidth frequencies. These results are generated at constant root mean square (r.m.s). values of seat suspension working space and compared with conventional passive seat suspension. Reducing the active system actuator bandwidth frequency increases the magnitude of the driver vertical acceleration, power spectral density. As shown in Fig. 4, worthwhile improvements are obtained in the r.m.s. values of driver vertical acceleration for active seat suspension at vehicle speed of 20 m/s on a minor road with roughness coefficient of 3×10^{-6} . The use of a 3 Hz actuator bandwidth in the active suspension gives a 42% improvement in the r.m.s. values when it is compared with passive system. Using higher actuator bandwidth frequency does not show noticeable improvement. The following generated results are calculated using a constant actuator bandwidth of 3 Hz which is competitive to the higher values as recommended by (12).

The effect of the driver seat damping coefficient on the driver vertical acceleration and seat suspension working space is illustrated in Figure 5 for passive and active systems. Increasing the damping coefficient slightly affects the active suspension performance. This is due to the system actuator operating instead of the damper. In the passive seat suspension, increasing the damping coefficient significantly affects the driver vertical acceleration and seat suspension working space.

The proposed active seat suspension system is examined for different vehicle speeds and its performance is compared with the conventional passive system in Figure 6. As shown, increasing the vehicle speed affects the driver vertical acceleration and suspension working space in both systems. The active seat suspension shows a better performance for all vehicle operating speeds. Also, the proposed active seat suspension system is examined for the vehicle moving on different road surface conditions as introduced in Figures 7 and 8. These results show the power spectral densities of driver vertical acceleration and power spectral densities of suspension working space with road roughness of 0.5×10^{-6} (smooth road) and 5×10^{-6} (rough road). Table 1 shows the r.m.s. of driver vertical acceleration and seat suspension working space for smooth, medium (3×10^{-6}) and rough road surfaces.

Table 1 R.m.s of driver vertical acceleration and seat suspension working space at vehicle speed of 20 m/s.

Road roughness coefficient	Seat suspension type	Driver vertical acceleration, m/s^2	Seat suspension working space, cm
0.5×10^{-6}	Passive	0.68	1.25
	Active	0.42	0.82
3×10^{-6}	Passive	1.68	3.06
	Active	1.03	2.00
5×10^{-6}	Passive	2.16	3.94
	Active	1.33	2.95

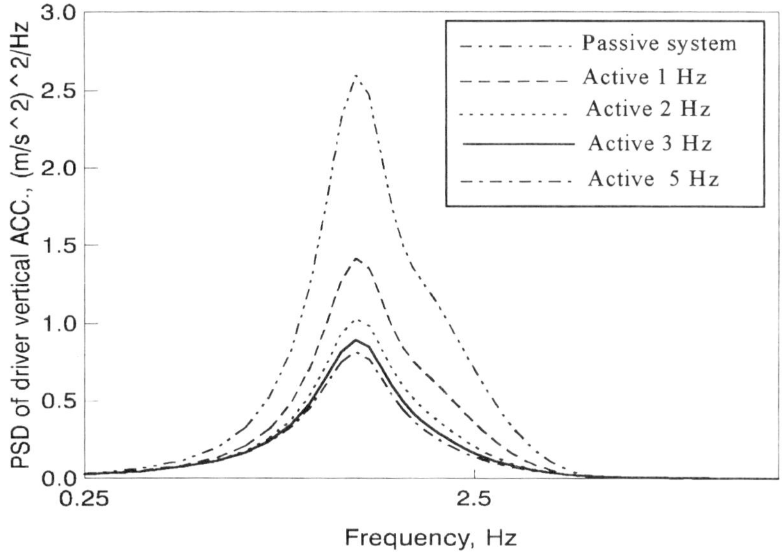

Figure 3 Power spectral density of the driver vertical acceleration for active seat suspension with different actuator bandwidth frequency compared with passive system at constant r.m.s of seat suspension working space 2 cm and 20 m/s vehicle speed.

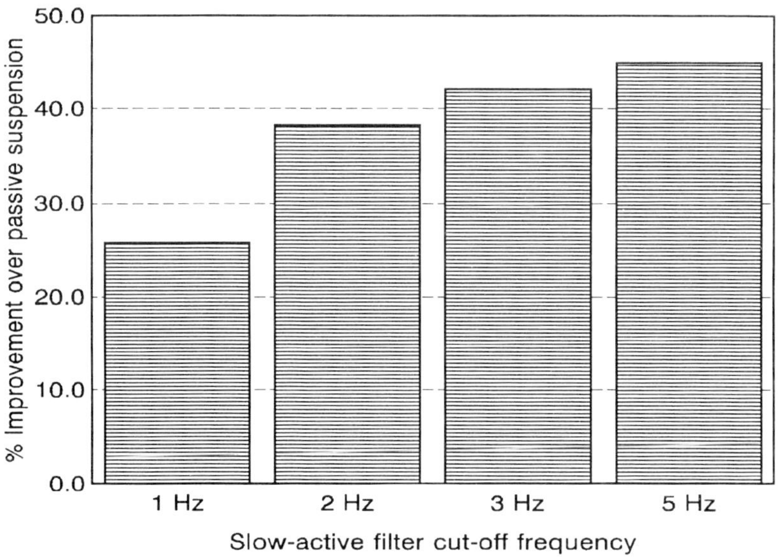

Figure 4 Improvement of driver vertical acceleration for limited bandwidth active system over passive system at constant r.m.s of suspension working space 2 cm and 20 m/s vehicle speed.

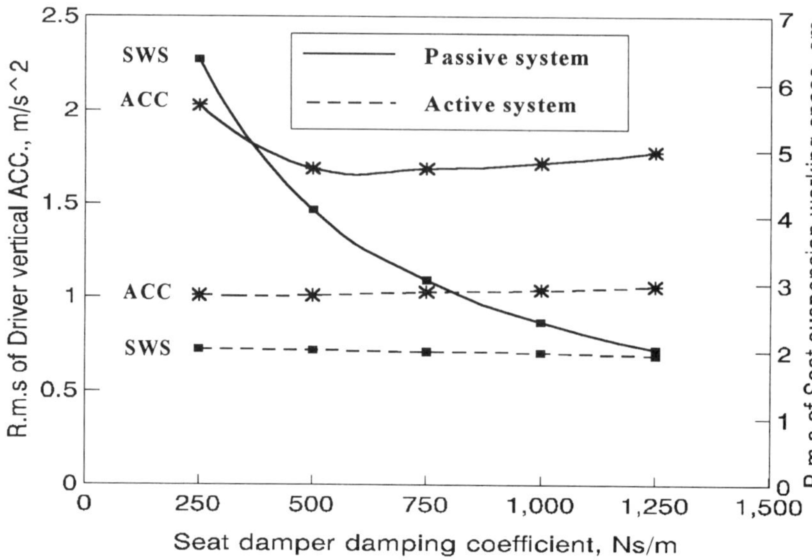

Figure 5 Effect of the driver seat damping coefficient on the driver vertical acceleration and seat suspension working space at vehicle speed of 20 m/s on road surface coefficient of 3×10^{-6}.

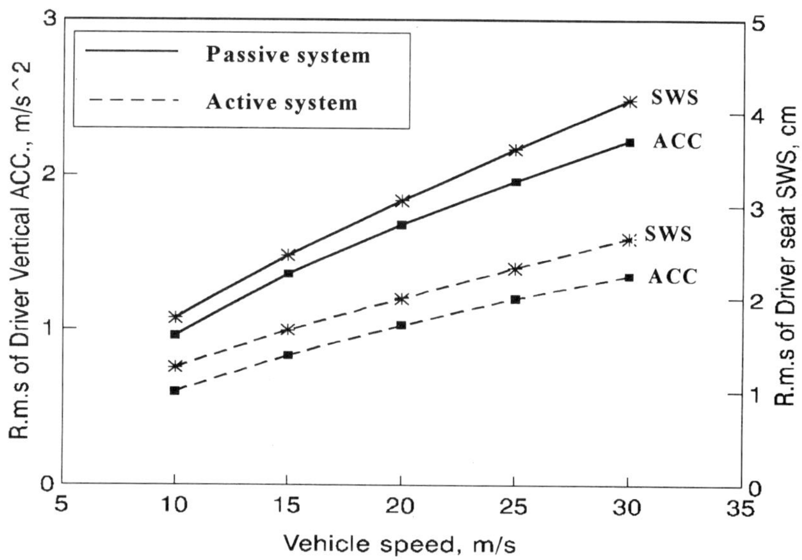

Figure 6 Effect of vehicle speed on the driver vertical acceleration and seat suspension working space for active and passive system on road surface roughness coefficient of 3×10^{-6}.

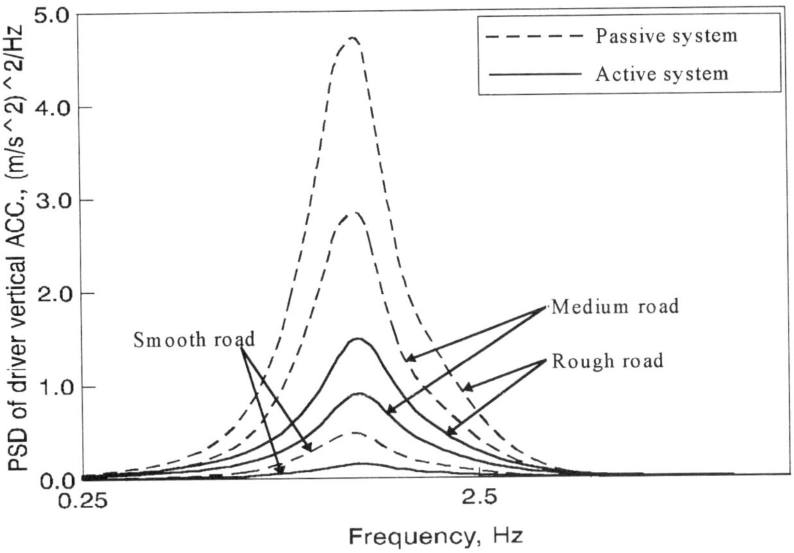

Figure 7 Power spectral density of driver vertical acceleration for active and passive seat suspension at vehicle speed of 20 m/s on road surface roughness coefficient of 3×10^{-6}.

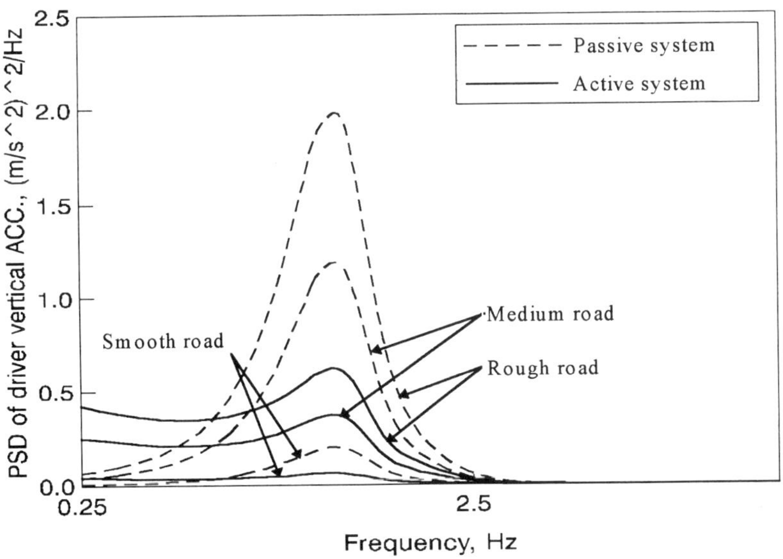

Figure 8 Power spectral density of seat suspension working space for active and passive systems at vehicle speed of 20 m/s on road surface roughness coefficient of 3×10^{-6}.

7. Conclusions

In this paper, a limited bandwidth, active seat suspension for a tractor semi-trailer vehicle has been modeled using eight degrees of freedom system. The model analysis were examined for different operating conditions and compared with conventional passive seat suspension. From the generated results, the following conclusions were made :

- The limited bandwidth active seat suspension is recommended for commercial vehicle seat suspension designs.
- The active system actuator with 3 Hz bandwidth analogue controller satisfies the driver ride comfortability requirements.
- For vehicle speed of 20 m/s and road roughness coefficient of $3x10^{-6}$, the active system shows a 42% improvement in the r.m.s. of the driver vertical acceleration in comparison with conventional passive system at constant r.m.s. suspension working space of 2 cm.

8. Refernces

(1) El -Madany, M.M. "Design and optimization of truck suspensions using covariance analysis", Computers & Structures, vol.28(3), !988, pp.241-246.

(2)Blue, D. ad Kulakowski, B. "Computer optimization of heavy truck suspension parameters", Proceedings of The 4th International Symposium on Heavy Vehicle Weights and Dimensions, 1995, pp.517-523.

(3)El -Madany, M.M., Samaha, M.E. "On the optimum ride control of a stochastic model of a tractor semi-trailer vehicle", Journal of Sound and Vibration, vol. 156(2), 1992, pp. 269-281.

(4)Wilkinson, P.A. and Crolla, D.A. "Development of sub-optimally controlled suspensions for articulated truck to reduce vibration and road wear", XXIV FISITA Congress, London, 1992, pp. 239-245.

(5)Huisman, R., Veldpaus, F., Heck, J. and Kok, J. "Application of a preview controlled active suspension to a nonlinear 2-D truck model", AVEC 94, 1994, pp. 293-304.

(6)Lee, G., Jong, L. and Gi-Tae, K. "Improving ride quality on the cab suspension of aheavy duty truck", Commercial Vehicle and Highway Dynamic, SAE Special vol. 1201, 1996, pp.11-15.

(7)El -Madany, M.M. "Design of an active suspension for a heavy duty truck using optimal control theory", Computers & Structures, vol.31(3), 1989, pp. 385-393.

(8)Stein, G.J. and Ballo, I., "Active vibration control system for the driver seat for off-road vehicles", Vehicle System Dynamics, vol.20, 1991, pp.57-78.

(9)Amirouche, F., Palkovics, L. and Woodrooffe, J., "Optimal driver seat suspension design for heavy trucks", Heavy Vehicle Systems, Int. J. of Vehicle Design, vol.2(1), 1995, pp.18-45.

(10) Hac, A., "Suspension optimization of the 2-DOF vehicle model using a stocastic optimal control technique", Journal of Sound and Vibration, vol.100(3), 1985, pp.343-357.

(11) Sunder Vaduri and E. Harry Law " Develoment of a simulation for assessment of ride quality of tractor semi-trailers", SAE Paper No. 932940, Proceeding of Heavy Vehicle Dynamics and Stability, 1993.

(12) Sharp, R. S.and Crolla, D. A., "Road vehicle suspension system design - a review", Vehicle System Dynamic, vol.16, 1987, pp.167-192.

C521/033/98

Close coupled catalyst solution and its effects on exhaust noise control

A LOREA and **V SACCÀ**
Magneti Marelli, Exhaust System Division, Venaria Reale Torino, Italy

Synopsis

The legislation for vehicle exhaust gas emission (ECE III/ECE IV) is becoming more and more strict. In order to comply with the legislation, the catalyst has to be mounted close to the engine; this can be done by adding a pre-catalyst to the exhaust line, or, if possible, by positioning the catalyst immediatly downstream of the manifold confluence (close coupled catalyst).

In this paper we present mono-dimensional acoustic and fluidynamic calculation carried out in order to evaluate the effect of the close coupled catalyst solution on exhaust noise control and engine performance.

Acoustic calculation shows that this particular type of layout allows a reduction of the tail pipe noise for the engine orders (2nd,4th,6th), with an improvement of the passenger compartment comfort and noise quality.

Moreover,fluidynamic calculation shows that, with a correct tuning of the manifold pipes length, it is also possible to improve the engine performances.

1. INTRODUCTION

Nowadays, the time required for the development and production of vehicle components is shorter than before; in the future, the "time to market" (time lasting from the design to the production phase) will have to be reduced even more to match the vehicle development time. Indeed, we have to take into account that the life of a vehicle model has decreased during the last years, mainly because of an increased competitiveness between car makers.Car component suppliers, and therefore exhaust systems suppliers, have to reduce the duration of design and experimental phases.In order to reach this goal, the exhaust system development has changed from an experimental approach to a predictive one, especially in these last years (1). This has been possible owing to the use of analytical and computational methodologies.

Magneti Marelli, is using more predictive tools, over a wider range of problems. The calculation codes reduce the time to market rapidly and continuously as each one is implemented, with important benefits in quality and cost of exhaust systems.

The wide use of predictive methodologies in acoustic , vibrational and fluid dynamic calculations, and the employement of more and more rigorous experimental procedures is a necessary condition to the competitiveness of the product, although not yet sufficient. In the early design phases, indeed, it is necessary to incorporate innovation as well in order to design an exhaust system with components which simplify the product, reduce the cost, while mantaining functional and qualitative performances.

In this paper an exhaust system with the catalytic converter placed closer to the exhaust manifold is described; this layout has been studied using the principles expounded above. The incentive to implement new predictive capabilities arose from the new and more restrictive gas emission legislation (ECE IV). With the application of analytical and predictive methods the acoustic and vibrational design has become more effective,with a general profit for the driver comfort (noise level and quality, level of the vibrations transmitted to the car-body).

2. EXHAUST SYSTEMS LAYOUT AND FUNCTIONS

An exhaust system is normally divided in three parts: a downpipe, from the engine manifold to the decoupling element, that has the purpose to reduce the vibrations transmitted to the other system components; a central part that incorporates the catalyst; a rear part consisting of two or three silencers, according to the engine type and car body layout (fig.1).

The main assignment of an exhaust system is obviously to "manage" the exhaust gas; the working functions finalized to the satisfaction of the customer issues are:

- to transport the exhaust gases from the engine to the rear part of the car;
- to manage noise, reducing its level and improving its quality;
- to reduce the polluting engine emissions through the exhaust gas conversion;
- to interface to the car system, taking under control the level of the vibrations transmitted to the car body and the heat transmitted towards either the cabin or the external environment.

Obviously, the exhaust line must not cause too much backpressure wich would affect the engine and vehicle performance.

The traditional architecture of an exhaust system consists of a catalyst positioned under floor, at a certain distance from the engine. This distance is set in order to comply with the legislative prescriptions, in terms of gaseous emissions.

Cars with start production in 1999 / 2000, now in their development phase, will meet the new legislative levels set by the ECE Phase III, with predisposition to comply to the future and more strict legislation (ECE Phase IV). Insofar as it is necessary to modify the architecture of the front and central parts of the exhaust system, advancing the catalyst towards the engine will reduce light off and reduce the size of the catalyst underneath the floor,or even eliminate it.

For obvious reasons of cost and weight it is convenient, when it is possible, to design a catalyst positioned as close as possible to the engine, in the engine compartment (Close Coupled Catalyst, CCC hereafter) (fig.2). This kind of exhaust system layout, with the catalyst positioned immediately downstream the exhaust manifold, implicates some very interesting considerations, that are described in this paper, with particular stress on the acoustic aspect.

An exhaust system prepared with a Close Coupled Catalyst reduces the catalyst light off time in comparison with an under floor catalytic converter; this improves the exhaust gas emission levels. Moreover, from the acoustic point of view the CCC layout simplifies the exhaust line, reducing the number or capacity of the silencers. This results in a reduction of costs and weight of the whole exhaust system.On the other hand, the Close Coupled Catalyst is submitted to more elevated temperatures, being so close to the exhaust manifold; and its flow distribution is worse in comparison to the under floor catalyst. This gives rise to the need of optimizing the CCC design in terms of reliability.

3. EXHAUST MANIFOLD WITH CLOSE COUPLED CATALYST

As already mentioned, the future European legislative limits on the engine emissions (normative ECE Phase III / Phase IV) will be more restrictive than the present ones. In particular, the limits will be critical for the engine start up phase which is responsible for more 80% of the legislative cycle emissions.

To meet these limits , it is necessary to develop a system tipology able to reduce the initial phase of light -off of the catalyst, which in turn requires that the catalyst temperature should reach operating temperature as fast as possible so that the reactions converting the harmful chemical elements start as soon as possible in the monolith. This can be realized by decreasing the thermal exchange between the exhaust gas and the external environment and placing the catalyst closer to the engine, through a compact collector of the type 4 in 1 (for 4 cylinders engines). In this way we reduce the pipe lengths and the surface of thermal exchange, in comparison to a traditional manifold configuration of the type 4 in 2 in 1.

Besides the control of the gas emission during the legislative cycle, it is necessary to guarantee time durability of the monolith as well. Considering the fluid dynamic phenomena that take place in the exhaust system, the monolith is exposed to oscillations of temperature and pressure of the exhaust gases; this causes thermal and mechanical stress. Moreover, the monolith is invested by a flow with an uneven distribution hence the consumption of the open surface will be patchy, possibly leading to a reduced conversion efficiency and emission of harmful chemical compounds.

The CCC positioning, closer to the engine, increases the amplitude of the oscillations. It is therefore necessary to guarantee on the monolith a good uniformity of the flow, that involves a more uniform mechanical consumption, a reduced thermal gradient, great thermostructural stability and a great conversion efficiency.

The flow uniformity on the monolith of a catalyst joined to a 4 in 1 exhaust manifold can be obtained in the following ways:

- with a traditional solution, lengthening the inlet catalyst cone. In this case the monolith "sees" a single inlet (the inlet cone) and the fluid dynamic optimization allows to minimize the bad distribution of the flow evenly over the monolith;

- with a direct connection between the manifold and the monolith through an inlet cone of reduced size, to obtain a global coverage of the section of the monolith by the flow coming from all of the four cylinders.The monolith sees four inlets (the four primary pipes of the manifold). The optimal solution from the fluid dynamic point of view is to match on the monolith the single gas traces. The overlap of the four flows coming onto the monolith from the cylinders can be visualized to optmize the whole collector and catalyst system , in order to evaluate from the fluidynamic point of view the whole collector and catalyst system. It allows to identify a large or small overlap of the distributions of local massflow of the exhaust gases;

The optimization has been performed through Computational Fluid Dynamics calculations (CFD) using the Star-CD calculation code.(2).

4. EXHAUST SYSTEM ACOUSTIC DESIGN

The principal mechanisms of noise generation involved in an exhaust system (3) are:

- opening of the exhaust valve with consequent expulsion of the exhaust gases; it gives rise to tailpipe noise with two main components: the *breathing noise* and the *flow noise* ;the former is responsible for the low-frequency exhaust noise component. In fact the *breathing noise* is produced from the engine firing frequency with relative superior harmonics (4th and 6th) (4). The *breathing noise* is controlled by optimizing the silencers capacity, the position of the silencers and the geometry of the pipelines. The *flow noise*, measured at the tailpipe as well, is due to the jet noise produced from the impact of the high speed gases (in the tailpipe) with the still surrounding air. The discontinuities along the exhaust line give rise to flow noise as well. *Flow noise* is controlled reducing the outflow velocity of the gases, for instance increasing the tailpipe section and taking particular care in the definition of the geometry of section variation along the exhaust line.

- acoustic, or gas pressure excitation due to pressure fluctuations, produced at the opening of the exhaust valve is mainly responsible for the *shell noise*, because it excites the vibration modes of the different components, the downpipe, the catalyst and the silencers is very important for noise quality. This component is controlled by the mufflers geometry and insulation and reducing the amplitude of the exciting source.

- vibration excitation; this noise source may be from a gas pressure fluctuations as described above or by mechanical vibration transmitted from the engine. This component is controlled by decoupling the pipeline from the engine, through the insertion of flexible elements. The study of the vibrational behaviour (frequencies,modal shapes etc.) through calculation and experimental setup optimizes the position of the pipeline hangers and static and dynamics characteristcs of the rubber elements which connect the exhaust system to the car body.

5. ACOUSTIC AND FLUID-DYNAMIC CALCULATION

5.1. The acoustic calculation code

Magneti Marelli uses, in the design phase, the acoustic calculation code SILLOS, developed in collaboration with the Fiat Research Center.

The code takes into account the geometry of the whole exhaust line, from the exhaust valves down to the tailpipe. The model of the exhaust system is built up from a database of components (manifolds, catalystic converters, silencers, etc) which is continuously updated.The code is one-dimensional and it is based on the transfer matrix algorithm (5).

The computation code requires as input an exciting acoustic source, simulating the noise generated from the engine; this source is given in terms of mass-flow and temperature of the gas at several points of the line. Such a source is built by different ways, according to the engine data available:

- through monodimensional fluid dynamics code (6), based on a model of the engine using data furnished by the customer. The same program is used for assessing the impact of the geometry of the exhaust line on the engine performance This allows our Division to improve the design of the exhaust manifold;

- when the complete model of the motor is not available, the program SILLOS is able to generate a simplified noise source , on the basis of a series of readily available engine data. In this case, however, it is not possible to obtain engine performance evaluation.

The calculation code SILLOS has a user friendly interface, and it is installed on all the workstations in the designing area at Magneti Marelli.

The output provided by the program is a set of information that, correctly combined, optimizes the geometry of the exhaust system (diameters and lengths of the pipes, position, volumes and inside schemes of the mufflers):

- acoustic quantities of the exhaust line and its components (Transmission Loss, Insertion Loss, Noise Reduction);

- acoustic modes shape for the resonant frequencies of the system, i.e. distribution of pressure maxima and nodes along the exhaust line;

- levels of sound pressure at the tailpipe for the engine orders (mainly 2nd, 4th, 6th engine order).

An example of the capabilities of the calculation codes is provided by the study of an exhaust line with Close Coupled Catalyst geometry, in comparison with a traditional layout,with an under - floor catalyst .

5.2 Results of the acoustic calculations

As we will see in more detail subsequently, the results of the calculation regarding the exhaust line with Close Coupled Catalyst underline that this type of geometry improves the acoustic performance of the exhaust system.

The principal positive effects of geometry CCC can be summarised according to the following:

- it is well known that the variations of cross-section, that is expansions, in an exhaust system contribute to the noise reduction. In the CCC geometry, the expansion ratio between the cross-sections of the primary exhaust pipes and the inlet catalyst cone, is greater than that for the under-floor catalyst. A confirmation of this can be seen in figure 3; in which is plotted the course of the instant gaseous pressure in front of the first silencer, versus the crank angle. This quantity (see point 4) represents the acoustic and mechanical excitation produced by the engine. The maximum values of instantaneous pressure are smaller in the case of the Close Coupled Catalyst layout type; moreover the excursion between a minimum and maximum is smaller, as well.

- in the traditional architectures, the front line that goes from the engine flange to the first expansion (the catalyst), has quarter wave resonances, defined by its length (ca 1 m), with typical values around 100 Hz. This resonance gives rise to a peak in the tailpipe pressure level around the 3000 RPM for the second engine order. In the case of the CCC architecture, this typical length is very short and it coincides with the length of the primary pipes of the manifold (ca 0.1 m); this moves the quarter wave resonance to higher frequencies, that don't influence the engine orders in the range of interest (1500-6000 RPM);

- we compared the acoustic performances, on the same engine, of a serial production exhaust line with an under-floor catalyst, front and rear mufflers, and a CCC layout. Our computation showed that the CCC solution is favorable for the positioning and the sizing of the mufflers. This layout, in fact, allows an architecture with only one rear muffler (fig 4). In this case the pipe between the two expansions represented by the catalyst and by the rear muffler is much longer than for the layout with the under floor catalyst. This produces an appreciable reduction of the sound pressure levels for the second engine order , at low regimes of engine rotation (1500-3000 RPM); this is the range in which noise is particularly critical .These considerations are confirmed by the results collected in figures 5a) and 5b). In these figures the sound pressure level at the tailpipe for the 2nd engine order of the CCC line is compared with the corresponding serial production line. Both theoretical and experimental data (see in particular the circled areas in both figures) show a sensitive noise level improvement for the CCC line. On the other hand, the worsening that arises at 4500 RPM and last up to 6000 RPM is not critical; in fact, this is a range in which we are usually below the acoustic targets.

  Owing to the decreased intensity of the acoustic source, the levels of the superior engine harmonics (4th and 6th), decreased in comparison to the traditional solution. It is well known that these components contribute to determine the' noise quality ' of the exhaust noise (7).

- it has been verified that an opportune sizing of the primary pipes of the exhaust manifold can influence positively the 4th and 6th engine orders; fig. 6a and 6b show the comparison between two solutions, with symmetrical and asymmetrical manifold primary branches. There is a sensitive improvement for the 4th engine order, in the case of asymmetrical primary branches.

Further benefits to the inside comfort are due to the better behaviour of the line from the vibrational point of view; as we have seen the acoustic exciting source is less critical in the CCC configuration; therefore the shell noise produced by the components of the line (see par. 4) results reduced, and the noise quality improved.

Finally, the use of a CCC geometry allows the employment of softer supporting rubbers; as the catalyst is supported by the engine and the weight that the rubber and hangers must sustain is small in comparison with a traditional geometry (under floor). The use of hangers with values of static and dynamic stiffness which are much less than to those used traditionally, reduces the forces transmitted to the floor with obvious benefits for the inside comfort.

5.3. Engine performance calculation

Fluid dynamic computations have been carried out in order to verify the influence of the CCC layout on the performance of the engine, as used for the acoustic calculations. The results show that with an appropriate tuning of the length and diameter of the manifold primary pipes, it is possible to improve the torque curve,in comparison with the normal production 4-2-1 manifold, especially in the range between 2500 and 3500 RPM (see fig.7). In this range the normal production manifold geometry gives rise to a depletion in the torque curve, which is very bad for the engine performance.This is a very interesting result, at least for an engine of medium capacity and power, as the engine we considered in our calculations.The trend showed by these results has been confirmed by the experimental data.

6. CONCLUSIONS

In this paper we have presented some results of an analytical predictive study on the acoustic and performance behaviour of an exhaust system with a Close Coupled Catalyst layout. The results are interesting from different points of view; firstly, good correspondence between theoretical and experimental data has given a confirmation of the validity of the predictive approach in the planning of an exhaust system.

Secondly they show the positive effects that the CCC layout can have on the acoustic and vibrational performances of the exhaust line have been explained and underlined.

In future it could be interesting to investigate more deeply the link between the acoustic and engine behaviour of the engine-exhaust line system and the geometry of the CCC, with particular stress on the optimization of the manifold primary pipes geometry and catalyst inlet cone shape.

7. REFERENCES

(1) Lorea, A. and Barbera, E. "Designing and Testing for Exhaust System Development", ATA International Seminar, 19-20 October 1995, "Innovative Testing Procedures and Facilities for Engine Development in True Road-Like Conditions"

(2) CRF Report "Internal fluidynamic study of close coupled catalyst", n.ro 00018, 1997

(3) Bocchi, "Motori a 4 tempi"

(4) Mascellini, R. et al: "Fluidacoustic design of engine exhaust systems", 4th international Conference, Florence ATA ' 94 "New designs frontiers for more efficient, reliable and ecological vehicles".

(5) Munjal, "Acoustic of ducts and mufflers", Wiley, NY 1982

(6) Benson, R.S "The thermodynamics and gas dynamics of internal combustion engines", Vol 1, Clarendon Press, Oxford , 1982

(7) Zwickler, E. "Psychoacoustic", Springer-Verlag ISBN 3-540-52600-5

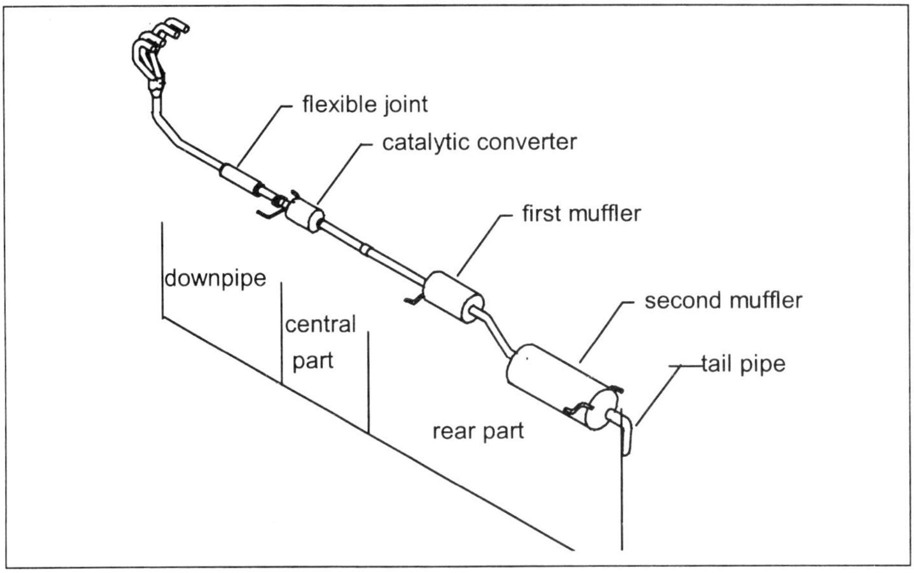

Fig.1: Exhaust system layout

Fig 2: Close Coupled Catalyst

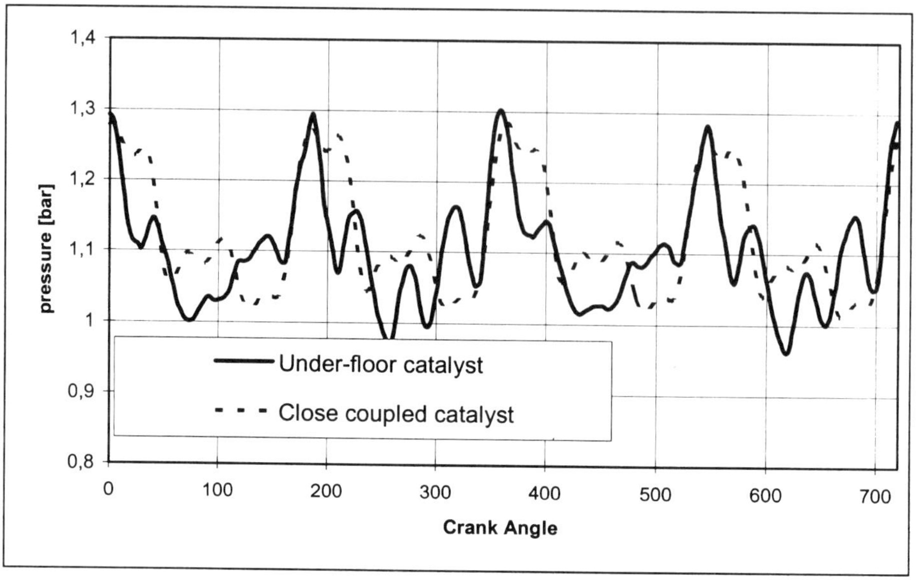

Fig.3: Instant pressure in front of the first silencer

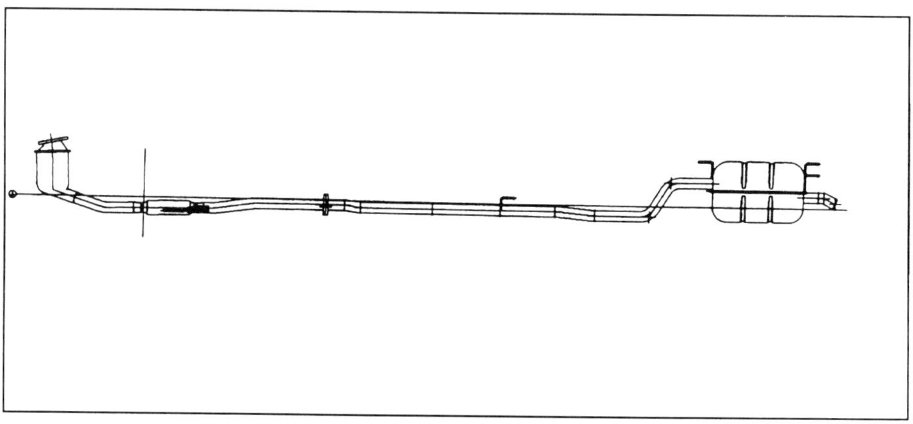

Fig.4: Exhaust line with close coupled catalyst and one rear muffler

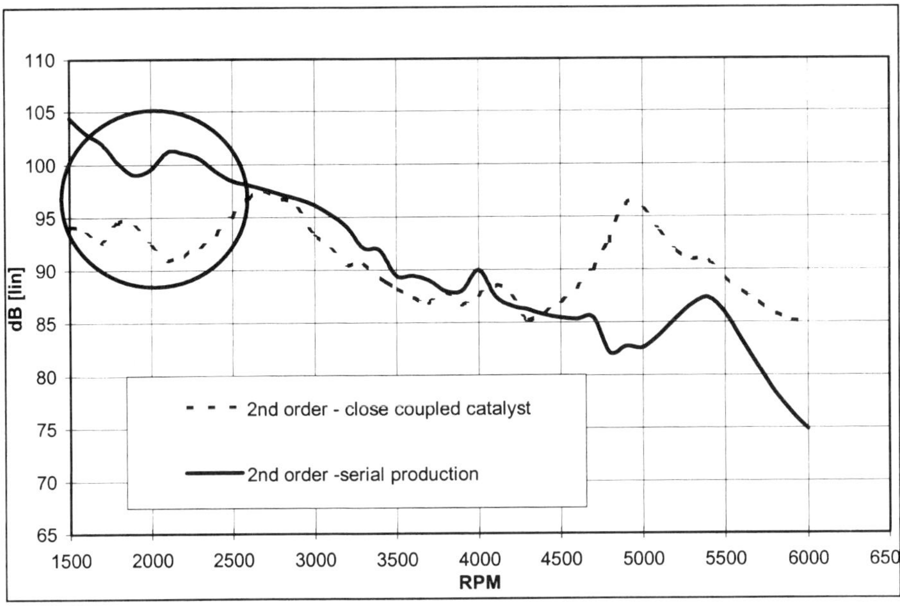

Fig.5a: Acoustic calculation for second order tailpipenoise

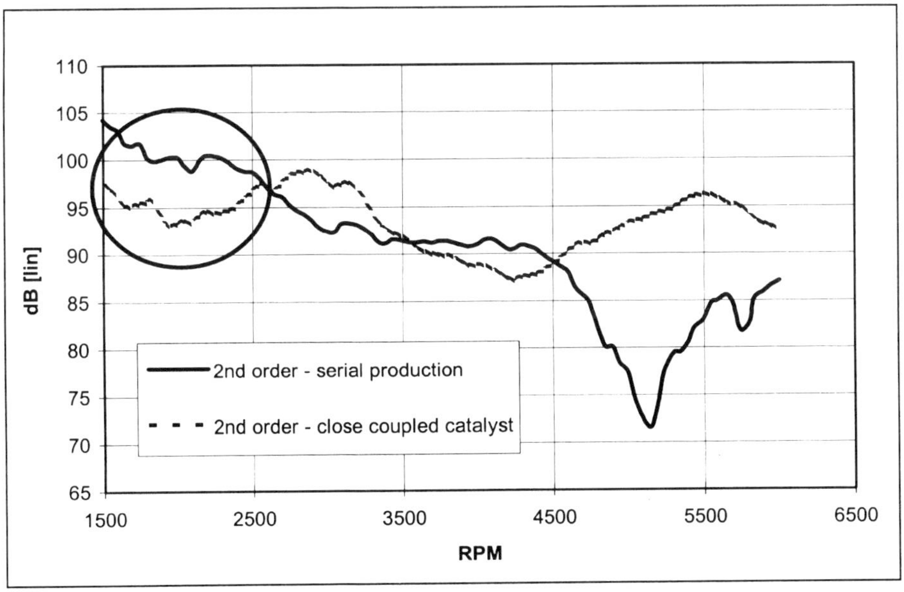

Fig.5b: Experimental results for second order tailpipe noise

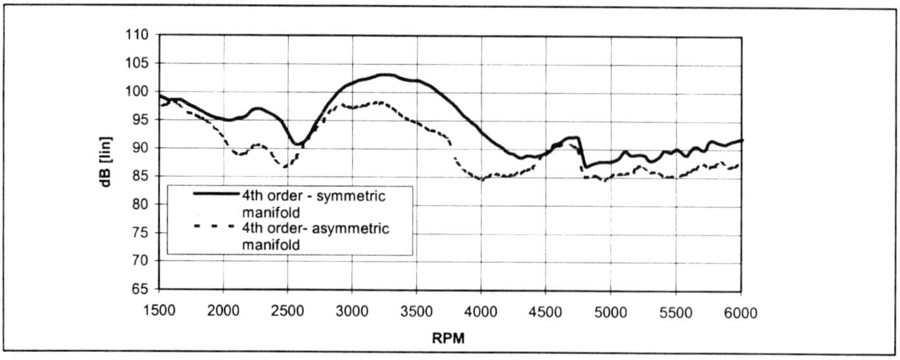

Fig.6a: 4th order tailpipe noise with manifold pipes optimization

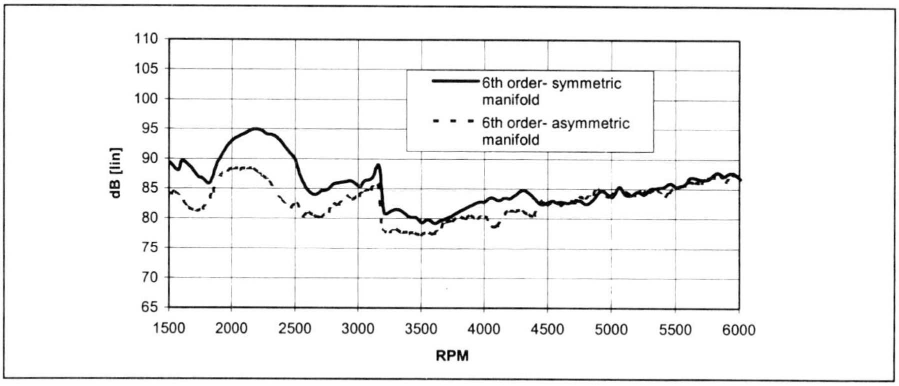

Fig.6b: 6th order tailpipe noise with manifold pipes optimization

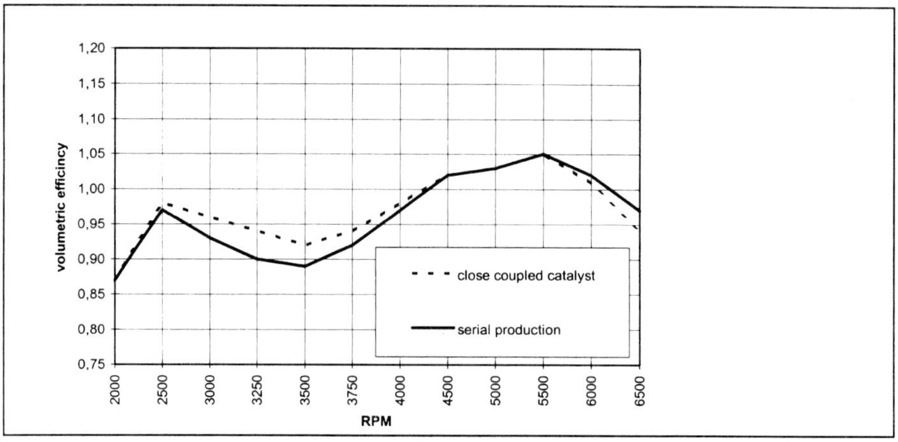

Fig.7:Engine performance evaluation

C521/035/98

Computer aided development of exhaust systems

K WOLFF, M SCHNEIDER, and **CH SCHERNUS**
FEV Motorentechnik GmbH & Co. KG, Aachen, Germany

ABSTRACT

Since exhaust systems contribute significantly to both, vehicle interior noise and pass-by-noise, they must be developed with respect to tailpipe and shell noise right from the beginning. In addition, the system vibration behavior influences the structure-borne noise transmission via the hangers into the vehicle with consequences in terms of noise quality and comfort as well as durability.

This paper describes a time and cost effective method of analytical treatment of planned exhaust systems during the concept phase. The method is based on Finite Element Analysis (FEA) and Computational Fluid Dynamics (CFD).

The exhaust system concepts and technical potentials can be evaluated already in the early stage of vehicle design, when layout and package are still not finalized. So, the simultaneous optimization with regard to NVH, durability and performance facilitates better results and also significant time saving during the optimization process.

1. INTRODUCTION

To keep pace with the continuously accelerating development processes of passenger cars, CAE tools are increasingly introduced into the design process. In the beginning, CAE analysis started with the more scientific approach of explaining real world problems; intensive and time consuming studies of boundary conditions were required. Later, with experience growing, the models could be simplified and boundary conditions carried over from earlier experiences. Thus, the explanation of complex problems arrived in time to assist the designer in focusing his development efforts. The next step, once a prototype and a model existed, was to extrapolate the technical benefits of the development-"step behind". For most technical CAE applications, this is still the state-of-the-art. Complex system modeling still needs calibration against absolute numbers from measurements of prototype hardware. Nevertheless, the target of the analytical approach is to predict the merits of design improvements on a purely computational base. For exhaust system development this paper shows how far technical progress allows a purely analytical approach today.

2. Gas-Dynamic-Calculation

Engine thermodynamics and acoustics used to be developed sequentially. Each department, one after the other, pursued the related targets. Therefore, development results were often quite contradictory.

The target values of exhaust and noise emissions, fuel consumption, package and performance will increasingly conflict in the future. "Simultaneous Engineering" is necessary to fulfill all these requirements equally. That means that computational and experimental efforts have to be carried out in cooperation. Suitable compromises must be developed to obtain the targets of design, thermodynamics and acoustics, as shown here for the development of intake and exhaust manifold.

This necessity to cooperate demands a computer code that can be applied to simulations of gas exchange, working process and acoustics and, therefore, will support cooperation of the specialists from the different disciplines.

In the past, one dimensional CFD codes were only used for gas exchange calculations. Over the past few years, however, these codes were expanded with acoustic tools to predict tailpipe noise. Through growing experience, the results of the calculations have become so accurate that the codes can be implemented in the development of exhaust systems. Nevertheless experimental tailpipe noise optimisation work is still important.

Figure 1 shows a CFD model which computes the gas flow from the intake orifice of the air cleaner up to the tailpipe. The exhaust system consists of a catalyst, an absorption front muffler and a reflection rear muffler (1).

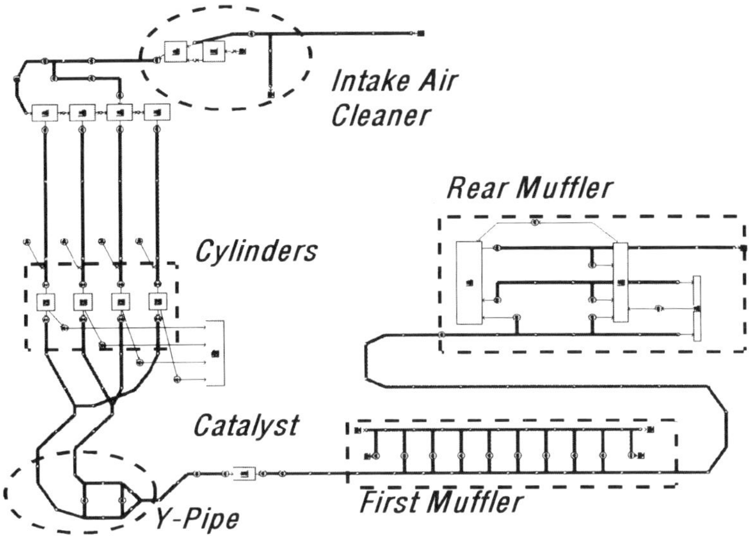

Figure 1: 1-Dimensional CFD-Model from Intake to Tailpipe

The comparison between calculated and measured tailpipe noise is shown in Figure 2. This is an example for a 2.0 litre 4 cylinder engine with a rated power of 74 kW. The predicted narrow band sound pressure levels of integer engine orders correlate with the measured data in the lower to medium speed range and up to 7th order within 2 dB. At higher speeds and higher orders, the sound pressure is dominated by flow-noise. The calculation does not include this noise share, so that the calculated level is lower than the measured one. The less significant noise levels at non-integer orders (i.e. 0.5, 1.5, ...) of the experiment tend to be higher than those calculated. This can be caused by deviations of the AFR (Air Fuel Ratio), by combustion rate difference from cylinder to cylinder and by cyclic variations that were neglected in the CFD model.

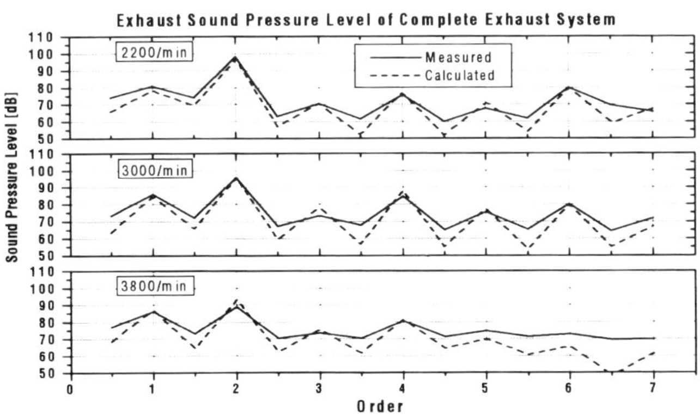

Figure 2: Comparison of Calculated and Measured Taiplpipe Noise

With today's modeling techniques, the tailpipe noise potential of different muffler designs can be determined. Thus, helpful recommendations for quick experimental muffler system development are obtained.

3. Shell Noise Radiation

For modern exhaust systems, the radiated shell noise contributes about 50% of the overall noise at drive-by test (2). The shell noise radiation of an exhaust system originates mainly from the manifold, the catalyst and the mufflers. Acoustic radiation analysis can concentrate on these components. The excitation mechanisms are due to gas pressure as well as vibration transfer from the cylinder head. The dominance varies significantly from engine to engine, so analytical prediction has to be focused on the very item since different frequency ranges are involved.

A FE model, as coarse as the frequency range allows, serves for Forced Response Analysis. The excitation is taken from FEV's database or, for the lower frequencies, from gas-dynamics of combustion force calculation.

In most cases, the problem of muffler shell noise radiation covers frequencies from .1 to 4 kHz (3). The effect of design modifications can be quantified analytically to determine the acoustic potential. The example shown in Figure 3 illustrates the effect on surface velocity of an additional baffle in the muffler (see marker). The front muffler -freely suspended- was excited virtually by forces at the tube entrance in a wide frequency range. This simulation stands for a vibration excitation from the cylinder head. In the frequency range between 1600 Hz and 2000 Hz, a noticeable improvement occurs without drawback in other frequency spans which results in a significant surface velocity level decrease.

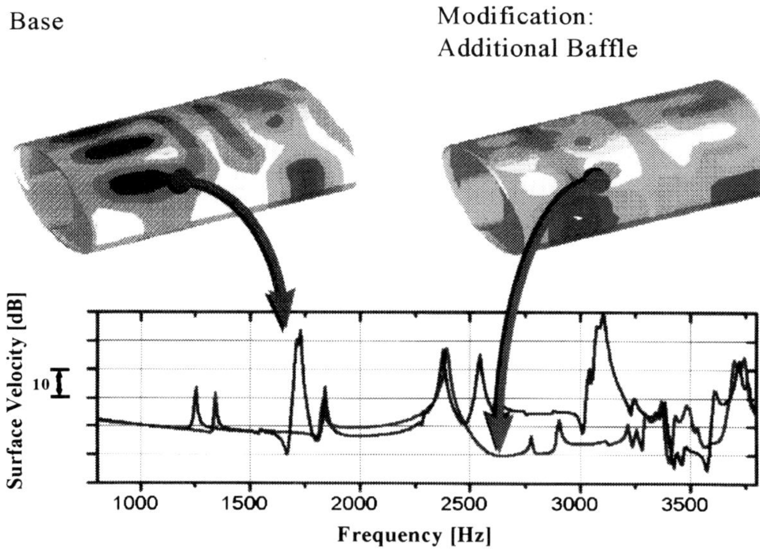

Figure 3: Surface Velocity

4. Global Exhaust System Vibration

Based upon the global dynamic behavior of the exhaust system, additionally an important compartment noise source arises. This is due to modal vibrations of the exhaust system structure which are transmitted into the car body via the hangers, especially in case of resonance with engine orders (4). Frequently this also causes durability problems.

In the past, the vibration effects of the exhaust system on vehicle interior/pass-by noise and also on durability used to be investigated experimentally after completion of first test hardware. Due to this procedure, too often problems occurred just before the start of series production, thus frequently causing time and cost intensive late production changes.

Meanwhile analysis tools such as the FEA (Finite Element Analysis) help to predict and solve these problems. Due to their high standard, they allow the development of complex dynamic systems at an early development stage. With FEA the dynamic behavior and durability of thin shelled structures can be calculated. Based on drawings or CAD files, FE models can be set up within a few days to dynamically layout and optimize the system structure during the concept phase before any hardware exists, thereby avoiding many time and cost intensive development iterations.

Since the basic modes of an exhaust system exhibit the highest amplitudes, they require special attention. As they are low frequent, they cover the range of 2nd, 3rd... engine order, depending on the engine, and resonance effects are quite likely. To this extent, a good understanding is achieved even with relatively rough FE models of the exhaust system. Pipes, mufflers and the manifold should be represented by shell elements. Typically the length of the shell elements is in the range of 20 to 40 mm, independent of the component beyond the manifold. Along the circumference about 12 elements are a good choice for the tubes. Figure 4 shows a rough meshed model of a exhaust system sufficient to serve the specific purpose. The flexible exhaust system model is coupled to a rigid body model of the powertrain which is originally mounted. The exhaust system is suspended at its rubber hangers.

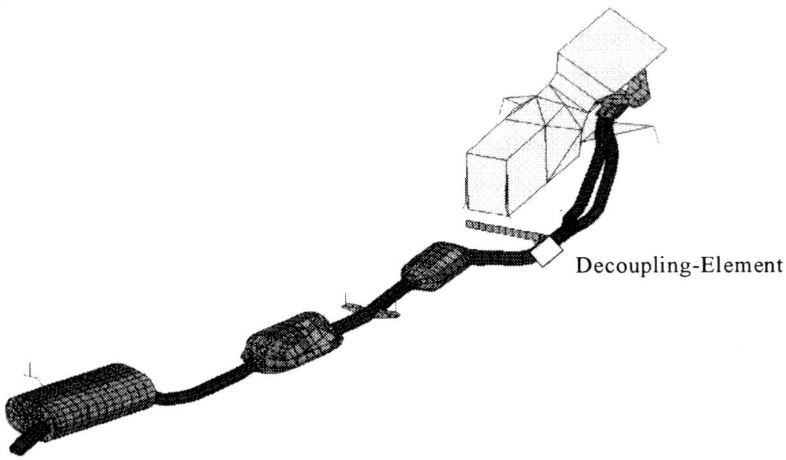

Figure 4: FE-Model for Dynamic Calculation

With the input of powertrain vibrations, excited by the gas and mass forces of the critical engine order(s), the dynamic forced response is calculated. Computational running mode analysis, if based on a realistic simulation code, facilitates the optimization of the following items:

- **dynamic load** of the exhaust system
- **exhaust hanger positions**
- **decoupling element** concerning position and properties
- **idle shake**
- **transient deflection** (package clearances)

As an example Figure 5 reveals the accuracy of displacement calculation of a 4 cyl. SI engine. The measured and simulated 2.order displacements at 2 positions along the exhaust system, that is front and rear muffler, are shown versus engine speed. In the diagrams the amplitudes of the resultant vector of the 3D-displacement values are depicted. At the interesting resonance peaks the calculated displacement is in agreement with the measured data within 20%. Taking into account that the reproducibility tolerance of the measurement is about 10% or even higher the reliability of the calculation is fairly good.

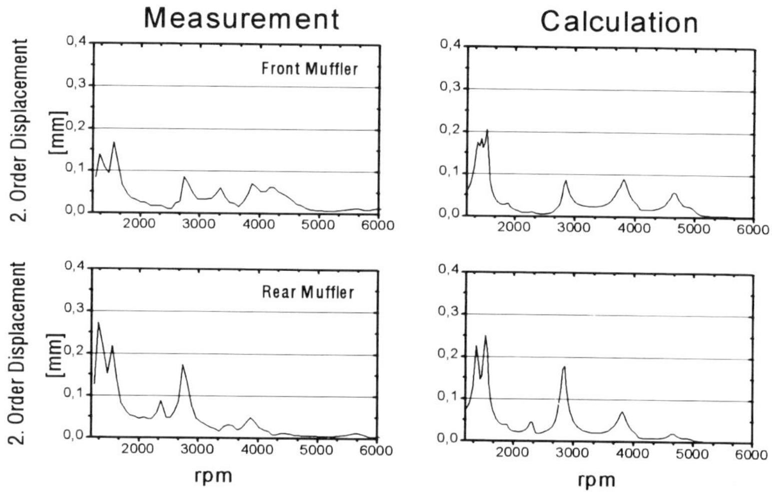

Figure 5: Comparison of Calculated and Measured Running Mode Analysis

The quality of the dynamic exhaust system model enables realistic stress predictions as well. The stress distribution is calculated at resonance speeds for critical components, such as muffler, catalyst and manifold. Contrary to modal analysis for running mode analysis different effects have to be considered in addition. In this context the vibration damping of the assembled components which is frequency dependent must be carefully assessed. Due to the impact of temperature on material properties, in the first step a reasonable temperature distribution along the exhaust system must be established. According to these boundary conditions the very data especially for the Young`s modulus are selected.

Figure 6 shows an example for the excellent standard of the prediction. Very high stresses with durability problems are shown at the rear muffler at a discrete resonance speed. The durability test confirmed the predicted results.

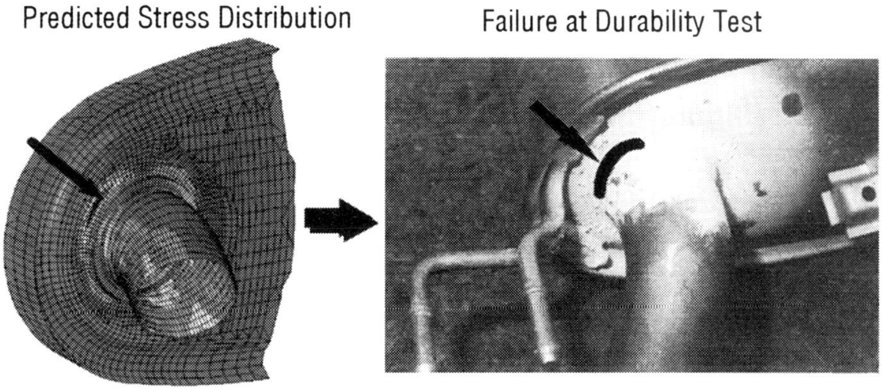

Figure 6: Calculation of Dynamic Load

The exhaust system may significantly contribute to idle shake comfort concerns (5). At resonance the system vibrations are transmitted via the exhaust hangers into the car body, causing noticeable vibrations amplitude at the steering wheel, seats and the instrument panel. Also peaks of the interior noise may be generated. If a hanger position is set close to an antinode of critical exhaust system modes, these phenomena become critical.

Figure 7 explains the three-step procedure of finding the best hanger position taking into account the vibration excitation of the exhaust system and the transfer function "body to interior". Based on the running mode shapes, the critical vertical amplitudes are summarized and displayed in colour for the areas of potential hanger positions. The summation covered all critical frequencies from 25 to 250 Hz. Above the excitation plots, the vehicle body sensitivity is depicted in terms of the transfer function "body to interior noise". With shakers applied to various points of the vehicle floor, the transfer function right to the passenger's ear was measured and displayed geometrically. The interior noise level is the product of the excitation and the transfer function which is shown in the upper maps as a function of the hanger position. With this information, the best hanger position can be determined. Thus by means of realistic simulation a good solution is found at a very early design stage, when possibilities concerning layout and package are still less restricted.

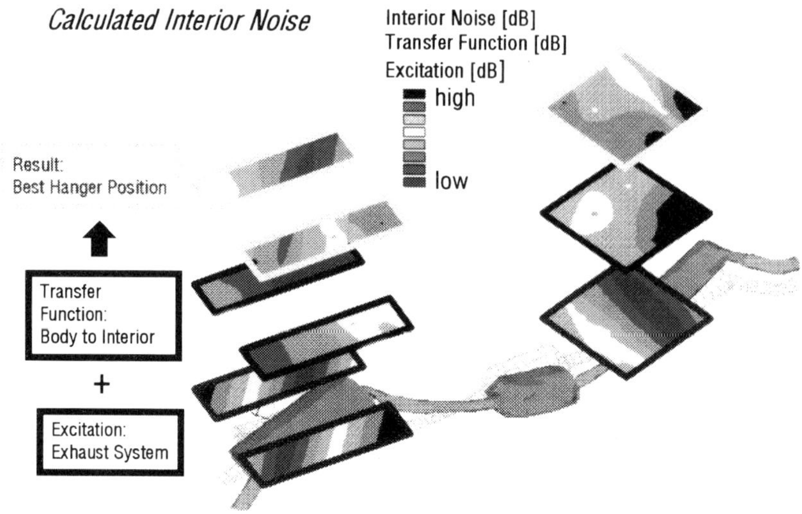

Figure 7: Best Hanger Position

For transverse engine applications, decoupling elements are used to compensate the roll angle and also to reduce the engine excitation. Over the past few years, FEV has collected a database to simulate non linear elements like that. Depending on the stiffness of the element the model varies from a single spring damper unit to a string of those units. Apart from the acoustical benefits, the decoupling elements help to solve dynamic load problems. The best position and properties, such as stiffness and damping of a decoupling element, will be worked out analytically, thus saving time and cost intensive experimental work.

The results of an analytical comparison between an exhaust system with and without decoupling element is shown in Figure 8. The diagram shows the displacement at one hanger for the 2nd order. The decoupling element has already been optimized. The benefit of the decoupling element is a displacement reduction of about 60%.

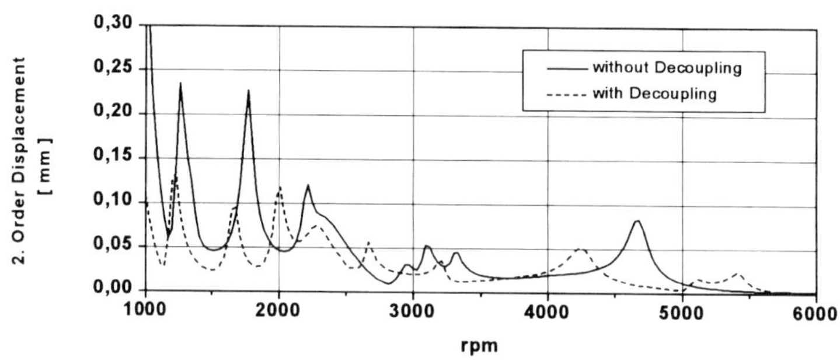

Figure 8: Influence of Decoupling Element

5. CONCLUSIONS

Exhaust systems may be critical to pass-by noise legislation and to interior noise comfort. For the development of exhaust system components, NVH, durability and performance are important criteria. For system optimisation in a time and cost saving manner and at an early design stage, FEM (Finite Element Method) and CFD (Computational Fluid Dynamic) are useful tools. FEM and CFD facilitate an early analysis and optimization concerning tailpipe noise, radiated shell noise, structure vibration behavior and dynamic stress load.

With today's CFD modeling techniques, the tailpipe noise potential of different muffler designs can be quantified. Thus, helpful recommendations for quick experimental muffler system development are obtained.

Shell noise is predicted based upon the dynamic FE model and measured or calculated information of the gas and structure excitation. Zones of high excitation or structural weakness are identified. Appropriate measures are worked out by means of locally modified FE-Models.

Global vibration of the exhaust system leads to comfort problems as well as durability problems. Calculation can quantify the excitation of the car underbody as well as the dynamic stress load.

References

(1) Schernus, Ch.; Lang, O. Computational and Experimental Investigation of the Acoustic and Gas Exchange Behavior of a Four Cylinder SI Engine; GT-Power Users Conference, Fellbach, 1997

(2) Lehringer, F., Kattge, D. The Radiation of Sound from Silencing Exhaust System; Automobiltechnische Zeitschrift 1987

(3) Kohichi, Muto A new device to reduce Exhaust Pipe shell noise; SAE Technical Paper Series 870631

(4) Hanus Müller Verbesserung des Innengeräusches von Kraftfahrzeugen mit Hilfe von Elastomerbauteilen; Haus der Technik, Fahrzeugakustik, Dezember 1996

(5) Schedl Moderne Analyse- und Berechnungsverfahren in der Motorakustik; Haus der Technik, Essen, Geräuschminderung bei Kraftfahrzeugen, März 1992

Analysis

C521/009/98

A study of disc brake squeal propensity using a parametric finite element model

Y S LEE, P C BROOKS, D C BARTON, and **D A CROLLA**
University of Leeds, UK

An integrated approach to the analysis of disc brake squeal noise is presented which is based upon a parametric finite element representation of the disc-pad assembly. The fundamental stages in the procedure are outlined and illustrated through an example analysis. Results from parametric analyses, typical of those routinely performed, are presented and the predictions shown to agree with reported behaviour. The complex modal analysis, used to determine the onset of squeal, permits individual unstable modes to be studied in detail by focusing on the modal motion of individual nodes in the contact region. This permits the engineer to obtain an understanding of the physics of the underlying squeal phenomenon which has been hitherto difficult to obtain.

1. INTRODUCTION

The dynamic instability of a disc brake can be viewed as a self-excited oscillation brought about by friction interaction at the rotor-stator interface. It is most frequently manifest in squeal noise and vibration. Behaviour of this type represents a serious problem in all mechanisms which are subject to relative motion at interfaces. Disc brake squeal is inherently fugitive in nature meaning that it is difficult to control or eliminate. The driving force for the research outlined herein derives from the high cost to the industry from brake noise related warranty claims coupled to potential environmental noise legislation.

There are two significant considerations which must be borne in mind when addressing the brake squeal problem. These are the mechanics of the friction interface and the incomplete unsymmetric contact which occurs over the frictional rubbing surface when braking loads are imposed on the system. Assessment of the system dynamics can be performed through the extraction of the complex eigenvalues and eigenvectors. The key point in achieving the goal of squeal elimination is directly related to an understanding of the frictional interface behaviour under realistic dynamic braking conditions. This then is the challenge of the current research which has explored ways in which to realistically represent the friction induced contact mechanics which govern the accuracy of the prediction of the disc brake instability. The finite element method presents an appropriate technique since it permits the flexibility of the component parts to be properly represented.

Tirovic and Day [1] performed a three dimensional finite element analysis of a floating caliper disc brake system under conditions of zero and non-zero interface friction and predicted the interface contact area and pressure distributions. The extent of contact and therefore the distribution of pressure was found to be critically dependent on the level of interface friction. Contact over the interface was found to reduce at the trailing end and build up towards the leading end as a result of interface drag forces. The notable work of Liles [2] reports an analysis of brake instability using a sophisticated finite element model. This employed equally stiff springs over the entire area of each interface to connect the adjacent nodes on the rotor and pad linings, thereby allowing variable normal forces at the interfaces. Constraint equations were used to maintain compatibility of displacements of the interacting components. The imposition of frictional forces upon the system gives rise to additional off-diagonal frictional coupling terms in the stiffness matrix which becomes unsymmetric and so leads to a complex eigenvalue problem. The solution to this problem yields an instability measure which can be used to evaluate the squeal propensity. However, no allowance was made for the actual variation in contact area and pressure distribution under braking conditions.

This paper presents an overview of an analytical technique which combines each of the above mentioned methodologies using conventional FE software tools to achieve the desired aim of an integrated finite element approach to disc brake squeal. The availability of a parametric design language has allowed the model to be described in terms which enable differing geometries and loading conditions to be easily generated and their effect on squeal performance evaluated.

2. MODEL AND METHODOLOGY

A foundation brake comprises of a number of individual components and clearly the overall noise performance of the system will in some way be influenced by the degree of interaction which takes place between one element and its neighbours. Since it is the purpose of this research to develop an understanding of the friction interface mechanics on the squeal propensity of the brake, the finite element model is made up of the rotor and the two pads of friction material. Connection to other components is achieved through appropriate constraints and boundary conditions. The caliper, which for this particular brake is of a floating design, is assumed to behave as an inertial mass and so is not explicitly defined within the model. It is recognised that this component is likely to exert some control over the system by virtue of its compliance.

2.1. The finite element model

The baseline finite element model is shown in Figure 1. Each of the components is modelled using 8 noded three dimensional solid elements. In total, there are 2016 elements with 3416 nodes in the structural model. When the nonlinear contact analysis is conducted, three dimensional psuedo nonlinear interface elements are used to connect across the entire frictional interface. These are replaced by linear interface elements which act between the contacting interface, as predicted by the nonlinear analysis, during the modal analysis.

The pads maybe constrained at leading and/or trailing abutments in the plane of the disc whilst the disc is fully constrained around the inner ring of the top hat. Piston loading is applied centrally to the inboard pad backplate whilst the outboard pad backplate is loaded symmetrically about its radial centre line by two fingers which form part of the caliper paw structure.

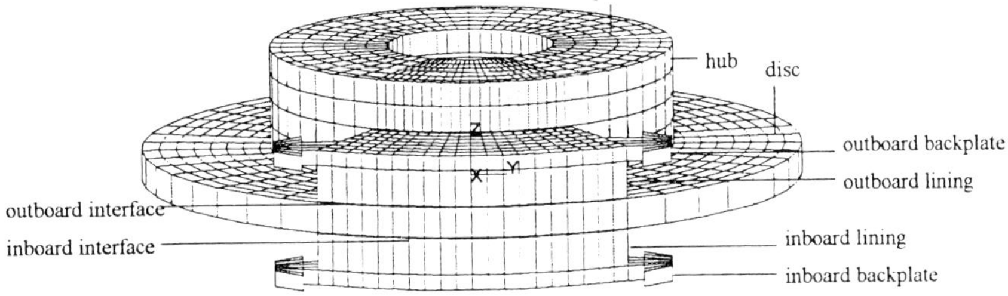

Figure 1: Baseline finite element model

The parametric nature of the model permits control over the following quantities, thereby allowing the analyst to quickly and efficiently undertake parametric sensitivity studies:

- The geometry of the three components, including the provision of a radial slot and its position within each pad of friction material.
- The loading condition from the piston and paw.
- The friction coefficient at the interface.
- Abutment constraints (between the caliper and pads).
- The magnitude of the contact stiffness between the lining and the disc.
- Material properties of all components.
- Mesh density.

Some of the essential parameters which control the setup of the baseline model are given in Table 1.

2.2. Analysis technique

The method of analysis essentially has two parts. The first permits calculation of the interfacial contact behaviour and the second part takes a linearised version of the same system and extracts the complex eigenvalues and vectors.

Parameter name	Definition	Parameter name	Definition
Abutment mechanism	Combined	Slot	None
Caliper loading	Piston - Paw	Backplate elasticity	209 GPa
Lining elasticity	1 GPa	Disc elasticity	220 GPa
Backplate density	7800 kgm^{-3}	Lining density	3000 kgm^{-3}
Disc density	8000 kgm^{-3}	Backplate thickness	5 mm
Lining thickness	12 mm	Disc thickness	10 mm
Piston diameter	42 mm	Piston location	Central
Contact stiffness	1.2 GNm^{-1}	Interface friction	0.4
Actuation pressure	0.8 MPa	Pad arc	56°
Inside rubbing radius	76 mm	Outside rubbing radius	118 mm

Table 1: Parameters for baseline model

The nonlinear contact analysis is used to determine the contact boundary, normal contact force and friction force distribution under normal braking conditions. The nonlinear behaviour within the structural model is associated with the contact elements used to determine the extent of the contact patch and the magnitude of the interfacial forces. The key to linearisation lies in the replacement of only those nonlinear elements which are closed as a result of the applied braking loads with equivalent linear interface elements. This process is illustrated in Figure 2.

Since the solution to the contact problem is unique, a linear interface element must be capable of replicating those normal and tangential loads predicted by the nonlinear model. The interface element chosen is one whose geometry is undefined but whose elastic kinematic response can be specified by a stiffness. The resulting unsymmetric stiffness matrix relates the displacement of two nodes, each having 6 degrees of freedom, and is formulated by combining Coulomb's law with elastic theory. Thus each pair of contacting nodes has associated with it a unique stiffness matrix which can transfer both normal and in-plane forces.

Once linearised, the modal analysis of the coupled brake system is based on that discussed in [2] and relies upon the theory given in [3]. The natural frequency and measure of instability are derived from the two parts of the complex eigenvalues. The software solves the eigenvalue problem using the Lanczos algorithm [4] which uses the full mass and stiffness matrices. The real part of the complex eigenvalues returned by the algorithm is the measure of modal frequency whilst the imaginary part quantifies the degree of instability, a positive imaginary part indicating an unstable mode of vibration.

The propensity of the system to squeal is best represented by a single number as against a set of eigenvalues, each having its own real and imaginary component. This work uses the same method developed in [2] to calculate such a measure. It is derived from the standard deviation of all the positive instability measures of those eigenvalues within a pre-determined frequency range from the mean value of zero. System instabilities are brought about by the presence of interface friction forces and so the same measure of standard deviation has been applied to the distribution of friction forces over the interfaces and the results compared to those derived from the system eigenvalues.

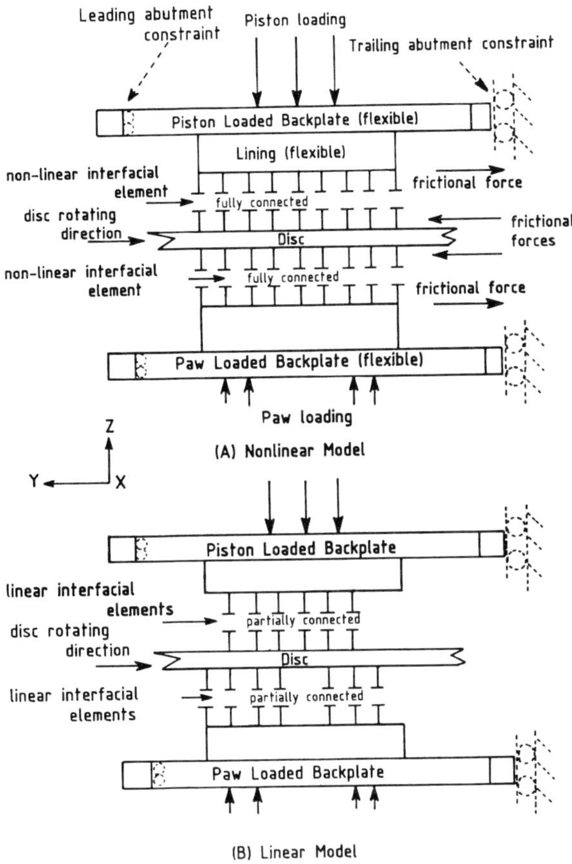

Figure 2: Idealisation of the disc brake system

3. SIMULATION RESULTS

A set of results is presented for the baseline model commencing with the interface pressure distributions, through the eigenvalue plot and associated instability measurement and ending with a discussion of the modeshapes associated with a stable and unstable mode. This is then followed by two sets of results derived from a series of parametric sensitivity studies in which the geometry of the system is first modified and then the load case altered.

3.1. Baseline system performance

Figures 3 and 4 show 3-dimensional plots of typical interfacial contact force distributions over the pad rubbing surfaces which have been imposed on a rectangular mapping plane. These normal loads are the nodal forces generated by the structural finite element model when subject to the brake actuation loads for a given simulation. The horizontal axis defines the circumferential direction and the remaining in-plane axis represents the radial direction. The vertical axis represents the magnitude of the normal contact force in

Newtons. The effect of in-plane frictional forces is clearly seen since the apparent centre of pressure in each plot is skewed towards the leading end of each interface. The piston loaded interface exhibits the worst contact performance with 21.4% of the interface losing contact around the trailing end. In contrast, the paw loaded interface is far more evenly loaded, with only a marginal loss in contact, 1.2%, at the corners of the trailing end, predicted by the analysis.

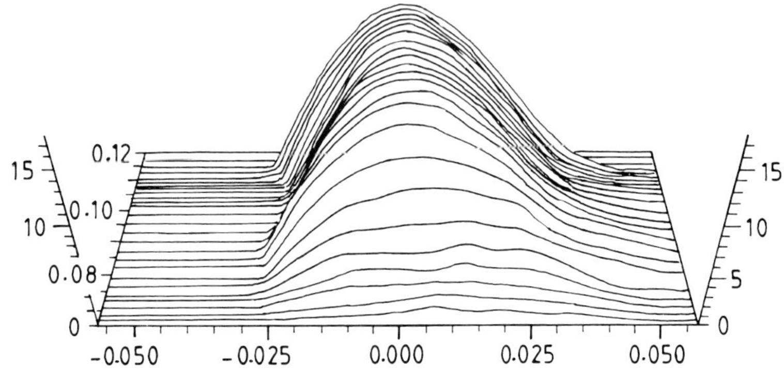

Figure 3: Contact force distribution over the piston loaded interface

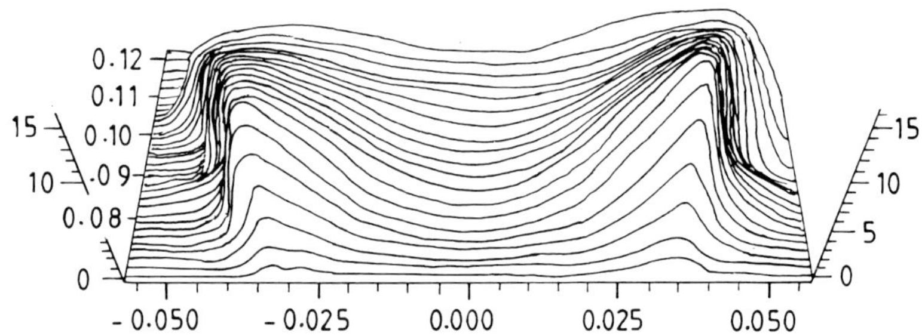

Figure 4: Contact force distribution over the paw loaded interface

The system eigenvalues, derived from the linearised model, are shown in Figure 5. In addition to the large number of stable normal modes, characterised by zero instability measure, a significant number of unstable complex modes are predicted each having a positive instability measure. The unstable modes lie mostly in the frequency range above 6 kHz and the standard deviation of the instability measure of the model up to 14 kHz is 199.8 s^{-1}.

Associated with each eigenvalue is a unique eigenvector which characterises the motion of the mode as a set of relative amplitudes together with phases if the mode is complex. The conventional approach to the study of mode shapes derived from a finite element analysis involves looking at contour plots onto which are mapped the relative displacements

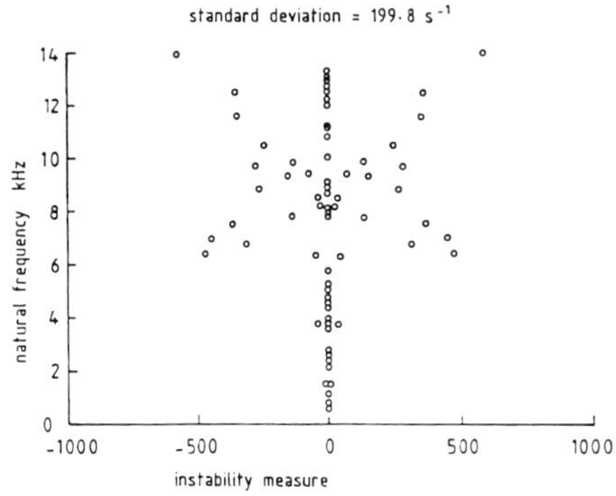

Figure 5: Eigenvalue plot for baseline model

occuring in the direction of interest. This is satisfactory for a normal mode but rather limiting when studying a complex mode as the important phase information becomes almost impossible to visualise. Nonetheless the spatial phasing of the real and imaginary parts of a typical complex mode when viewed through contour plots are in accord with the experimental results of Lang and Newcomb [5].

In order to explore the movement at the disc-pad interface during a complete cycle of oscillation, the real and imaginary parts of the relevant eigenvector, as predicted by the finite element model, are processed to yield a normalised amplitude and phase for each node in the contact region. These are then plotted enabling visualisation of the key interface motion of the two lining surfaces together with that of the disc, over a complete cycle. This highlights fundamental differences in the motion of stable and unstable modes and is accomplished by plotting the motion of the components along a circumferential arc at a specified rubbing radius. By looking at several plots, each taken at different rubbing radii, a picture of the true three dimensional motion can be quickly assertained.

Figure 6 shows such a plot for a stable normal mode, along the central circumferential arc of the rubbing surface, which has a frequency of 5775 Hz. The abscissa represents the circumferential position of the interface whilst the ordinate represents the normalised amplitude in the axial direction calculated throughout a half cycle of oscillation. The normal mode is characterised by nodal lines which connect stationary positions where the displacement remains zero throughout the whole oscillation. The pad and disc surfaces demonstrate in-phase motion within the contact region whilst those non-contacting areas exhibit out of phase displacements. This separation is most noticable on the outboard lining surface. If the same eigenvector elements are plotted instead on an argand diagram, then all the nodal displacements are either in or out of phase.

Repeating the above procedure for a typical unstable complex mode leads to an argand diagram in which the nodal displacements occur at random phase with respect to one another. A modeshape associated with an unstable complex mode is illustrated in Figure 7 which has a frequency of 6471 Hz and an instability measure of 475.9 s^{-1}. As a consequence of this random phasing, the complex mode shape is characterised by having no clearly defined nodal lines within the region of contact. This phenomenon is argued

[6, 7] to provide the necessary energy required to stimulate the divergent response of the unstable mode. In common with the normal mode, structural compliance is maintained in those areas in contact whilst large differences in axial displacement are evident in those regions which separate. The motion of the two pads and disc are in phase at the leading end of the assembly whilst at the trailing end the outboard lining is executing out of phase motion with respect to the rotor and inboard lining. A physical interpretation of this behaviour suggests that the outboard pad impacts the rotor at the modal frequency of 6471 Hz.

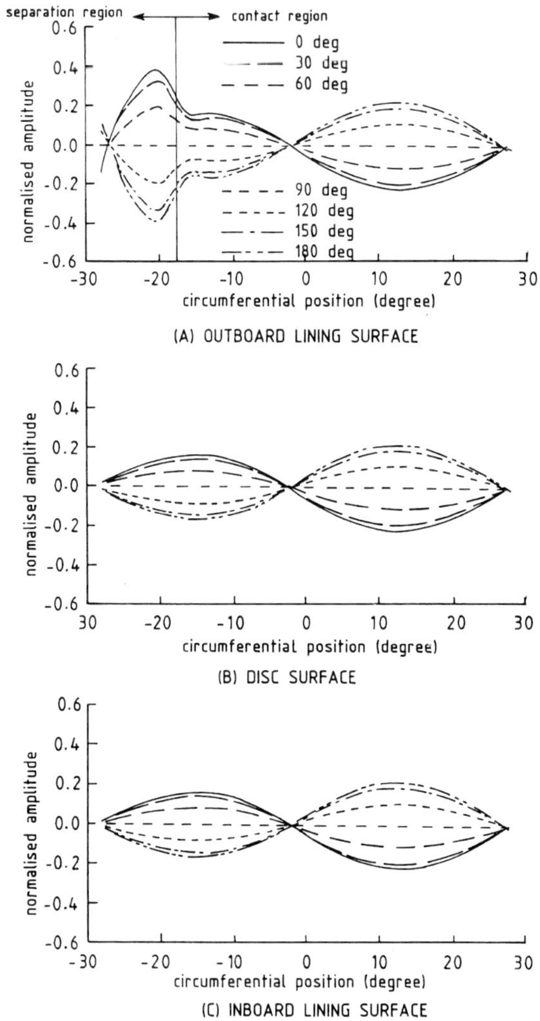

Figure 6: Stable normal mode

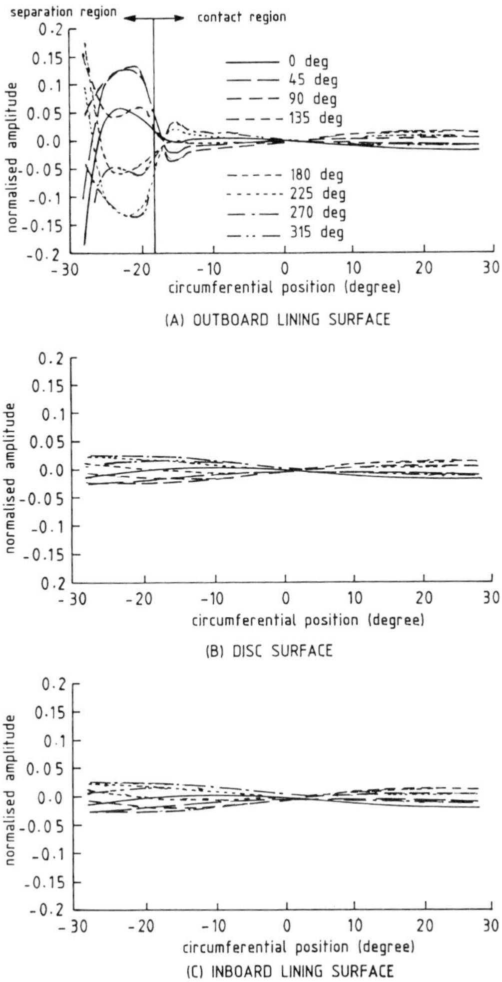

Figure 7: Unstable complex mode

3.2. Typical parametric analyses

3.2.1. Backplate thickness

The thickness of the friction material backplate is a parameter which can be readily varied by the designer. It is also a parameter which has been demonstrated [8] to influence the ability of the pad to evenly distribute the normal caliper loads. As the thickness of the backplate is reduced then the interface presure distribution becomes less uniform. The results, shown in Figure 8, demonstrate that the squeal propensity of the brake is also similarly affected by reduction in backplate thickness. When the thickness is reduced by 50% then the analysis predicts a 34% increase in squeal propensity. The standard deviation of the frictional force is also plotted in Figure 8 and this follows the same trend.

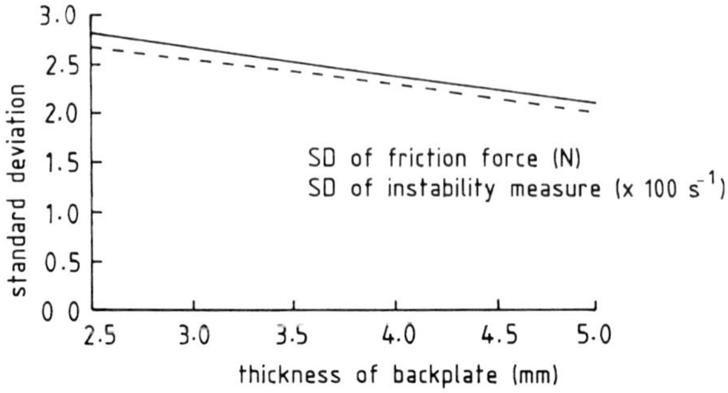

Figure 8: Effect of backplate thickness

3.2.2. Pad abutment

Brake torque is reacted between the caliper and pad backplate by abutments located at each end of the backplate. There are three generic designs: A leading, trailing and combined abutment. In the latter the brake force is notionally shared between the leading and trailing ends of the backplate. Each of these cases can be represented within the model and so the effect of abutment condition on squeal may be readily evaluated. The results, Table 2, indicate that a disc brake incorporating a trailing end abutment is least likely to give rise to squeal, even though this configuration is associated with the least uniform interface pressure distribution. This agrees with the findings of holographic experimental studies [9].

Abutment condition	**Standard deviation Instability measure s^{-1}**
Leading	175.6
Trailing	145.0
Combined (Baseline)	199.8

Table 2: Effect of abutment condition

4. CONCLUSIONS

An integrated approach to the analysis of disc brake squeal has been presented which relies upon a three stage finite element formulisation of the problem involving a non-linear contact analysis, system linearisation and modal analysis. The finite element model is parametric enabling efficient quantification of the effect of geometric and loading parameters on squeal.

A baseline set of results has demonstrated the use of the technique and this has also led to the proposal of a squeal mechanism. Further results, taken from a sensitivity study, show the usefulness of the parametric approach and the results have been found to be in agreement with those of other workers in the field.

The technique has proven to be robust and the potential to extend the method to a complete foundation brake exists. The current model does not take into account disc thickness variation or temperature related effects which are present in a real brake. The study of these aspects is the focus of ongoing research.

REFERENCES

1. Tirovic M, Day A J. Disc brake interface pressure distributions. Proc IMechE, Vol 205, p137-146, 1991.
2. Liles G D. Analysis of disc brake squeal using finite element methods. SAE Technical Paper Series, SAE891150, 1989.
3. ANSYS User's Manual for Revision 5.0, Volume 4. Swanson Analysis Systems Inc, PO Box 65, Johnson Road, Houston, USA.
4. Rajakumar C, Rogers C R. The Lanczos algorithm applied to unsymmetric generalised eigenvalue problem. Int Journal for Numerical Methods in Engineering, Vol 32, p1009-1026, 1991.
5. Lang A M, Newcomb T P. The vibration characteristics of squealing brakes. FISITA Congress, Paper 905170, 1990.
6. Ibrahim R A. Friction induced vibration, chatter, squeal and chaos: Parts 1 and 2 - Mechanics of Friction. Trans ASME DE-Vol 49, p107-138, 1992.
7. Ghesquiere H, Castel L. High frequency vibrational coupling between an automobile brake disc and pads. AUTOTECH'91, IMechE Paper C427/11/021, 1991.
8. Tirovic M, Day A J. Disc brake interface pressure distributions. Proc IMechE, Vol 205, p137-146, 1991.
9. Fieldhouse J D, Newcomb T P. An experimental investigation into disc brake noise. Proc Int Conf Braking of Road Vehicles, Paper C444/0016/93, IMechE, London, 1993.

C541/014/98

The investigation of sound fields and their radiating surfaces using acousto–optic holographic modal analysis

C H BUCKBERRY BSc, MTech, PhD, MIMechE and **N C TAYLOR** BSc
Rover GroupLimited, Warwick, UK

SYNOPSIS

Sophisticated full-field measurement techniques are now available commercially, and researchers can now readily study both the surface vibration and the resulting propagation of acoustic energy from that same surface. This can be done using two types of Holography. Using optical wavelengths, Holographic interferometry can study the surface motion and provide measurements of both vibration amplitude and phase. Similarly, by transferring to the domain of acoustic wavelengths, the radiated sound field can also be mapped under steady-state conditions using the process known as Nearfield Acoustic Holography.

This paper describes preliminary work into understanding how the surface and its sound field are related to each other under some simple test conditions. This work has also allowed the authors to understand how these new measurement tools can best be used together and how this new data can be used to supplement that provided by more traditional techniques such as that acquired by accelerometers.

1. INTRODUCTION

By virtue of the fact that full-field measurements provide an image of an object's behaviour, they provide a simple but significant advantage over discrete point based measurements. However, as these techniques and the data that they provide are still relatively new much has yet to be learnt about how best to utilise them in combination when studying acoustic related phenomena. Rover has two full-field systems that are used daily for their own specific application areas. They are Holographic Interferometry (HI) and Nearfield Acoustic Holography (NAH)[1]. Traditionally, these two systems have been used separately to identify resonant mode shapes and measure pass-by-noise levels respectively. However, recent advances with an in-house developed Television(TV) based HI system has permitted new types of data to be extracted. This has consequently allowed the two systems to be used together to study how sound fields and surface vibrations are actually related to each other. Specifically, the aim was to observe what happens to the sound fields when travelling waves are present on the vibrating surface.

1.1 Holographic Interferometry.

During the last decade, HI has become a widely used tool in the automotive industry for the study of surface vibrations and deformations. By using a laser light source, usually operating at a wavelength in the visible region of the electromagnetic spectrum, a two-beam interferometer can be constructed, providing that the bandwidth of the laser is sufficiently narrow. Such a device will then record the interference pattern that results at the superposition of the two beams. The system is usually so arranged that the path length of one of the beams is altered by the motion of an object, and as a result the motion of the object whilst exposed to the light is recorded. The recording is then a fringe map that defines contours of equal displacement.

Traditionally, this recording would have been made upon a glass plate, which would have been previously coated with a photographic emulsion. Rover has helped pioneer a modified version of the technique that utilises a CCD camera instead, thereby permitting the recording of 'Holograms' at a rate of 25/second rather than approximately 1 every 15 minutes with the glass plate as a recording medium. This system is known as TV Holography[2] although its more traditional name is Electronic Speckle Pattern Interferometry (ESPI). A typical system configuration and result are shown in figures 1 and 2 respectively.

1.2 Near-field acoustic holography.

By scanning an array of microphones to measure the acoustic intensity at a number of points in a measurement plane, it is then possible to reconstruct the whole sound field within a volume utilising diffraction theory. The Spatial Transformation of Sound Fields [3] is the Bruel & Kjaer trademark for a measurement system of this type that can, from a grid of pressure cross-spectra over a planar surface, allow a detailed 3 dimensional map of the sound field to be calculated. The Helmholtz integral equation is used to calculate the resultant sound in the far-field and Near-field Acoustic Holography is used to calculate active and reactive intensity in the near-field. The active intensity vector calculations are particularly useful in showing the propagation from the near-field to far-field and are used here to show the acoustic radiation from the surface of the disc.

1.3 A combined approach.

Previous researchers have also made a considerable effort to extend the application of HI by reproducing its output in a manner more akin to that of classical modal analysis using accelerometers. Hyodo et al[4], performed a full optical modal analysis, i.e. combining a number of holograms of an object excited at a single frequency (taking the vibration amplitude term only), to see if they can be used to reconstruct a hologram of the same object undergoing excitation at all frequencies. Imai[5] showed that it was possible to predict the sound pressure level at a point above the surface, again using only the measured vibration amplitude term and making assumptions about the vibration phase. Brown[6] utilised measured values of vibration amplitude and phase for the first time to predict sound pressure at a point close to the object. More recently Freymann[7] has shown that the data density can be increased to the point of gathering amplitude and phase data "Hertz by Hertz", in order that true modal data can be acquired rather than doing so at discrete frequencies.

At Rover the possibilities of how best to use the HI with NAH in order to correlate their time-averaged spatial maps of sound pressure and surface vibrations, were considered. We have attempted to show the possible alternatives in figure 3. Additionally, it is shown in figure 3 where data could also be cross correlated between theoretical and experimental data.

2. AIMS

In this piece of work the aim is to show that with a simple object, the sound pressure level map back-projected to the surface from the scan plane does correlate with the map of vibration amplitude, for simple sinusoidal motion of the surface.

With the TV Holography system, vibration phase maps that possess phases other than 180 degrees between anti-nodes, are frequently measured. These results occur for a number of reasons, both material and structural. This is indicative of a significant level of damping and therefore requires the presence of a damping coefficient in any numerical description of the object's motion. Such maps therefore provide a more complete and generalised description of the motion of the object and hence a better means of correlating Finite Element models to experimental data than vibration amplitude maps. Finite Element models are typically based upon a linear solution to a set of second order differential equations. This is achieved by assuming that the coefficient of damping is zero and that the velocity term can be cancelled out. As a consequence the only vibration phase values that can exist in a linear numerical solution are -90 and 90 degrees. This piece of work seeks to observe that the presence of vibration phases other than -90 or 90 degrees as measured by TV Holography, manifested themselves in the sound field.

3. EXPERIMENTAL ARRANGEMENT

The object chosen for these experiments was a vented brake disc. From earlier investigations into brake squeal, the vibration behaviour of discs is now well understood and we know that the modal density is low and that therefore the resonant modes are very widely spaced. This reduces the possibility of simultaneously exciting more than one mode.

Secondly, all the vibrating areas of the disc's surface lie on a single geometric plane of the object, thereby alleviating any requirement to measure the object's form in order to measure the vibration amplitude relative to a line drawn normal to the surface. There is a central aperture for the wheel hub, but relative to the wavelengths found in the kiloHertz region, this was thought unlikely to produce any diffraction effects.

Thirdly, it is very easy to produce travelling waves on the surface of the disc. This is because the material damping is minimal and the shape is highly symmetric, allowing orthogonal modes, within a few Hertz of each other at each resonance to be excited easily. The orthogonal modes are generated by a lack of absolute symmetry and are fixed in space by the suspension holding the disc and the point of excitation.

The two sets of measurements required for HI and NAH were taken simultaneously. This prevented any frequency and amplitude variation occurring between the two and allowed the HI to monitor any frequency drift that might occur during the experiment. The experimental layout is shown schematically in figure 4.

The microphone arrangement in the scan array relative to the brake disc is shown in figure 4. The microphones were spaced 0.05m apart. However, by scanning the sound field twice a final measurement spacing of 0.025m was obtained. A reference signal was taken from the signal used to drive the piezo-electric crystal that was bonded to the disc to provide the source of excitation. At this grid spacing spatial aliasing will occur above (343/0.025) = 13720Hz, 343 being the value taken for the speed of sound in air in metres per second. With a grid size of 25 x 26 points the measurement will be valid down to a lower frequency limit of (343/(24x0.025) = 572Hz.This is the lowest frequency at which one complete wavelength will be captured within the smallest of the scan dimensions (total x-axis traverse distance). Note that $343ms^{-1}$ quoted above, is taken as the speed of sound in air in metres/second. The distance between the scan plane and the source surface needs to be of the same order as, but slightly greater than, the grid spacing, this was therefore set to 0.035m from and parallel to the disc surface.

The holographic data was imaged onto a CCD array and then digitised to 512x512x8 bit data. Rover's proprietary software then analysed the data to generate maps of absolute vibration amplitude and phase using optical phase measurement techniques described in reference [8].

The object itself was mounted in manner consistent with a 'free-free' boundary condition. Resonant modes of the disc were then identified by observing the live response on a TV monitor to a frequency varying sinusoidal excitation over the range 1 to 10kHz.

4. RESULTS

Figure 5 shows the vibration amplitude and phase maps for the resonance centered about 3918Hz. This frequency was chosen because it most clearly showed the coupling that can occur between two orthogonal modes. Figure 6 shows the corresponding active acoustic intensities at the surface. It can be seen that for each anti-node on the vibration amplitude map there is a corresponding set of concentric sound intensity contours. No vibration amplitudes or acoustic intensity levels have been given because we are only correlating spatial maps at this stage.

As the orthogonal mode rotates between sub-figures 5a and 5c so do the sound intensity contours in figures 6a and 6c. Note that the vibration phase map in figure 5a clearly shows six consecutive areas in anti-phase to each other i.e. 90 then –90degrees. Figure 5c shows this less clearly because the lower frequency orthogonal mode is still being partly excited.

More importantly, when both orthogonal modes are simultaneously excited, figure 5b, it can be seen that a set of three continuous phase ramps is generated and that there is no nodal line delineating the nominal anti-nodes in the amplitude map. In the corresponding acoustic intensity map, figure 6b, the travelling surface wave results in a time-averaged ring shaped pattern like that of the vibration amplitude map.

This simple experiment shows that whenever possible a combination of the two techniques leads to a better understanding of what type of vibration generates a particular sound field, and removes some of the ambiguity of making time-averaged spatial measurements.

5. CONCLUSIONS

The two techniques have shown that vibration phases do manifest themselves in the acoustic field and that as a consequence theoretical models should accurately predict the vibration phase map, for which TV Holography is an appropriate tool to measure it with. There also now exists a possibility that by using the combination of the two techniques it maybe possible to separate the structural and airborne sources of noise. This would be very advantageous when measuring the transmissibility of vehicle bulkheads.

In the immediate future, more work will be done to measure how the different effects of damping manifests itself on the resulting vibration phase maps and identify the different sources of damping. We are also writing the data into other appropriate file formats in order that the TV Holography data can be exported to other software packages such as Sysnoise™. This would then allow correlation between HI and NAH at planes other than just that immediately above the surface of the object. Finally, the same files will be accessed in a slightly different manner in order to allow classical modal parameters to be obtained in order that Finite Element models can be more accurately defined and allow a correlation between practice and theory.

Ultimately we hope to measure and separately identify, structural and acoustic sources on objects that have a complex surface form.

6. REFERENCES

1. Maynard, J., Williams, E., and Lee, Y., "Near-field acoustic holography. I: Theory of generalised holography and the development of NAH", J. Acoust. Soc. Am., Vol. 78, (4), p1395, 1985.
2. Davies, J. and Buckberry, C., "Television Holography and its applications", Chapter 8, Optical Methods in Engineering Metrology Ed. D.C. Williams, pub. Chapman & Hall.
3. Hald, J., "STSF. A unique technique for scan-based near-field acoustic holography with restrictions on coherence", Brüel & Kjær, Technical Review No.1, 1989.
4. Hyodo, Y. and Konomi, T., "Mode analyses using laser holography", Jnl. Japan. Soc. Auto. Engnrs, Vol. 7, (3), p44, 1986.
5. Imai, M., Suzuki, S., Sugiura, N., and Saito, H., "Radiation efficiency of engine structures analysed by holographic interferometry and boundary element calculation", Soc. Auto. Engnrs, SAE 861411, 1986.
6. Brown, G., "Fringe analysis for automotive applications", Fringe Analysis 92, Leeds, Institute of Physics, 1992.
7. Freyman, R. and Honsberg, W., "Holographic Modal Analysis", LASER 95, Munich, 20-23 June 95.
8. D.P.Towers, C.H.Buckberry, B.C.Stockley, M.P.Jones, "Measurement of Complex Vibration Modes and Surface Form - A Combined System", invited paper to Measurement Science & Technology, Volume 6, pp.1242-1249, 1995.

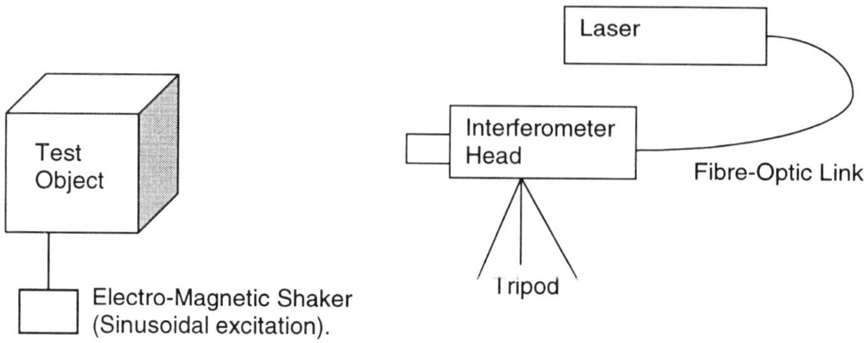

Figure 1: Typical layout of a TV based holographic interferometer.

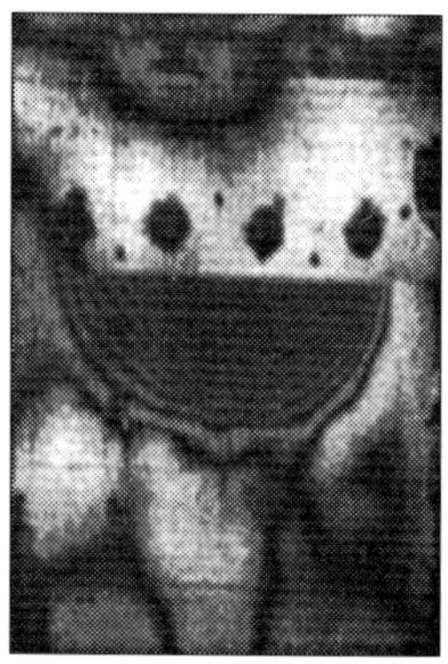

Figure 2: Typical vibration fringe map from a TV Holography system.

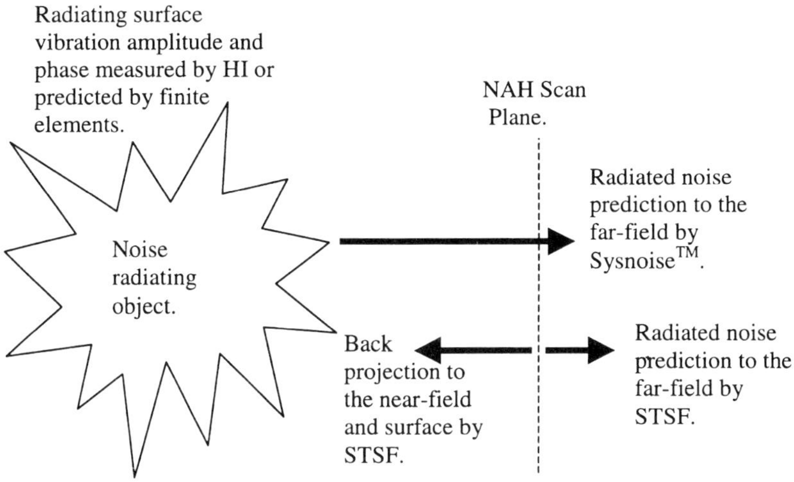

Figure 3: Interaction between full-field measurements and computational methods.

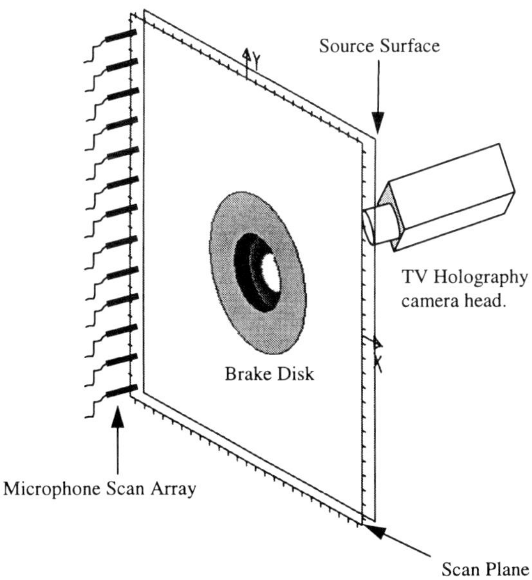

Figure 4: Schematic diagram of the experimental arrangement.

Amplitude
(Node = Black, Amplitude ∝Grey Level)

Phase
(Black = -180deg, White = 180deg.)

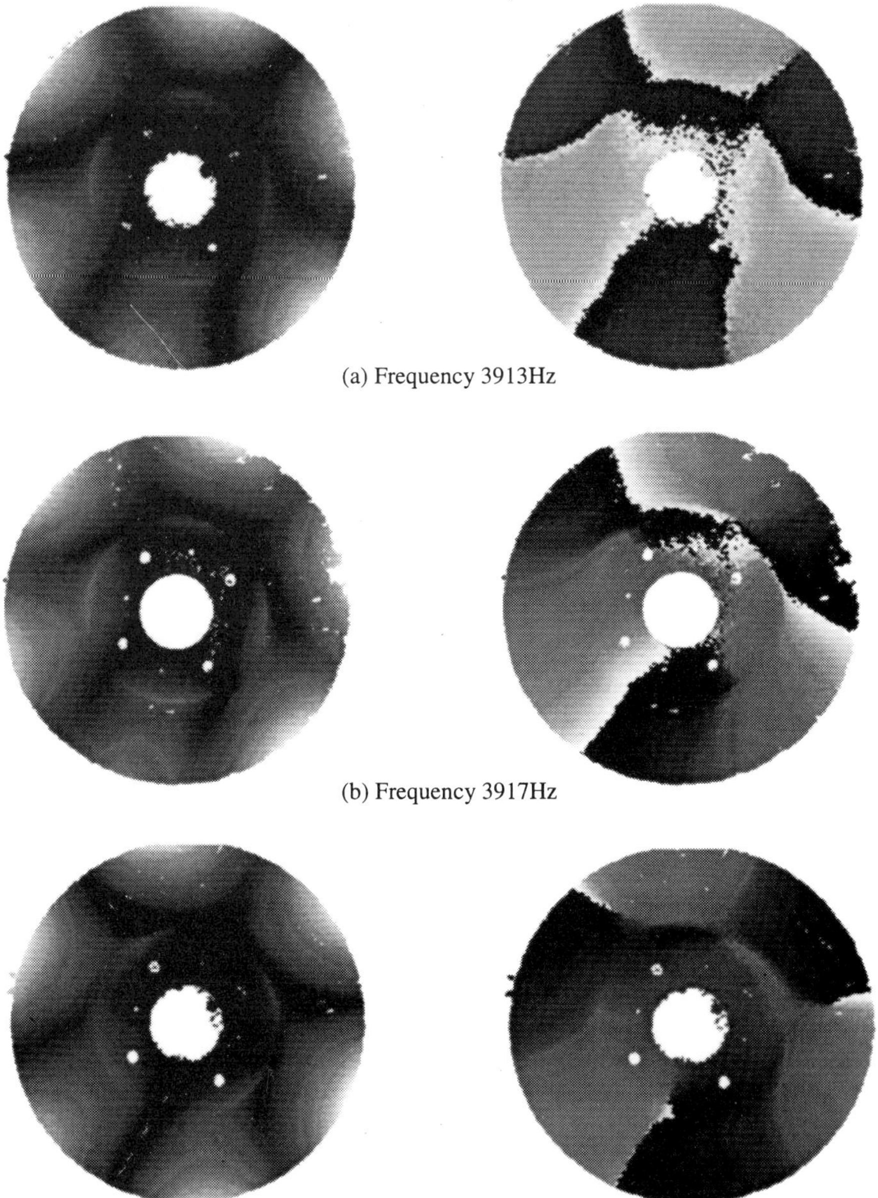

(a) Frequency 3913Hz

(b) Frequency 3917Hz

(c) Frequency 3923Hz

Figure 5: Vibration maps of amplitude and phase for the disk resonance centered on 3918Hz.

(a) Frequency
3913Hz.

(b) Frequency
3917Hz.

(c) Frequency
3923Hz.

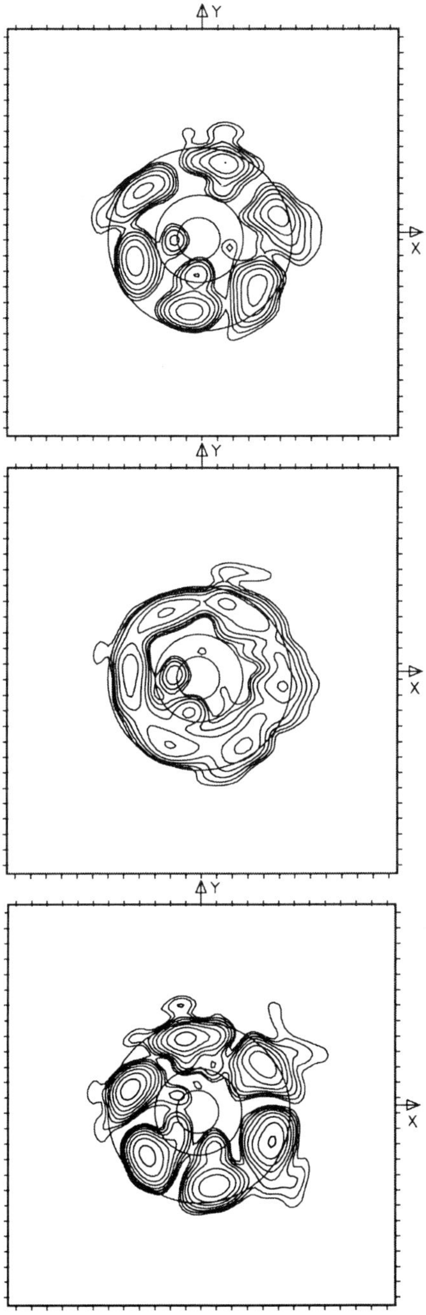

Figure 6: Active intensity on the surface of the disk. Minimum threshold of 79dB.

C521/019/98

The use of indirect coupled boundary element analysis in vehicle NVH refinement

J A HARGREAVES BSc, MSc, AFIMA
Ove Arup and Partners, Coventry, UK

SYNOPSIS

NVH refinement is one of the major aspects of automotive design alongside crashworthiness, occupant safety and durability. Considerable effort is invested in the refinement process which is a combination of predictive methods in the concept and detail design phases and experimental methods once prototype hardware becomes available.

Market competition has lead vehicle manufacturers to strive to reduce and control the levels of interior noise and to avoid cabin noise 'booms' in the frequency range below 200Hz. One way of approaching this issue is to refine the vehicle body design to achieve target levels of noise sensitivity so that the interior noise is controlled for a number of possible power-trains force inputs.

Reduction of interior noise is often accomplished by a combination of trim design for acoustic absorbence and measures to identify the controlling aspects of the vehicle body structure with respect to interior noise and make various design changes accordingly.

This paper discusses how the indirect boundary element method can be used to predict the noise sensitivity of a vehicle body and also how the controlling factors of the body structure might be identified and improved using panel contribution and modal participation analysis.

The approach taken here is to use a modal model of the vehicle body coupled to a boundary element model of the cabin and boot airspace and to apply unit forces in the three translational degrees-of-freedom at the engine mounts and damper towers. Noise transfer functions are then predicted between the force input locations and the occupant ear positions.

1. INTRODUCTION

There have been many examples of vehicle interior noise analysis over the last twenty years or so in which a modal model of the body structure has been coupled to a modal model of the cabin airspace (1),(2) & (3).

There are also many different types of NVH analysis that could be undertaken on a vehicle structure depending on the frequency range. Statistical energy analysis (SEA) is used for the mid-to-high frequency range and involves parametric studies on the vehicle structure in which the effects of the trim and panels on interior noise can be assessed. Other methods, such as ray tracing, might be used to help design the trim package for a particular vehicle again in the mid to high frequency range or more specifically for source and receiver applications such as audio design. Finite element methods (FEM), which transform the physical matrices of the system into modal models of the vehicle body, and modal/boundary element models of the interior airspace are applicable to the low-frequency range where noise 'booms' occur.

Nowadays it is common for vehicle manufacturers to focus on the acoustic sensitivity of the body structure so that the interior noise performance is improved by controlling the various noise transfer functions or NTFs. NTFs may be typically measured at the engine mount, suspension and exhaust hanger locations and would also be subject to a target depending on the class of car. In this way the vehicle interior noise might be acceptable for a variety and power-train and road inputs. The target setting process is normally accomplished using experimental measurements on one or more reference vehicles providing mass, stiffness, modal, NTF and forcing level information about the vehicle.

The total noise in a vehicle is then given by (4):

$$\mathbf{P} = \Sigma \mathbf{p}_i = \Sigma(\Delta \mathbf{x}_i \mathbf{k}_i)\mathbf{H}_i$$

p_i = contribution of ith route

Δx_i = relative displacement across i'th mount

k_i = dynamic stiffness of i'th mount

H_i = body acoustic sensitivity (P/F) at i'th attachment location = NTF

$(\Delta x_i k_i)$ = dynamic force applied to body at i'th attachment location

2. FINITE ELEMENT METHOD

The process of predicting body acoustic sensitivity is usually started by generating a finite element model of the body structure. The normal modes in the frequency range below 200Hz are then extracted to form a set of eigenvalues and eigenvectors which, with the addition of modal damping ratios, forms the modal model. The FE model itself would consist of the body-in-white (BIW), glazing and closures together with the various items of trim. Such models have become very detailed over the years since accurate prediction of the natural frequencies and mode shapes of automotive panels requires a fine mesh. A typical trimmed body model might have a mesh containing some 200,000 elements and over 800 modes have on occasions been extracted (5).

A modal model of the cavity airspace has also been in common use. The airspace is modelled using solid elements yielding some 10-20 modes in the sub-200Hz frequency range, depending on the cabin airspace geometry. A modal damping model has often been used for the acoustic model which effectively smears the damping over the entire model rather than representing the local absorption from the different trim items such as seats, headliner etc.(6). The wavelengths of the acoustic waves are much greater than those of the body structure in the frequency range below 200Hz and therefore a larger element edge length can be used and might be 50-100mm while the structural finite element model might have an element edge length of 10-20mm.

The next step in the analysis procedure consists of coupling the two modal models together, applying the loads and solving for pressure in the airspace. The coupling model depends on the code and in some cases only allows the vibrations of the structure to drive the airspace. Fully-coupled models take account of the feedback term from the airspace also.

The method is suitable for use during the concept and detailed phases of a vehicle development programme. In the concept phase the model will be relatively crude as design data is limited and consequently the low-frequency noise sensitivity is generally focussed upon.

During the detailed phase as panel surface data becomes available the model can be much more refined and the analysis frequency range extended. However, even at 100Hz the modal density of the structure is very high and so it remains a challenge to accurately predict NTFs in the 100-200Hz frequency range. A number of out-of-range structural modes also need to be included in the calculation to account for the residual term.

The main drawback with the acoustic FEM is the time required to generate the mesh although various auto-meshers are helping to overcome this.

The lack of a suitable absorption model in many codes is another disadvantage although alternatives such as acoustic modal damping and even arrays of viscous dampers have been used. Some codes, such as SYSNOISE, have a normal impedance/admittance boundary condition which is suitable for modelling the absorption of seats, headliner, carpet and parcel shelf covering is a representative manner using experimental impedance data.

It has advantages too, in that a modal model provides the engineer with more of a physical 'feel' for the acoustic behaviour of the cabin. The calculation of the acoustic normal modes is a standard feature of many finite element codes. The solution is also more efficient for a given frequency step.

3. BOUNDARY ELEMENT METHOD

The boundary element method(BEM) is suitable for acoustics applications in which there are exterior or interior fields. The direct collocation BEM is suitable for problems in which there is a single domain and the domain can be interior or exterior. The indirect variational BEM can be used on problems in which there are one or more domains which may or may not be open.

3.1 BEM Solution Domains

An introduction to the BEM theory for the direct and indirect approaches is given in (7). The indirect BEM has been used here for studying vehicle interior noise because the problem has several domains: the cabin, seats and boot airspace. In addition the prediction of body acoustic sensitivity necessitates a large number of loadcases which are efficiently handled in an indirect solution which leads to a symmetric system of matrices.

The BEM uses the same structural modal model as above but the cabin and boot airspace are modelled using 2D elements. A typical structural modal model is shown in Figure 1 and the boundary element model is shown in Figure 2.

The boundary element model can be created in a proprietary pre-processor and can be very rapidly created from the structural mesh. The element size would be comparable to that of the acoustic finite element mesh.

A field point mesh also needs to be generated to enable occupant ear sound pressure level (SPL) functions to be calculated and also to visualise the pressure field in the cabin and boot airspace so that the presence of an acoustic mode can be detected. See Figure 3.

3.2 BEM absorption models

A surface impedance model is used to represent the acoustic absorption of the seats, headliner, carpets and parcel shelf covering. The indirect BEM model usually deals with admittance which is simply the inverse of impedance. The admittance is applied at the element level and can be applied to either the positive, negative or both sides of the element. This feature allows trim items such as the parcel shelf to be represented in a variety of ways depending on the design.

Admittance = (Velocity in the direction of the element normal)/Pressure

In the case of a structural parcel shelf then a structural modal model is combined with a surface impedance model applied to both sides of those elements on the parcel shelf. A non-structural parcel shelf is, in effect, an acoustic screen and so is only represented by an acoustic model. A transfer relation is available in SYSNOISE that allows pressures and velocities on both side of the screen to be related.

Whichever model is used the impedance values must be experimentally determined using impedance tube measurements.

3.3 BEM models

The coupled indirect BEM used here requires two models: a so-called surrogate structural model (SSM) and an acoustic mode.

The structural modal model must be transferred onto the SSM and the weighted search algorithm maps the eigenvector component of each of the structural nodes onto the nodes on the boundary element model. The number of nodes in the structural model is generally much larger than that of the boundary element model and so the modal model is considerably reduced by this process. The modes are then visually checked using a graphical user interface (GUI).

4. FEM v BEM

In this section an outline of the main advantages and disadvantages of each method is provided

4.1 Advantages of FEM

- The banded nature of the stiffness matrix for FEM problems lead to a rapid solution time when using a coupled modal model of both the airspace and structure.

4.2 Disadvantages of FEM

- It is often difficult to generate the 3D mesh efficiently and quickly.
- A large number of acoustic modes are required to predict response in a given frequency range in order that the residual term is accounted for. This is also important because the coupling between structure and fluid can significantly reduce the frequency of certain acoustic modes.
- The modal damping model is not physically representative and some form of acoustic modal testing would be needed to derive modal damping factors.
- The FEM is susceptible to mesh discretization errors

4.3 Advantages of BEM

- The mesh is very easy to create both efficiently and in a very short time.
- The BEM is not a modal model and does not therefore require a large number of modes to be calculated.

4.4 Disadvantages of BEM

- The solution is very computationally expensive

5. SOLUTION OF INDIRECT BEM PROBLEM ON A CRAY J916 COMPUTER

The problem size is determined by the number of elements in the acoustic model, the number of structural modes, the number of frequencies and the number of loadcases. The prediction of body acoustic sensitivity in this case involves a 2000 element acoustic model, 531 structural modes, 91 frequencies and 24 load cases. This is quite a large problem in terms of boundary elements and so needs to be run in parallel on the computer. The objective was to reduce the elapsed run time such that the solution would be completed overnight.

In order to achieve this four processors were used and the frequency-level parallelism implemented. This does require a merging of the results after completion of the solution.

The future trend in computer hardware appears to indicate more common use of multi-processor machines with a large (> 100) number of processors. This would certainly help overcome some of the computational shortcomings of a BEM solution and lead to manageable run times even for a relatively small frequency step.

The memory for indirect BEM solutions needs to be high enough to invoke the in-core solver which in this case was 132 Mwords (1056 Mbytes) at the request level or 33 Mwords at the process level.

6. Analysis And Presentation Of Results

The body-acoustic sensitivity of NTFs are simply displayed on a magnitude/phase plot as shown in Figure 4. The Cabin and boot airspace pressure distribution is shown in Figure 5.

These plots can then be examined for frequencies where the NTF is above target or shows the possibility of there being a noise boom. One such frequency occurs at 144Hz.

Panel contribution and modal participation analysis has been used to get some insight into the controlling factors for the NTF at this frequency.

The major panels that contribute to interior noise are defined as sets and again the best way of achieving this is through the use of a pre-processor. Typical panels are:

- Roof
- Front Floorpan
- Rear Floorpan
- Parcel Shelf
- Doors
- Firewall

In this case the contribution of each of the panels to noise at the occupant ear location and at specific frequencies is of interest. The vibrations of the panels contribute to the total response at a selected point as a vector sum. The contribution of one panel is therefore the projection of the vector component at the frequency concerned onto the vector of the total response, together with the relative phase angle.

It is possible for the panels to have a positive contribution, i.e. they increase noise at a certain frequency, or a negative contribution in which some degree of cancellation occurs.

The total pressure at a field point, P_t, is expressed as a vector $\mathbf{P_r + iP_i}$. The contribution vector is expressed as $\mathbf{C_r + iC_i}$. The contribution percentage is defined as

$$\mathbf{100*[C_rP_r + C_iP_i]/[P_r^2 + P_i^2]}$$

The Percentage can be positive, negative and/or greater than 100%.

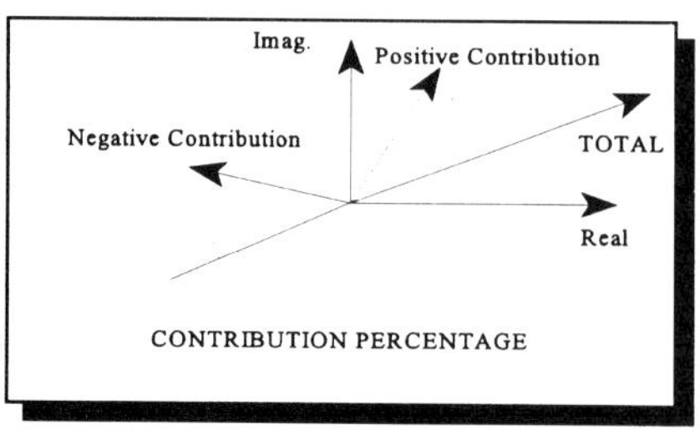

There are two possible types of contribution analysis:

- Total contribution
- Effective contribution

6.1 Total Contribution Analysis

Total contribution assumes that all of the panels are vibrating and influencing the interior noise and is therefore useful in determining where to apply or improve an acoustic treatment.

The geometrical influence of the various panels is obtained by calculating the contribution vectors which essentially assumes that all the panels are vibrating with unit amplitude and equal phase.

6.2 Effective Contribution Analysis

Effective contribution enables the influence of the structure to be addressed because of the assumption that a single panel is vibrating. The effective contribution is defined as the product of the contribution vectors and the normal surface velocity the panel.

The effective contributions for the various panels can be plotted on an amplitude phase diagram shown in Figure 6 for 144Hz.

6.3 Modal Participation Analysis

Modal participation analysis is typically used in addition to panel contribution analysis to rank the important structural modes at a particular frequency. It is also possible to select a single mode and look at the influence of this mode across the entire analysis frequency range but this is not the best way of looking at modal participation in a vehicle application.

The total displacement for a modal superposition solution is, U_{ij}, where

$$\mathbf{U_{ij} = \Sigma a_k \Phi_{ijk}}$$

i: Node number
j: Direction
k: Mode number
ϕ: Mode Shape (eigenvector)
a_k : Modal Participation Factor

A modal participation diagram, showing the modal participation factors v mode number at 144Hz, is shown in Figure 7.

6. SUMMARY

Although FEM has been in common use in vehicle interior noise analysis with a modal model of both the structure and the airspace the BEM can also be successfully used to tackle these analyses. The main drawback of this method is the creation of the acoustic mesh which is a volume fill operation as compared with a simple surface mesh generation required by BEM.

An example of the application of the BEM to vehicle body acoustic sensitivity has been presented here discussing the various steps needed in the pre/post-processing and solution of the model. A modal model of the structure is used in a fully-coupled body acoustic sensitivity calculation, predicting NTFs using engine mount and suspension point inputs.

A surface impedance model has been used here to model the absorption of the interior trim items such as seats, headliner, carpet, parcel shelf covering. It has the benefit of respresenting the spatial distribution of trim and also allows the analysis to provide performance information which can be fed back to the trim supplier to assist in the design process.

Several useful post-processing tools can be used to provide engineering insight into the results. The most important two are modal participation and effective panel contribution analysis and both have been illustrated here.

Although the BEM tends to be rather inefficient at the solution stage when compared to FEM it has other advantages with regard to accuracy and mesh generation.

7. REFERENCES

1. 'Structural-Acoustic Finite Element Analysis Of The Automobile Passenger Compartment' J.A. Wolf, D.J. Nefske & L.J. Howell

2. 'Application Of A Structural-Acoustic Diagnostic Technique To Reduce Boom Noise In A Passenger Vehicle' - C.A. Joachim, D.J. Nefske & J.A. Wolf Jr.

3. 'Automobile Interior Noise Reduction Using Finite Element Methods' - D.J. Nefske & L.J. Howell

4. 'Structural-Acoustic Analysis For The Prediction Of Vehicle Body Acoustic Sensitivities' B. Campbell, M. Abrishaman & W. Stokes

5. 'Automotive Modal Analysis Using CSA NASTRAN On High Performance Computers' P.J.Scharz, M. Venugopal & J.L. Bauer - High Performance Computing in Automotive Design, Engineering and Manufacturing - Paris, France October 7-10th 1996.

6. 'Prediction and Reduction of In-Vehicle Interior Noise Due to Road Irregularity And Other Inputs'
J. Bretl. - SAE P-221, Autotechnologies 1989, Monte Carlo, 23-25 January 1989.

7. 'Acoustic Radiation Of Power-train Components Using A Parallel Version Of SYSNOISE' L.Cremers & P. Guisset - High Performance Computing in Automotive Design, Engineering and Manufacturing - Paris, France October 7-10th 1996.

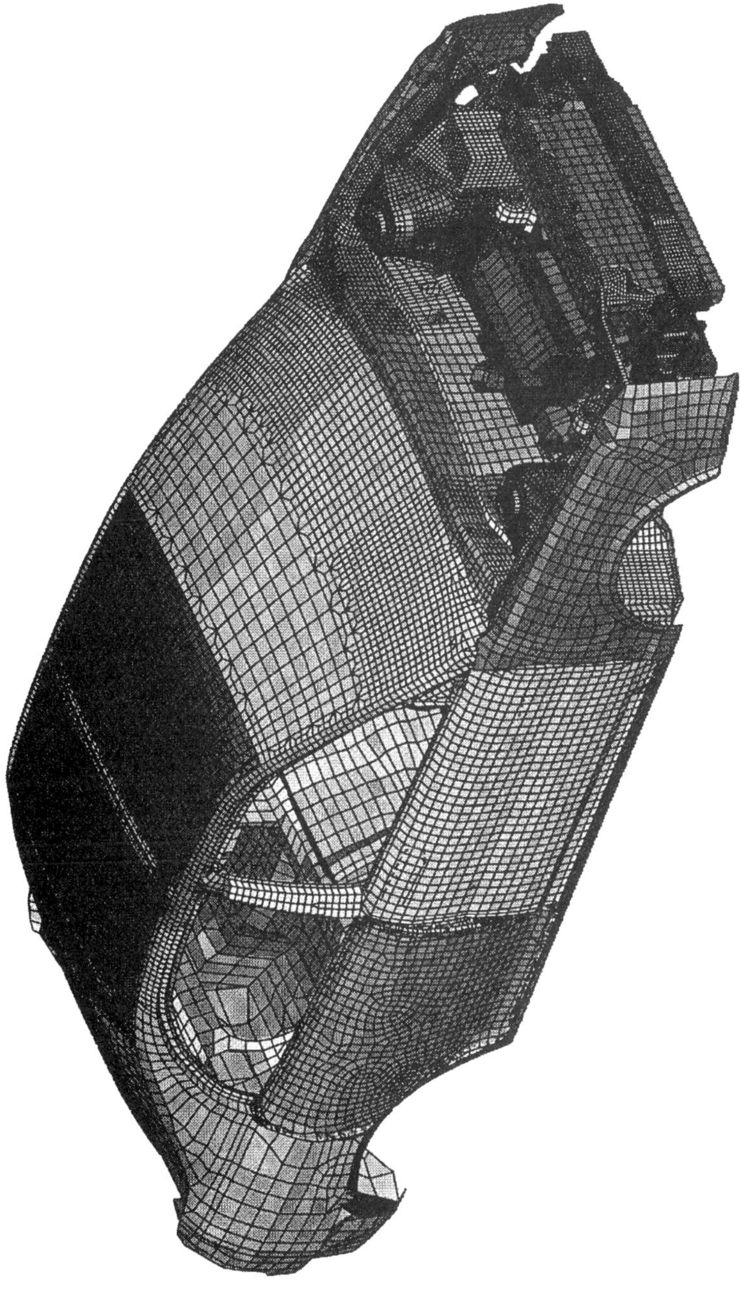

Figure1: Vehicle Body FE Model

Figure2: Vehicle Cabin Acoustic Model

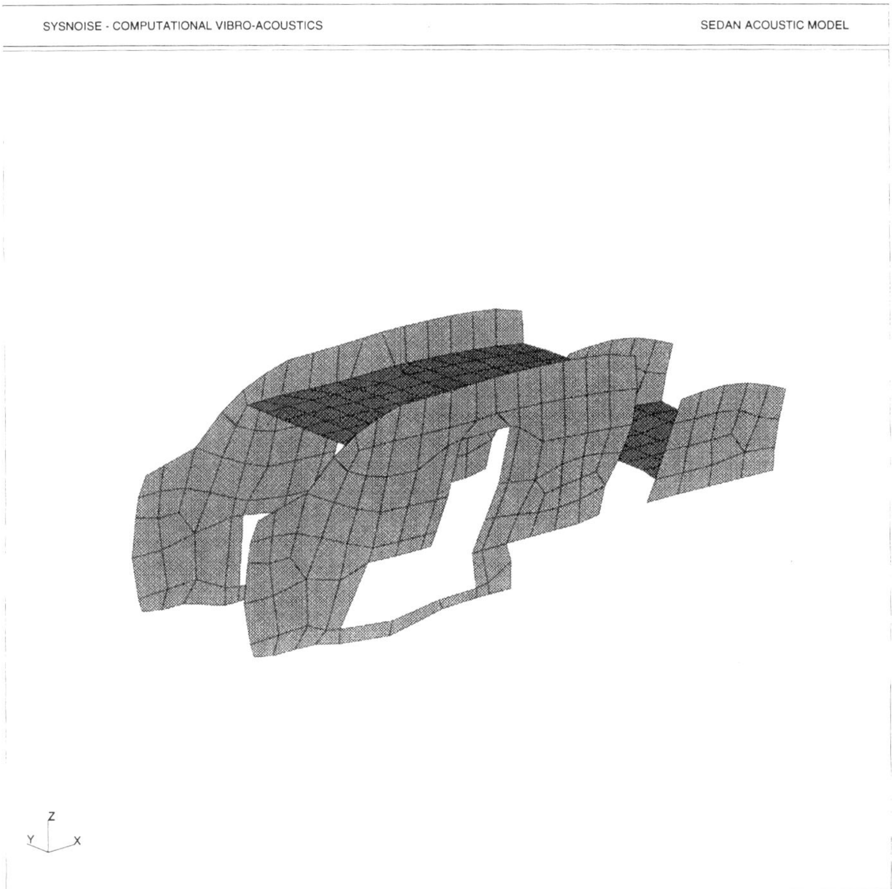

Figure3: Vehicle Cabin Field Point Mesh

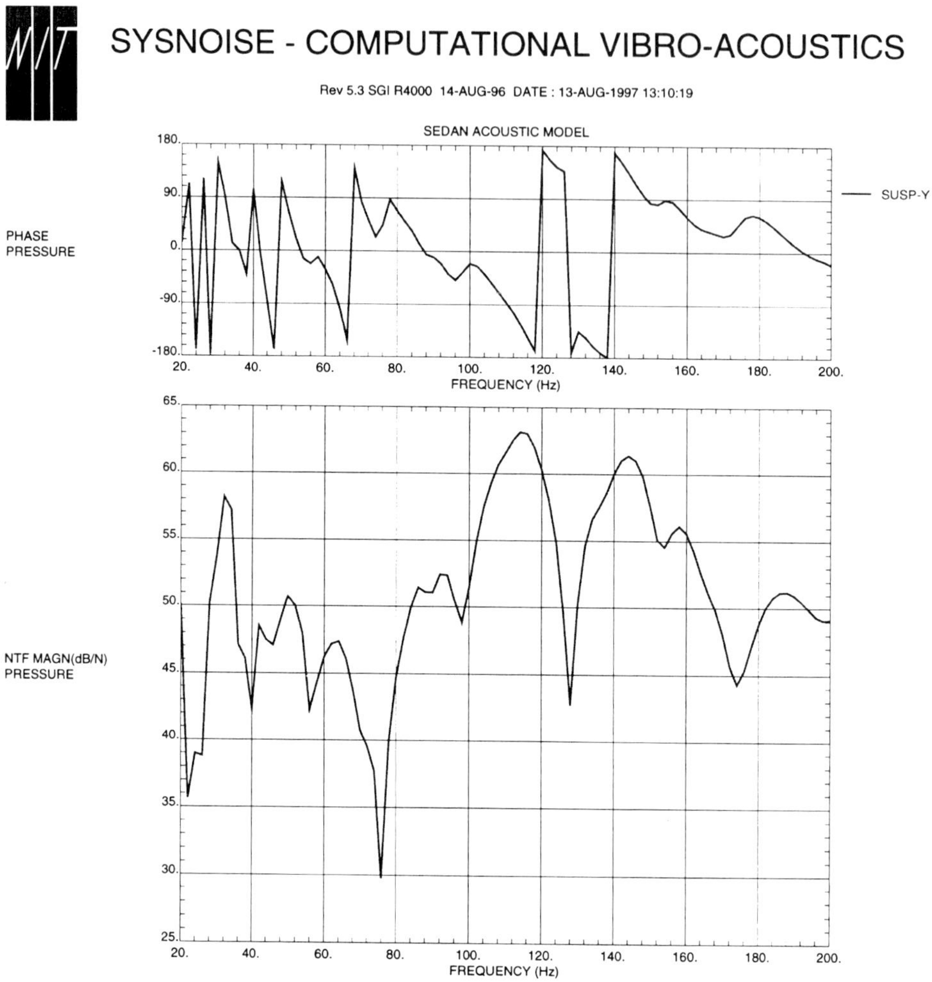

Figure 4: Vehicle Body NTF

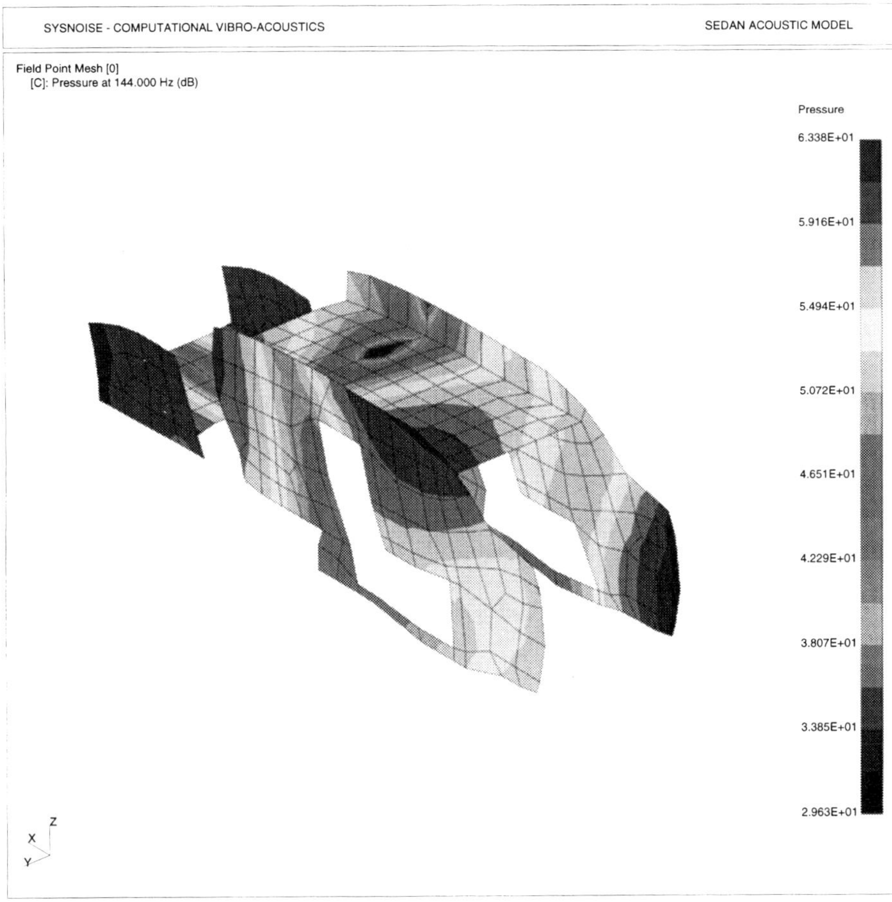

Figure 5: Interior Pressure Distribution

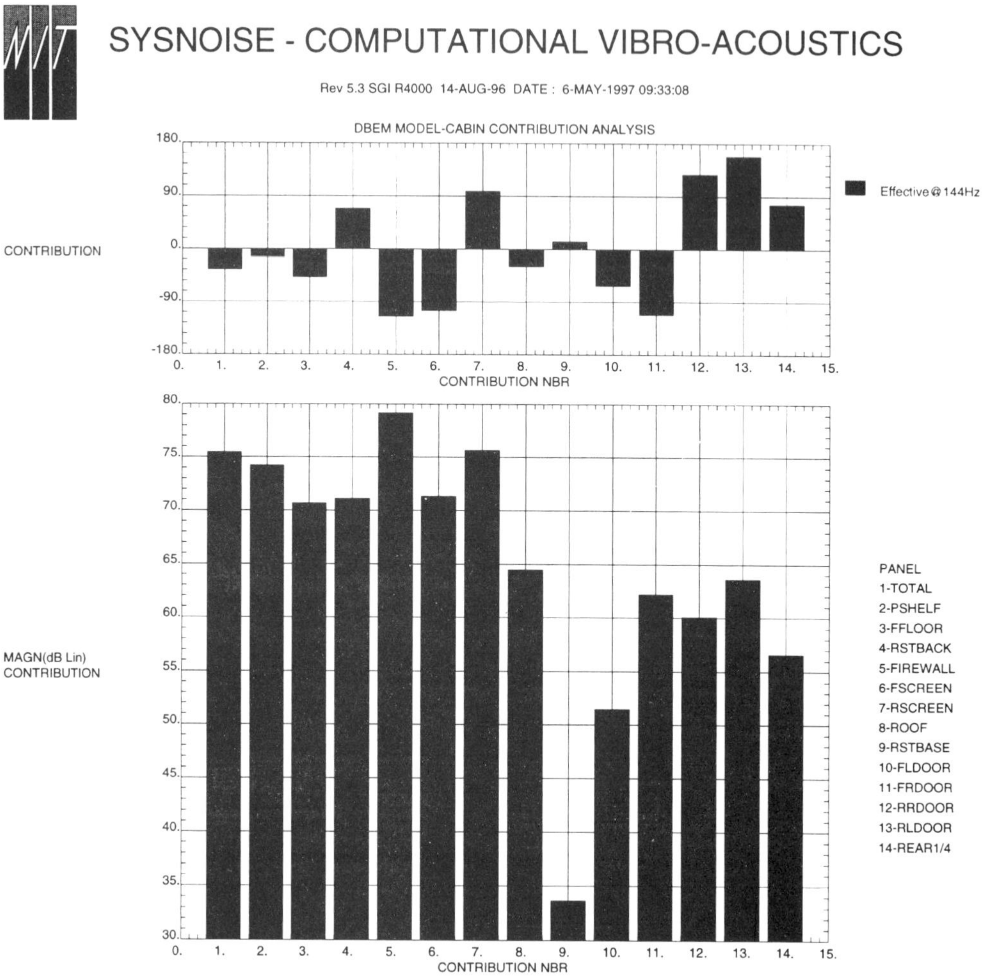

Figure 6: Panel Contribution Analysis

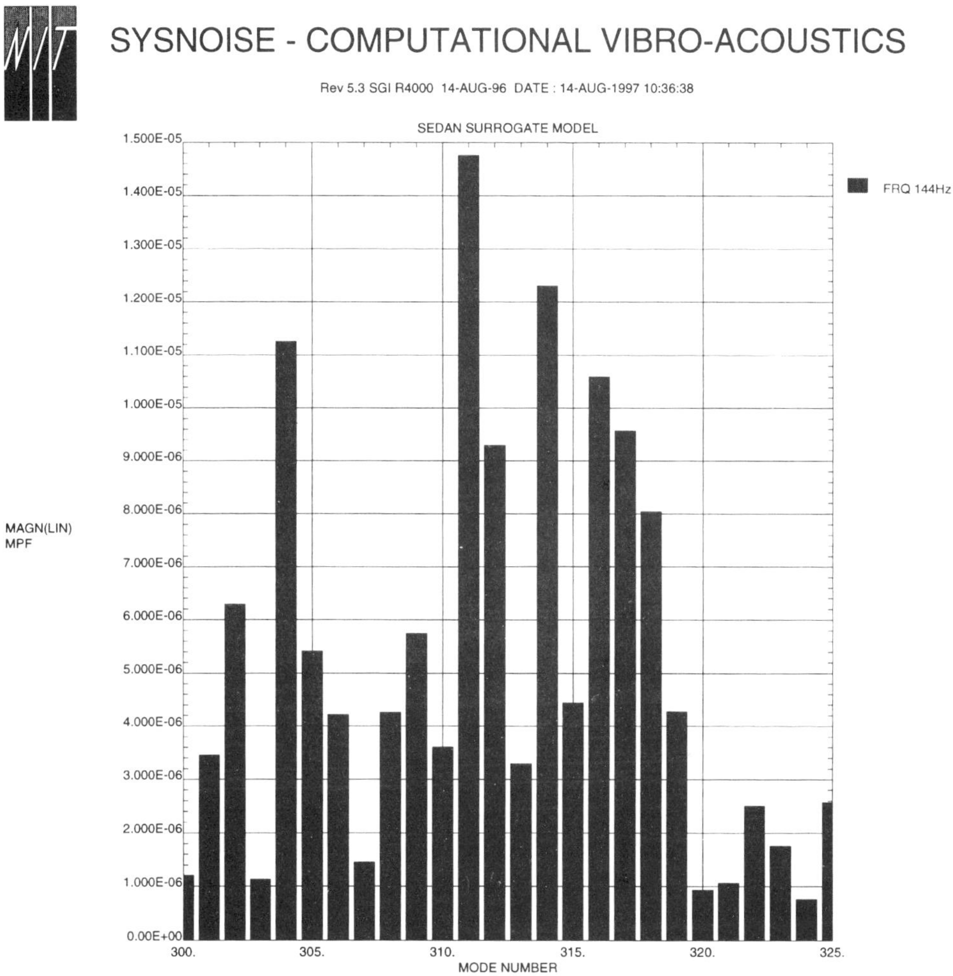

Figure 7: Modal Participation Analysis

C521/026/98

Development of coupled structure - acoustic FE models for vehicle refinement

W B TURNER BSc, MPhil, AFIMA
MIRA, Nuneaton, UK
F M TURGAY BSc, MSc, AMIMechE
Rover Group Limited, Warwick, UK

SYNOPSIS

This paper presents computer modelling work done in support of the NVH development of an experimental vehicle. Previous work in this field is reviewed, and modelling methods are applied within the constraints of a vehicle development programme.

To gain acceptance computer models must show that they can be applied to development issues equally as well as established test methods, and that they can reliably produce data representative of the intended vehicle. This pushed the analysis to consider the coupling of the body structure with the acoustic cavity and the effects of the interior trim.

1. INTRODUCTION

With the release of each new model, the automotive industry world-wide is pushing for reduced development cycle times and costs. However, with each new vehicle, manufacturers are also seeking increased enhancements and features in order to protect their share of a very competitive market. Just such a feature is the refinement of the vehicle.

With appropriate mathematical tools, the characteristics of a vehicle are explored and much development is done prior to production of expensive prototype hardware. When successful, this process reduces the number of development prototypes necessary, and consequently development time and costs are reduced.

Over the last decade much work has been done to put into place the tools and processes for computer simulation of vehicle NVH. As the mathematical tools are developed then so they need to be proved and accepted so that they may increase their contribution to vehicle development. Acceptance will only ultimately be gained when computer tools come out of the research laboratory and into the vehicle design and development office.

This paper presents the application of a finite element vehicle system model to a vehicle development programme. It extends the modelling process into the mid-frequency range to cover the full operating speed range of a diesel engine, and addresses the coupling of the structural and acoustic models such that structure borne interior noise is predicted.

2. NOISE VIBRATION & HARSHNESS IN MOTOR VEHICLES

The task of engineering a refined vehicle is to provide for improved travelling comfort by minimising unwanted vibration at the driver and passengers. To do this, the NVH engineer must understand the source of vibration and its route to the occupants.

In the mid frequency range of 80 to 200Hz, the flexural dynamics of a vehicle structure become significant. This requires more complicated mathematics to describe vibrational behaviour. Many limitations are imposed upon the available mathematical description of the vehicle. These may come from an inability to describe a physical phenomenon adequately, or to achieve a sufficiently accurate solution in a useful time.

3. REFINEMENT DEVELOPMENT TOOLS

A number of tools and processes are available to the NVH engineer. Many are routinely used with confidence and a knowledge of its own practical limitations. However, no single tool or process will solve the wide range of issues presented to the refinement engineer.

3.1 Experimental Tools

Test based processes include experimental modal analysis to understand structural and/or acoustic resonances, noise route tracking to understand structure borne vibration paths, and signature analysis to understand dominant sources of vibration.

For each of these however, the NVH engineer needs a vehicle structure to measure. Indeed, at the early stages of vehicle development, prototype vehicles may not necessarily be sufficiently representative of the final, volume vehicles. Measurements from early prototypes have the risk of misleading the development engineer.

The use of computer based calculations removes, to an extent, the need to have a physical structure to develop. Such calculations must, however, be upon the expected performance of the pure design. Fundamental dynamic characteristics only may be assessed and developed. Small allowances may be made for manufacture and build qualities, and these may only be reasonably included from quantified experience.

3.2 Vehicle System Models

The vibration path from source to occupant may not necessarily be direct. Engine vibration may be transmitted through the driveline, axles and primary suspension before it becomes incident at the body structure and ultimately the occupant.

For these reasons, if there is to be increased use and successful application of "computer prototypes", all aspects of the vehicle, as an assembly of dynamic components, need to be adequately considered. These include the definition of vibration source, calculation of its transmission path and, finally, its incidence at the occupants of the vehicle.

Finite element analysis of structures is now a mature technique. Much work has gone into the analysis of the acoustic cavity, and recent developments now enable this work to realise relevant application outside of the research laboratory.

4. REVIEW OF FE METHODS APPLIED TO ACOUSTIC CALCULATIONS

Finite element analysis is established and accepted for both static and dynamic structural analysis. The method has more recently, but less widely, been applied to the calculation of resonant frequencies of fluid volumes bounded by rigid surfaces (1).

Calculation of rigid wall cavity modes is in itself useful information to the NVH engineer, but ultimately we need to adequately predict acoustic response to real structural excitations.

The influence and interaction of the bounding structure and enclosed cavity of a vehicle has also been investigated (2),(3). Change of basis to modal co-ordinates is established as a necessary means of reducing the computational problem and improving analysis efficiency. This is fine until one wishes to couple the cavity to the structure.

Coupling of the cavity and bodyshell is considered through conservation of momentum at the fluid structure boundary. However, this results in non-symmetric matrices which do not lend themselves to standard finite element solution techniques. Consequently solution of the coupled problem is inefficient and results in a computationally expensive process.

The equations may be written, neglecting dissipation (4);

For the Structure $$[M_s]\{\mathbf{u}''\} + [K_s]\{\mathbf{u}\} = [A]\{\mathbf{p}\} + \{F_s\} \quad (1)$$

For the fluid $$[M_f]\{\mathbf{p}''\} + [K_f]\{\mathbf{p}\} = -[A]^T\{\mathbf{u}''\} \quad (2)$$

where [M] denotes mass matrix, [K] stiffness matrix. Subscripts s and f denote structure and fluid respectively. The basic unknowns of the equations are u, the structural displacement, and p the fluid pressure. $\{F_s\}$ signifies any externally applied force to the structure.

The coupling matrix [A] is a function of the interface area. In finite element notation its components are calculated by:

$$A_{ij} = \int_S N_j \, N_i \, ds \quad (3)$$

where Ni, Nj are structure and fluid element shape functions at the interface, and s is the interfacial surface area.

Significant savings in computational effort are made when the matrix equations (1) and (2) are transformed into the modal domain using the undamped invacuo structural modes, and undamped rigid wall acoustic modes as a change of basis. The resulting matrix equations are still not diagonal, but are at least significantly smaller and easier to solve.

The most commonly applied finite element code has been MSC/NASTRAN. This has provided the framework for the separate calculation of structural and fluid uncoupled modes, the boundary coupling terms and finally the coupled response solution(3),(5).

Each of these activities was a separate process. Manual intervention was needed to sort and transfer data from one process to the next. Whilst, therefore, it has in the past been possible to calculate the coupled behaviour of a vehicle structure and enclosed cavity, the process has not always been particularly straightforward or simple to apply outside of the research laboratory.

Limitations were previously imposed by the computational effort involved in achieving a reasonable dynamic representation of the body structure. Sung and Nefske (4) restricted their work to the investigation of sub 80Hz road noise as the modal density of the body became large, and modal clarity diminished.

Model reduction tools have been available for many years, but appear to find few applications outside the aerospace industry. Recent advances in the capability and efficiency of finite element codes have increased the available scope of coupled analyses. MSC/NASTRAN V69 now provides automatic calculation of coupled structure-fluid modes and forced response. Facilities are also provided for the inclusion of fluid damping and boundary impedance.

5. MODEL APPLICATION

To support the NVH development of a vehicle programme a computer model of the vehicle system was created All structural models were created from design information as physical hardware was not available at the time of model creation.

The final model was a comprehensive representation of the trimmed vehicle. A schematic of the model is shown in Figure 1. All flexural and dynamically active components were represented by finite element models and superelements.

The development and application of vehicle system models is referenced (6). This work also highlighted the use of superelement techniques to maintain analysis efficiency. Complex components of the system were modelled in detail such that their dynamic characteristics were captured with the required accuracy. Reduction of the detailed model to superelement form retained the pertinent dynamic features as a reduced dof set for inclusion in the system model.

In this way the model retained dynamic detail, yet remained an efficient development tool.

5.1 Excitation

To simulate the forced vibration behaviour of a driven vehicle, we must consider the engine as a main vibration source. As, in this application, the engine was an in-line four cylinder diesel, then the main causes of engine vibration are known to be unbalanced forces due to combustion pressures and reciprocating inertia's.

For an in-line four cylinder engine, both the primary sources described above are incident upon the crankshaft twice per engine revolution. The frequency of these sources are twice the frequency of crankshaft rotation, and as such are known as "engine second order" excitations, or "2E

The 2E vibrational forces acting at the crank were derived from measurements from a test bed engine. The measured data is shown in Figure 2. This figure shows that the 2E content is for the most part the major component of this engine's vibration.

Calculation of the engine disturbing forces from the velocity measurements was feasible by considering the engine to be "mass controlled". The engine running speed of 1000rpm to 4500rpm causes 2E forces from 34Hz to 150Hz. This range was well above the mounted modes of the engine.

5.2 Model Results

Figure 3 shows measured and predicted vertical acceleration at a front and rear engine mount. The black curves are model calculated and the grey curves are measured values from the motored vehicle. Overall, the correlation achieved gave confidence that the level and phasing of the excitation forces had been adequately defined and that the rigid body response of the engine unit was correct. This in turn implied that the forces applied through the engine mounts were representative.

The other transmission path from the engine is via the driveline. Figure 4 shows accelerations at two other entry points to the body. The levels and general form of the calculated accelerations (black curve) compare very well with those measured on the motored vehicle (grey curve).

Coupling between the body superelement and the cavity air was achieved on a 1:1 mapping between adjacent structural and fluid nodes. This required additional preparation time, but rewarded the effort with computational efficiencies.

Figure 5 shows the calculated sound pressure level (SPL) at a point representative of the driver's ear. The calculated values are plotted in black. The grey curves of Figure 5 are sound pressure levels measured with a microphone during a rolling road test of the vehicle.

Given the spread of the measured SPL values, the calculated SPL may be considered to be a reasonable fit. The general level and overall trend of the cabin noise are well represented over the engine operating range. Specific noise features up to @4000rpm would also appear to have been captured.

6. CONCLUDING DISCUSSION

The progress of the automotive industry towards reduced development times continues. Vehicle development by computer simulation is seen as a way to facilitate this, and many programmes to develop the necessary modelling tools and techniques are active.

This paper has presented work which explores the application of a computer model to vehicle NVH within the process of a true development programme. Important to such applications is the ability of the modelling process to provide results as reliable as those which may be obtained from measurement, to provide them within an acceptable time frame, and so generate a more efficient tool to assess design changes. This was achieved.

The scope of the model was extended to cover the full operating range of a diesel engine. Previously accepted limits were imposed by the need to balance finite element model size against computer capability and model analysis times. Developments in computer hardware and software are pushing back these boundaries.

Coupling of the acoustic cavity to the structure of the vehicle is also now more accessible, and provides a process improvement is over earlier methods. The quality of achievable interior noise results through coupled models is demonstrated. This provides a foundation for models to be used as NVH optimisation tools.

REFERENCES

1. Sung S H Automotive Application of Three Dimensional Acoustic Finite Elements. SAE 810397 1981 pp1 - 9

2. Nefske D J, Wolf J A, Howell J R. Structural Acoustic Finite Element Analysis of the Automotive Passenger Compartment: A Review of Current Practice Journal of Sound and V ibration1982, 80(2) pp247-266.

3. Sung S H, Nefske D J. A Coupled Structural-Acoustic Finite Element Model for Vehicle Interior Noise Analysis. Trans ASME. Journal of Vibration, Acoustics, Stress and Reliability in Design. Vol. 106 pp314-318

4. The MacNeal-Schwendler Corporation MSC/NASTRAN Users Manuals V69

5. Sung SH, Nefske D J. Component Mode Synthesis of a Vehicle Structural-Acoustic System Model. AIAA 1986 24(6), pp 1021-1026

6. Turner W B Vehicle Refinement using Superelement Techniques in MSC/NASTRAN. Proc. 1996 MSC UK Users Conference.

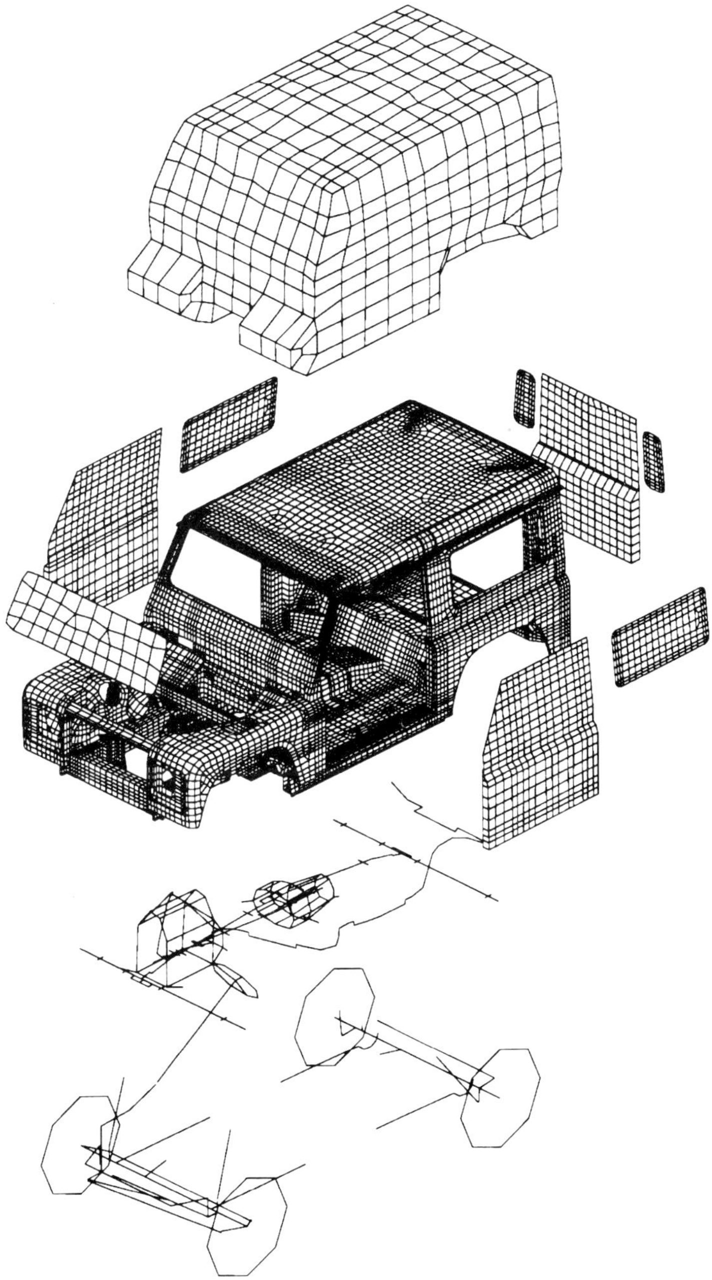

Figure 1. Components of Coupled System Model

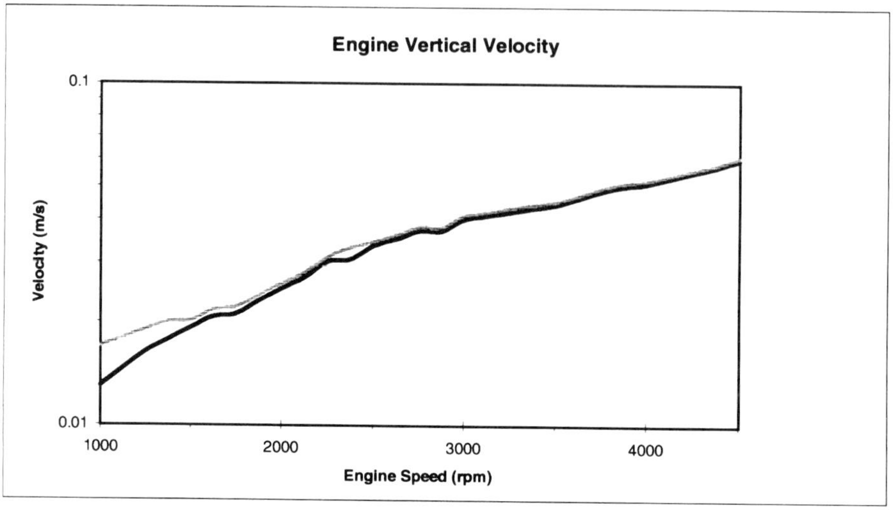

Figure 2a. Crankshaft Vertical Velocity

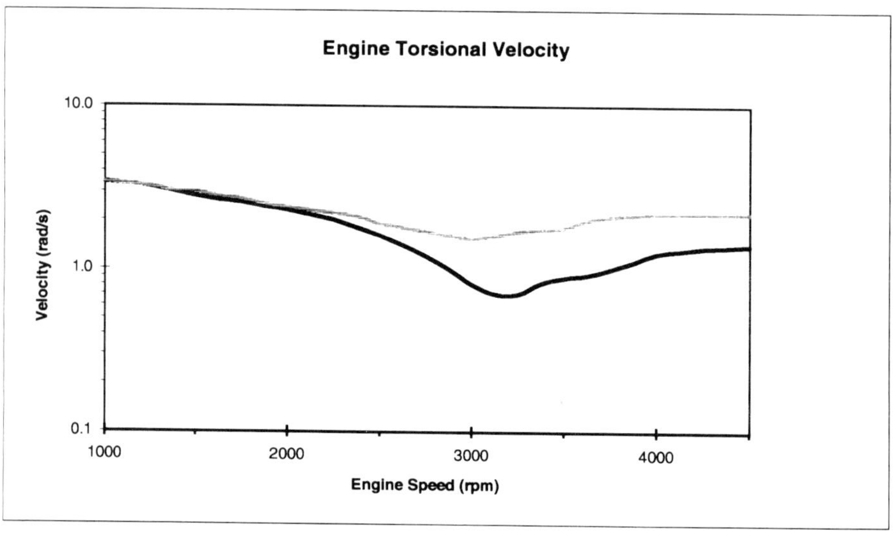

Figure 2b. Crankshaft Rotational Velocity

Figure 2 Engine Vibrational Velocities

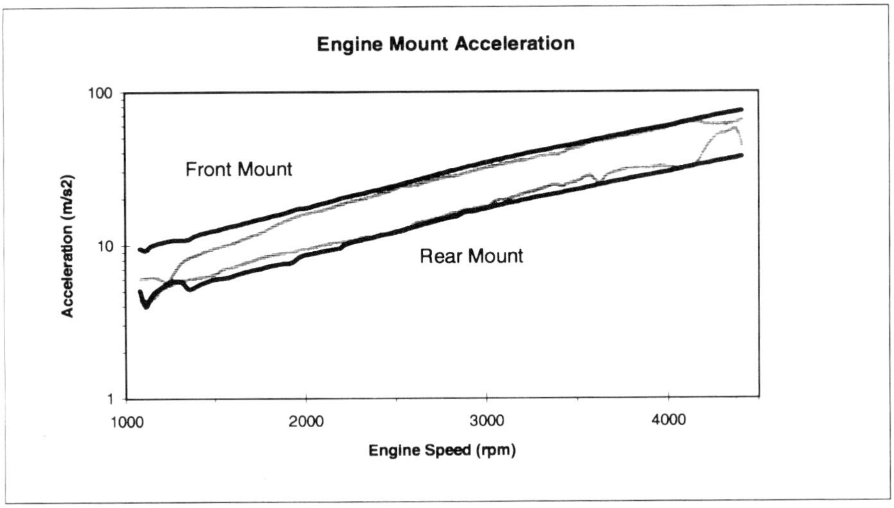

Figure 3. Predicted and Measured Engine Accelerations

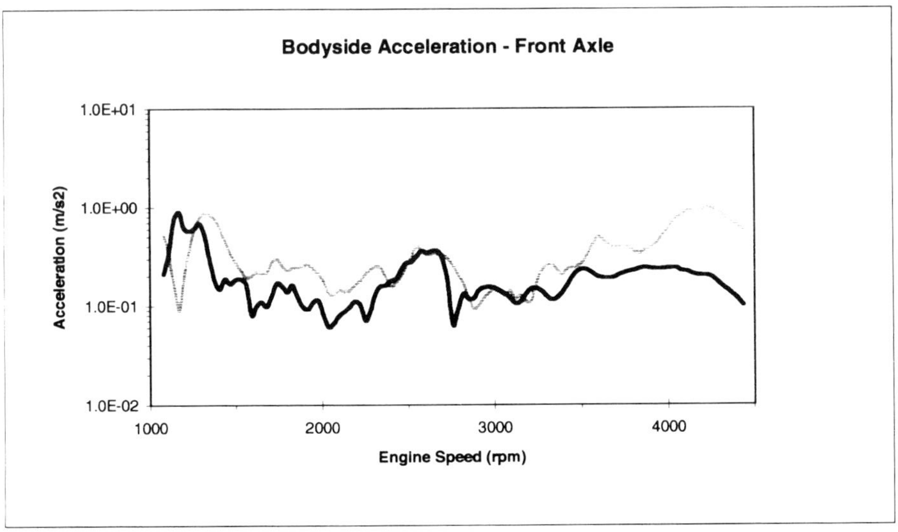

Figure 4. Accelerations of body at Front Axle Tie Bar

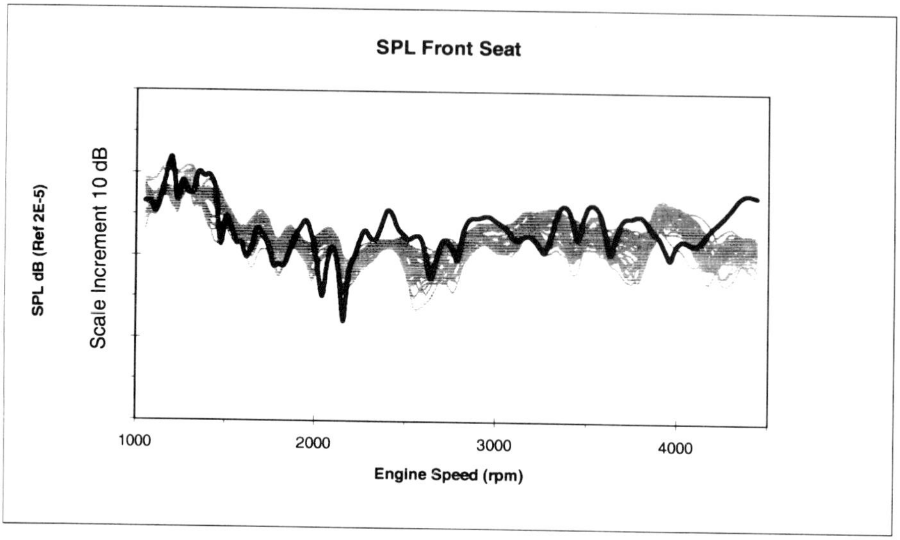

Figure 5. Predicted and Measured SPL at Front Seat

C521/036/98

Cylinder head noise reduction on a 4-cylinder 4-valve SI engine

N KRAUS and **E SCHÄFER**
FEV Motorentechnik GmbH & Co. KG, Aachen, Germany
M JOERRES and **T GRÜNERT**
Ford–Werke AG, Kölin, Germany

Due to decreasing legislative noise limits, the noise emitted by the cylinder head has to be analysed when optimising the NVH behaviour of an engine. This paper describes investigations concerning valve train excitation, vibration transfer behaviour and noise radiation of a cylinder head including test bed measurements and FEM calculations.

The results show, that the main excitation mechanisms are the valve opening and closing impacts. The examined cylinder head structure reveals a high mobility at the bearing pedestals, which is improved by additional inner walls. A laboratory type modification of the cylinder head leads to noise reductions of up to 2 dBA for the pure valve train noise.

1 INTRODUCTION

The requirement to optimise the vehicle noise and vibration is still increasing due to more stringent legislative noise limits and elevated customer awareness for high riding comfort. The engine as a primary source of vehicle noise has been topic of various NVH improvement work and will be in the future. To this end, all potential noise sources must be considered.

The cylinder head together with the valve train can be a significant contributor to the mechanical noise radiation. Due to the high frequency characteristic of the cylinder head noise it is often felt to be annoying.

This paper describes investigations concerning valve train excitation, vibration transfer behaviour of the cylinder head and noise radiation of a 4-cylinder 4-valve SI engine. Aim is to identify main excitation sources and to show the influence of structural cylinder head modifications on the noise emission.

The investigations include test bed measurements for quantification of the real excitation forces related to valve train parameters like valve clearances, further dynamic structural optimisation of the cylinder head by FEM calculations, impact hammer testing of the cylinder head and verification measurements on the test bed.

2 EXPERIMENTAL ANALYSIS OF VALVE TRAIN NOISE

Compared to conventional 2-valve designs, modern 4-valve engines tend to be more noisy. Among other things this is caused by the valve train. The valve train noise is mechanically induced and it is emitted by the surfaces of the cylinder head structure (and the attached components). Analysing the noise radiation of a cylinder head implies that both the excitation of the valve train and the dynamic structural behaviour of the cylinder head itself have to be examined. The analysis of the valve train noise of the 4-cylinder 4-valve SI engine under investigation was experimentally conducted. The dynamic structural analysis of the cylinder head was carried out analytically with Finite Element (FE) calculations based on the excitations evaluated empirically.

2.1 Status of Valve Train Noise

The 4-cylinder engine (valve actuation with mechanical tappets) was built up in a semi-anechoic test cell for the experimental acoustic investigations. The sound pressure of the emitted noise was measured with circumferentially located microphones at a distance of 1 m opposite the front, intake, exhaust and top side of the engine. For in-depth analyses of the noise generating mechanisms due to the valve train events, acceleration sensors were fixed on the cylinder head's outer surfaces, on camshaft bearing caps and on one combustion-chamber dome near the valve seats.

A comparison of the noise radiation of the total engine (fully equipped, fired condition) with the partial engine (pure valve train noise due to dismounted pistons and conrods, motored condition) reveals a relatively low share of the valve train noise on the total engine noise with no very pronounced directivity of the noise radiation. The results also show, that the cylinder head is relatively silent compared to competitors. Despite that the potential for further improvement was checked.

The detailed experimental analysis was performed on the partial engine in motored condition. The measurements were conducted with the engine in stable thermal status with heated cooling water and heated lubrication oil (80-90 °C).

2.2 Influence of Zero Clearance Tappets

When activating the valve train, several impulses and high forces of short duration appear due to kinematic processes. These are impacts of valves on the valve seat (valve closes), impacts of the camshaft journals on the bearings (valve opens) and mass forces on the camshaft and thus on the bearings during the phases of high valve acceleration and deceleration.

Beside these primary excitation mechanisms, the secondary movements of the valve train components due to clearances may generate noise, too. The tilting and the lateral movement of the tappet in the bore is an example for this kind of possible noise generation. In order to investigate the share of this mechanism on the valve train noise, zero clearance tappets were produced by graphite coating the tappet outer surface.

Regarding the sound pressure measurements with these modified tappets, no level shift can be detected. Also the frequency content does not differ significantly. Slight reductions of the acceleration levels on the outer surface of the cylinder head can be recognised, but they belong to local processes without implication for the acoustical far field. The influence of the zero clearance tappets on the acceleration levels on the bearing caps is very low, too. Event analyses of the time traces show predominating valve closing and valve opening impulses, but signal peaks belonging to secondary movements of the tappets cannot be identified. This points out, that the impacts and mass forces on the camshaft bearings and on the valve seats are the main excitation mechanisms for the valve train noise of the investigated engine.

2.3 Influence of Valve Clearance

Valve train noise usually becomes very loud if the valve clearance exceeds the height of the cam ramp. This is caused by stronger impacts of the cam lobes on the tappets. In order to investigate the influence of the valve clearances on the valve train noise, the valve clearances were varied between a minimal and a maximal clearance status within the production tolerances.

The sound pressure levels show only a slight tendency to higher levels for maximal valve clearance. The third-octave levels for maximal valve clearance are higher in the 3200 Hz to 5000 Hz frequency bands compared to minimal valve clearance, but these higher levels only lead to a slightly increased total level for the different speeds (**Figure 1**). This low influence of the valve clearance also can be detected in the acceleration levels on the cylinder head surface.

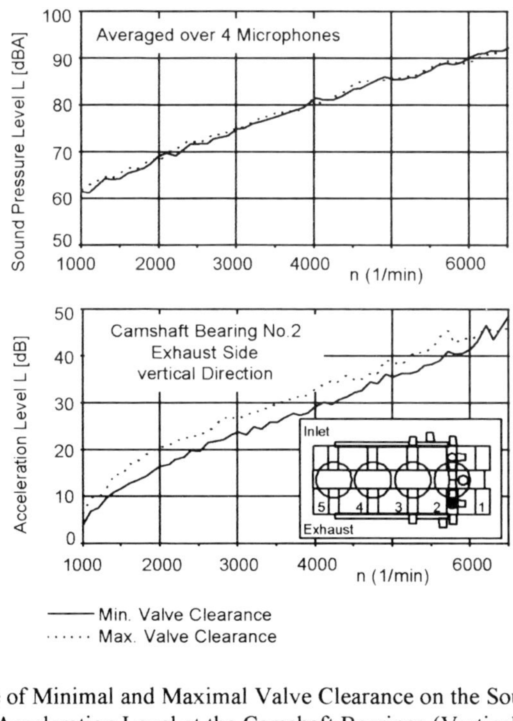

Figure 1: Influence of Minimal and Maximal Valve Clearance on the Sound Pressure Level and on the Acceleration Level at the Camshaft Bearings (Vertical Direction)

Contrary to that, about 3 dBA higher acceleration levels can be found on the camshaft bearing caps and on the valve seats. As an example, **Figure 1** also shows the acceleration levels measured in vertical direction on the camshaft bearings for minimal and maximal valve clearance. Clearly, the higher levels for maximal clearance can be recognised. Responsible for the different level curves are high frequency components of the vibration. These facts suggest, that some of the valves possessed larger clearances than the ramp heights. The reason could be that the valve clearances were controlled at room temperature (maximal valve clearance was slightly lower than ramp height), but during measurement, operational engine temperature conditions were simulated. Thus, the different coefficients of thermal expansion of the cylinder head and of the valves lead to an additional clearance.

As long as the cam ramps (constant opening and closing velocity) are adequately large, no significant influence of the valve clearance on the valve train noise can be expected.

2.4 Influence of Decreased Number of Valves

Further investigations were made in order to detect excitation mechanisms of minor importance. To this end, vibration and sound measurements were conducted at the cylinder head equipped by only half the number of valves. At each cylinder, one exhaust and one intake valve was dismounted. For all tests the intake and exhaust manifolds were replaced by sealed cover plates. Because of the dismounted pistons there is no compression and expansion of the air inside the cylinder. Thus the air flow noise is not relevant for these investigations. The halved number of valves should lead to a halved excitation. This yields theoretically towards a level decrease of 3 dBA $(10 \cdot \log(1/2) = -3)$.

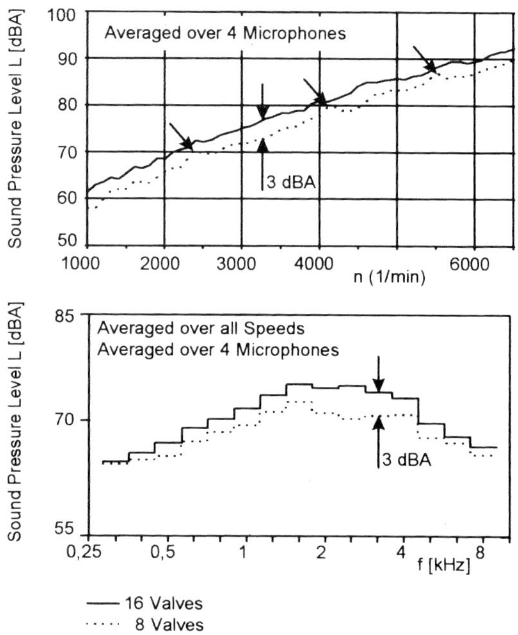

Figure 2: Influence of Halved Number of Valves on the Valve Train Noise

The results of sound pressure measurements with these lower excitations are displayed in **Figure 2**. They show in good approximation the 3 dBA shift of the level curves and of the characteristic third-octave spectra (the spectra are averaged over all speeds with the spectrum at each speed normalised by its total level).

When decreasing the excitations another effect can be recognised: the sound pressure level curve with 8 valves (**Figure 2**) shows slight singularities at 2400 rpm, 4100 rpm and 5500 rpm. In-depth analyses indicate that these singularities belong to the 19th, 38th and 57th engine order which are caused by the meshing orders of the timing belt. But this excitation is low compared to the impacts and mass forces of the valve train.

3 DYNAMIC STRUCTURAL ANALYSIS OF CYLINDER HEAD

As explained earlier, the excitation of the cylinder head of a motored engine is mainly effected by impacts and mass forces. Based on this information, the cylinder head structure itself will be analytically examined in the next step.

3.1 Simulation Method

Today Finite Element (FE) calculations are usually performed to optimise an engine component with respect to sound and vibration. Even large engine structures like block and cylinder head can be successfully treated. Eigenfrequencies of acoustical significance between 1 and 2 kHz are characteristic for engine blocks. Cylinder heads are stiffer, and typically frequencies up to 3 and 4 kHz dominate the noise.

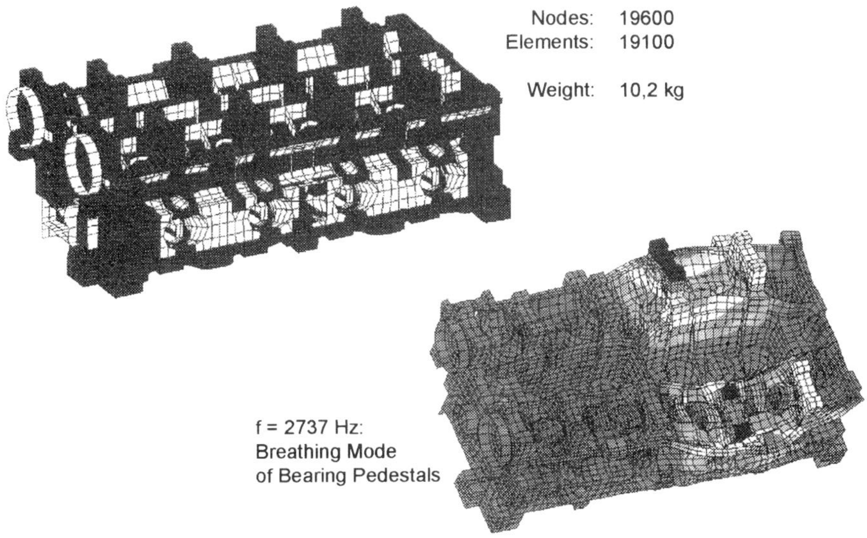

Figure 3: FE-Model of the Examined Cylinder Head with an Example of the Dominant Breathing Modes of the Bearing Pedestals

At FEV, a simulation method was developed which allows dynamic structural simulation based on design drawings. First, the component to be examined is set up as a FE model. The upper part of **Figure 3** shows the FE model of the investigated cylinder head which contains about 19600 nodes and 19100 elements (red colour: solid elements, yellow colour: thin shell elements). The material of the cylinder head is aluminium; the weight including camshaft bearing caps amounts to 10.2 kg.

Beside the mechanical excitations, the operated engine adds the combustion pressure excitation to the cylinder head. For an acoustical optimisation based on FEM and structural parameter variation, these excitations are simulated for Forced Response Analyses. In order to assess the excitation forces due to impacts and mass forces, accelerations were measured on 2 camshaft bearings and on 1 valve seat at the motored engine with maximum valve clearance (worst case) for a variety of engine speeds. Dividing the acceleration spectra by Input Transfer Functions (the cylinder head structure is hit with a hammer at the excitation locations and the acceleration measured nearby is divided by the hammer force) yields the excitation force spectra.

These excitation spectra and the spectrum of the combustion excitation (gained by cylinder pressure measurements) are loaded to the FE model separately. At first, the surface velocity as a response of the structure is calculated for various locations on the surface. In a further step, the intensity spectra of the velocities are surface area-weighted and summed up to the averaged spectrum of structure vibration which is used for assessment of structure modifications.

3.2 Present State Analysis

First of all, the vibration behaviour of the cylinder head is calculated. Comparison of this with the experimental modal analysis results shows, that the calculated eigenfrequencies differ less than 10% from the measured ones. This proves the high quality of the FE model for dynamic structure calculations. Also, calculated and measured transfer function pairs agree very well.

The calculated modal analysis determines the first global modes at 1113 Hz (1st torsion), 1892 Hz (1st vertical bending) and 2075 Hz (1st lateral bending). In the adjacent frequency range, breathing and bending modes of the bearing pedestals can be found. One example of mode shape is displayed in the lower part of **Figure 3**. The frequency range from about 3400 Hz to about 5100 Hz is characterised by axial bearing modes, but also higher order breathing and bending modes of the bearing pedestals can be found. Above 5500 Hz there are rotational bearing modes and skirt panel vibrations.

First of all, a selection of characteristic excitations has to be chosen in order to focus the calculation effort to the most important aspects. By loading the force spectra to the FE model at the camshaft bearings, at the valve seat and at the combustion-chamber dome, the importance of the different mode shapes for the acoustical behaviour of the cylinder head can be determined. The excitations with the highest influence on the surface velocity are those at the camshaft bearings in vertical direction, both intake and exhaust side, and the combustion excitation. To calculate worst case surface velocity, the excitation forces are restricted to these three forces at 6500 rpm.

The distribution of the surface velocities of the baseline cylinder head with high resolution is displayed in the upper part of **Figure 5**. It shows relatively low levels at the lower part of the cylinder head, represented by blue colour. Higher levels are represented by red colour and can be found in the upper part of the cylinder head. An in-depth analysis of the vibration behaviour of the cylinder head shows, that mainly the breathing and bending modes of the bearing pedestals (as already shown in **Figure 3**) and with minor importance the rotational bearing modes combined with skirt panel vibrations are responsible for these higher levels.

3.3 Structure Optimisation

For a dynamic structure optimisation of the cylinder head, a variety of potential modifications must be extracted from the modal analyses in order to obtain an optimum mass/stiffness ratio with a minimum surface velocity at noise radiating areas. In order to quantify the acoustical benefit of a modification, each surface velocity calculation is displayed in a standard way (see **Figure 4**) with narrow band spectra, third-octave band spectra for a better clarity and integral spectra which are calculated by summing up the levels at each frequency to a cumulative sum.

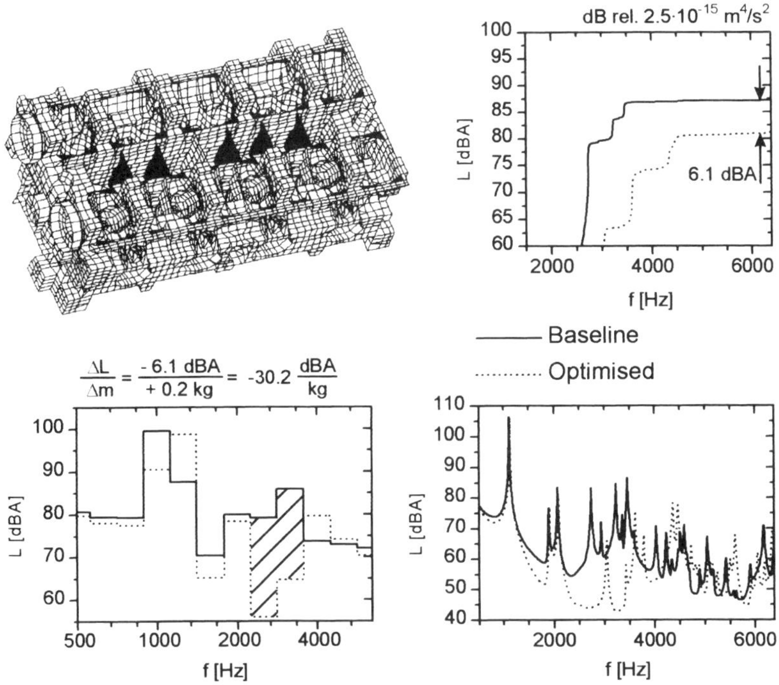

Figure 4: Calculated Surface Velocity of Optimised Cylinder Head

There exist two alternatives for optimising: stiffening of the entire structure and decoupling of the vibration transmission paths. Different modifications following the second way show, that this optimisation method does not lead to a significant level decrease for the examined cylinder head. Further analyses of the existing outer ribs demonstrate, that their influence on the acoustical behaviour of the cylinder head is very low. The calculation of a high number of variations shows that modifications like longitudinal or vertical ribs do not have the potential for a high acoustical benefit as they are not able to reduce the level of the dominant breathing and bending modes of the bearing pedestals. Only when stiffening both pedestals towards each other, high level reductions are achieved. One example is a link of the bearing caps on the exhaust side to the bearing caps on the intake side. This modification results in a decrease of the surface velocity level of about 4.3 dBA. But with this modification several changes of the valve cover would have to be carried out.

Inner walls between the pedestals are more efficient concerning level benefit and mass necessary, and they are simple modifications. A variety of walls with different heights were calculated, and in agreement with the designing engineers 5 inner walls could be realised. They reduce the breathing and bending modes of the bearing pedestals which results in a benefit of about 3.0 dBA. Additional triangular ribs at the camshaft bearings which influence the bearing modes decrease the surface velocity further to a complete level decrease of about 6.1 dBA. This result is shown in **Figure 4**. In the narrow-band spectra it can be seen that the peaks between 2500 Hz and 3500 Hz belonging to the breathing and bending modes of the bearing pedestals are strongly reduced. Thus the levels of the dominant 2500 Hz and 3200 Hz third-octave bands are significantly decreased. The benefit of about 6.1 dBA is indicated in the integral spectra at 6400 Hz.

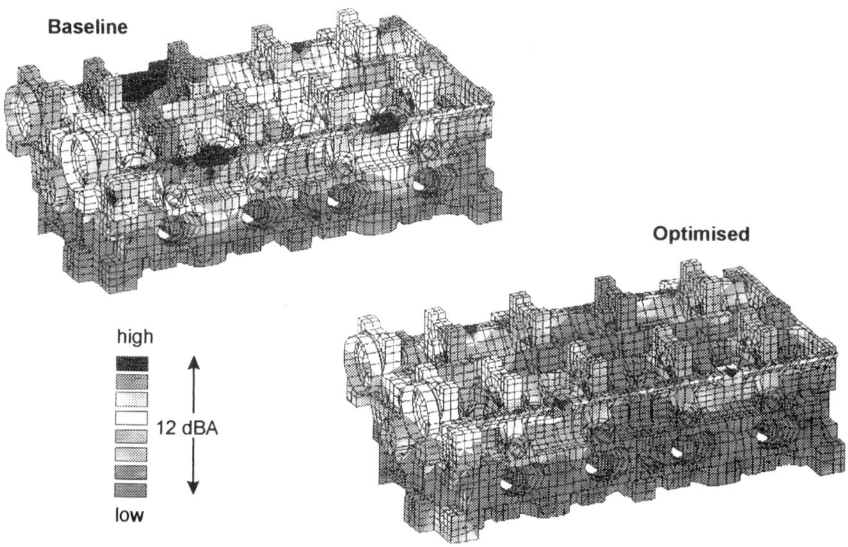

Figure 5: Calculated Surface Velocity Distribution of Optimised Cylinder Head

To visualise the calculated level reductions of this modification, the distribution of the surface velocity is shown in **Figure 5** compared to the baseline of the cylinder head. A large part of the baseline's high level area is reduced.

4 VALIDATION OF THE OPTIMISATION

An experimental Impulse Noise Analysis (INA) was carried out to evaluate the positive effect of the inner walls for the acoustical behaviour of the cylinder head. INA is a method where a component is excited with a wide-banded hammer impulse, and the surface velocity at several points is measured and summed up. As a laboratory type measure, five additional inner walls were glued into the cylinder head (to simplify matters without the triangular ribs at the bearings). The measurement of the surface velocity shows a very good correlation to the calculation.

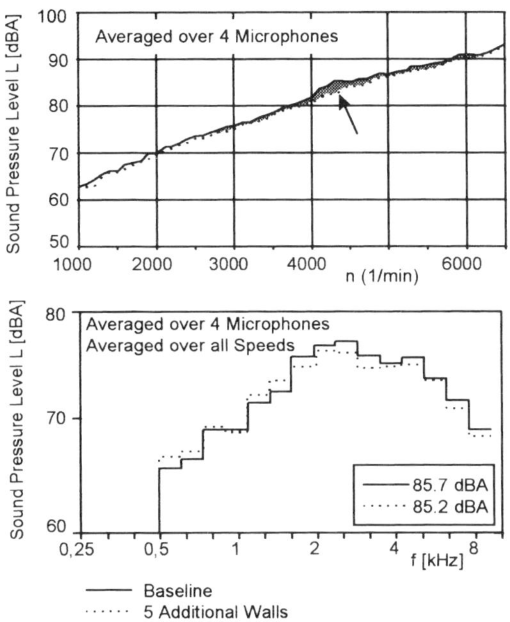

Figure 6: Influence of Additional Walls on Valve Train Noise

Of more interest are the sound pressure levels measured on the test bed with the partial engine in motored condition. **Figure 6** shows the valve train noise level averaged over all sides and the characteristic third-octave spectra (averaged over all sides and over all speeds). Especially in the speed range 4000 rpm to 5000 rpm which is relevant for the drive-by test, a level decrease of about 1-2 dBA can be recognised. The third-octave spectra show level reductions in the 2000 Hz to 3200 Hz third-octave bands which is in good correlation to the calculated surface velocity level decrease. So, the improvement of the cylinder head's acoustical behaviour by reducing the breathing and bending modes of the bearing pedestals can be confirmed on the test bed. Probably the additional triangular ribs at the bearings will decrease the sound pressure level of the already low-noise cylinder head further more.

5 SUMMARY

The experimental investigations of the 4-cylinder 4-valve cylinder head made clear that the main excitation mechanisms of the valve train noise belong to impacts of the camshaft journals on the bearings, impacts of valves on the valve seat and mass forces on the camshaft and on the bearings during the phases of high valve acceleration and deceleration. Secondary movements like tilting of the tappets have minor importance for the investigated cylinder head.

The dynamic structural analysis of the cylinder head conducted by the method of Finite Elements revealed level-dominant breathing and bending modes of the bearing pedestals. The improvement by 5 inner walls between the bearing pedestals of intake and exhaust camshaft were confirmed at the test bed. Further benefits can be assumed by additional triangular ribs at the bearings.

This result shows the good interaction of the tools „FEM Calculations“ and „Experiment“ by which a good cylinder head has been improved further more.

Tyre and Road

C521/011/98

Airborne road noise source definition

N C TAYLOR BSc
Rover GroupLimited, Warwick, UK
A P BRIDGEWATER B.Tech
Dunlop Tyres Limited, Dynamics Laboratory, Birmingham, UK

SYNOPSIS

This paper describes an investigation into defining airborne road noise in terms of source location and directionality, where possible relating this to classical descriptions of the noise generating mechanisms. All of the tyre noise measurements were made on a chassis dynamometer with the tyre under investigation being held in isolation and driven at a constant speed. Tests were conducted on the plain steel drum and a replica BS594 (coarse chip) surface fitted to the drum. Detailed maps of the sound radiation patterns were made at the side, front and rear of the contact patch over the frequency range of 200-2000Hz using a Brüel & Kjær Spatial Transformation of Sound Fields system.

1 INTRODUCTION

The object of this exercise was to characterise the tyre, particularly in relation to its tread pattern, as a noise source. The mechanisms of tyre noise generation are complex, but they may be characterised by a series of classical descriptions [1]. These are as follows:

The noise generation mechanism at the tyre to road interface may be broken down into three distinct regions:

i) At the leading edge the tread blocks strike the road surface and subsequently vibrate. The frequency of these impacts is the pattern repeat frequency. The block impacts lead to the production of a sound wave, the form of which has been shown empirically to be of sine form with some third harmonic influence [1].

On a smooth road the pattern repeat frequency of the tread blocks would be quite dominant although modern tread pattern design reduces this effect. The block impact effect on a rough road is similar, however there also exists some further influence from the road surface which breaks up the repeat effect.

ii) In the contact patch, area of the tyre in contact with the road surface at any given time, there is some movement of the tread blocks relative to the road which generates a slip noise. This mechanism will also be affected by road surface roughness as a more coarse surface tends to reduce the effect of slip.

The tread blocks that are in the contact patch are caused to vibrate by their impacts with the asperities in the road surface. This mechanism is greater on coarse road surfaces and this is partly the reason for the reduction of block impact and slip effects on coarse roads.

The groove organ pipe noise is the resonance of the air within the tread grooves in the contact patch. Grooves running circumferentially around the tyre are particularly significant because their length is such that the first mode of vibration of the air column is low enough to be coincident with other vibration mechanisms. This mechanism tends to be of low importance on coarse roads.

As the grooves progress through the contact patch they close up which results in the air which is ordinarily trapped within the tread grooves being compressed. This squeezing and then releasing of the air acts as a pump which generates noise, known as air pumping. The pressure increase in lateral grooves will create a noise source to the side of the contact patch. When traveling on a porous road surface the air, rather than being compressed within the tread grooves, is squeezed into the void spaces of the road. This is a major reason for the reduction in tyre - road interface noise when traveling on porous roads.

iii) At the trailing edge, the blocks vibrate after they have been released from the ground contact, this phenomenon is known as 'Block snap out'. The tangential forces which exist in the contact area apply significant deflections to tread blocks and it is as these forces are removed, and the block is released, that it has a tendency to vibrate.

Further mechanisms of noise generation exist and include:

i) Noise generated by the sidewalls of the tyre. The deflection of the sidewall is greatest in the region of the contact patch.

ii) The air cavity resonance is the resonance of the air within the tyre. The first circumferential mode of this air space is known to have a frequency of 200-300Hz, the actual frequency being controlled by tyre size and wheel dimensions. At higher frequencies, typically around 1100 - 1200 Hz, there is a lateral mode of the air cavity. Numerous researchers have investigated the cavity resonance [2].

iii)At the front and rear of the contact patch a 'horn' is formed by the tyre and road surfaces. It has been suggested that this may amplify noises produced at the front and rear of the contact patch within certain frequency bands. However, little evidence of this type of effect is cited in the literature.

Having discussed in some detail the classical descriptions of the tyre as a noise source, the objective of this work is to understand the source characteristics and to relate the measurements where possible to these classical mechanisms.

Initial investigations were made on a production tyre on both a smooth steel drum and a replica coarse road surface to BS 594. Measurements were made at three road speeds in an attempt to determine the basic excitation mechanisms. Consideration of these results showed that on the replica road surface no distinct features were discernible. On the plain steel drum the noise generation mechanisms of the standard tread pattern were shown to be difficult to separate. Figure 1 shows the radiated sound power spectra for these two road surface test conditions, and clearly demonstrates the importance of the road surface.

At this stage it was decided that the measurement should be simplified, which was achieved by using a single pitch lateral grooved tyre in place of a production tyre. This lead to a single fundamental excitation frequency that was a function of road speed rather than the many excitation frequencies of the standard production tyre. As the aim of this study was to characterise tread pattern effects as noise sources, the rest of this paper will describe the results obtained with the simple tread pattern on the plain steel drum. This tread pattern consisted of 63 equally spaced grooves, each 5 mm wide, cut laterally across the tread. This configuration of tread pattern gave a constant excitation frequency for a given rate of tyre rotation, and this is termed the fundamental frequency. For the tested tyre size, 195/65R15, at a pressure of $2.3x10^5$ Nm^{-2} and a load of 3000N the fundamental frequency was 440Hz at a road speed of 50km/hr.

2 METHOD

The noise measurements were made using the Brüel & Kjær Spatial Transformation of Sound Field (STSF) equipment [3]. The system was used to measure and calculate the sound field generated by the tyre rolling on a driven drum (diameter 1.52m). The STSF system assumes the ground plane to be flat and perfectly reflective. In order to achieve this the floor was extended to lie close to the tyre contact patch, which reduced the influence of the drum curvature.

Figure 2 shows the configuration of the measurement scan plane in relation to the tyre on the drum. The measurement scan plane was placed at a distance of 0.17 m from the surface of the tyre. This configuration allowed all of the sound energy that would radiate to the far-field, and as such would contribute to the pass-by noise level of the vehicle, to be captured. A calculation of the near-field acoustic holography allowed detailed sound intensity maps to be investigated at the surface of the tyre, enhancing the source identification.

Measurements were made at varying load and pressure conditions at 50, 55 and 60km/hr. These showed the shift in fundamental frequency as expected. However, a problem was encountered at the higher speed due to a resonance of the drum. The results described have been therefore limited to show the effect of tyre load and pressure at a road speed of 50km/hr.

3 RESULTS

The active intensity maps for the fundamental frequency are shown in Figure 3 as a function of load on the tyre.

At very low load (1000N) most of the sound energy is radiated from the contact patch. For this load case the dominant mechanism is air pumping from the side of the tread groove as described above. As the load on the tyre increases so the sound power from the rear of the contact patch increases. A possible explanation for this is that as the deformation of the tyre increases, so the recovery of the tyre to its undeformed shape as it releases from the contact patch is more severe. From the active intensity maps at the higher loads there appears to be a source behind the contact patch. Detailed investigations of the active intensity plots in vector form showed the source to be located at the tread face and tyre road interface. This source is highly directional, hence the appearance at the plane on the surface of the tyre sidewall (back calculated from the measurement plane).

The total radiated sound power at the fundamental frequency through the measurement plane was calculated and plotted against load. This showed a complex relationship. A measurement of the sound power for the 3000N $2.3x10^5Nm^{-2}$ load case was repeated three times and for the 3500N $2.3x10^5Nm^{-2}$ load case twice. The error in the load on the tyre was $^+/_-$ 100N and the error in the sound power measurements (based on experimental results) was $^+/_-$ 0.5dB (re 1pW). The sound power for the fundamental frequency as a function of load with the associated errors is shown in Figure 4.

Considering all of the measurement errors it is evident that a minimum exists on the curve at the 3500N load. This suggests that for this load case either the radiated sound power is reduced or that the sound power is radiated in a direction such that it does not pass through the scan plane. In order for the sound power not to pass through the measurement plane it would have to be radiated in a very tight pattern fore and aft of the contact patch. The air pumping mechanism is known to be very directional, but this tyre had no longitudinal grooves. Another possibility is that as the load changes so the contact patch shape and contact patch pressure distribution changes.

At this stage a relationship between contact patch shape and sound power radiated at the fundamental frequency was pursued. A number of contact patch shape parameters were considered and a relationship was found to exist between the contact patch length, width and radiated sound power at the fundamental frequency. An example of the contact patch shape as measured on the curved surface of the drum is shown in Figure 5.

The contact patch length and width for all of the load cases is shown in Table 1.

Table 1. Tabulated tyre parameters.

Tyre pressure (Nm^{-2})	Load (N)	Contact patch length (mm)	Contact patch width (mm)	Length/width
$1.3x10^5$	2000	80	128	0.63
$1.3x10^5$	2950	108	135	0.8
$2.3x10^5$	1000	46	90	0.51
$2.3x10^5$	1500	62	110	0.56
$2.3x10^5$	2000	74	125	0.59
$2.3x10^5$	2500	79	130	0.61
$2.3x10^5$	3000	96	135	0.71
$2.3x10^5$	3500	97	140	0.69
$2.3x10^5$	4000	100	137	0.73
$2.3x10^5$	4240	109	140	0.78
$2.3x10^5$	4700	117	140	0.84
$3.3x10^5$	350	32	66	0.48
$3.3x10^5$	3900	83	133	0.62

At a tyre pressure of $2.3x10^5Nm^{-2}$, as the load increases the contact patch length increases faster than the width. This results in the aspect ratio (length/width) increasing. However, at 3500N the width of the contact patch increases with little change in the length of the contact patch, i.e. the aspect ratio decreases. A plot of sound power of the fundamental frequency and the tyre contact patch aspect ratio versus load is shown in Figure 4. The sound power of the fundamental frequency versus contact patch aspect ratio gave a correlation coefficient of 0.9 for a linear least squares fit. This shows that there is a reasonable degree of correlation between sound power of the fundamental frequency and contact patch aspect ratio for this particular tyre and this test condition.

Figure 6 shows the sound for the fundamental frequency versus tyre contact patch aspect ratio for all of the conditions tested. The correlation coefficient was reduced to 0.8 for the linear least squares fit. Whilst correlating less well than for the $2.3x10^5$ Nm^{-2} tyre pressure conditions alone it is still considered to be of some significance. This suggests some relationship between contact patch shape and sound power. However, this relationship has only been shown for the simplified tread pattern and only at 50km/hr.

4 CONCLUSIONS

For the simplified laterally grooved tyre running at 50km/hr on a plain steel drum it has been shown:

- At low load and high tyre pressures air pumping is the dominant sound radiation mechanism at the fundamental frequency. As load increases block 'snap-out' and resultant tread face vibration dominates with possible influence of the 'horn' effect.

- A relationship between contact patch shape and radiated sound power at the fundamental frequency has been shown to exist. This relationship becomes more complex when the influence of pressure is considered. Further investigation is required on real production tyres.

- As load increases so the dominant source of sound power moves from the side of the contact patch to the leading and trailing edges. At high loads the dominant source is the tread face as it is released from the contact patch.

This project leaves a number of unanswered questions related to tyre road interface noise generation. These require extensive investigation before the conclusions given in this paper can be extended to real tyres and road surfaces.

REFERENCES

1. J. C. Walker, "Noise Generated at the Tyre-Road Interface", Ph.D. Thesis, The University of Aston, August 1981.
2. T. Sakata, H. Morimura, H. Ide, "Effects of Tire Cavity Resonance on Vehicle Road Noise", Tire Science and Technology, TSTCA, Vol. 18, No. 2, April-June, 1990, pp. 68-79.
3. J. Hald, "STSF A Unique Technique for Scan-Based Near-Field Acoustic Holography Without Restrictions on Coherence", Brüel & Kjær, Technical Review No. 1, 1989

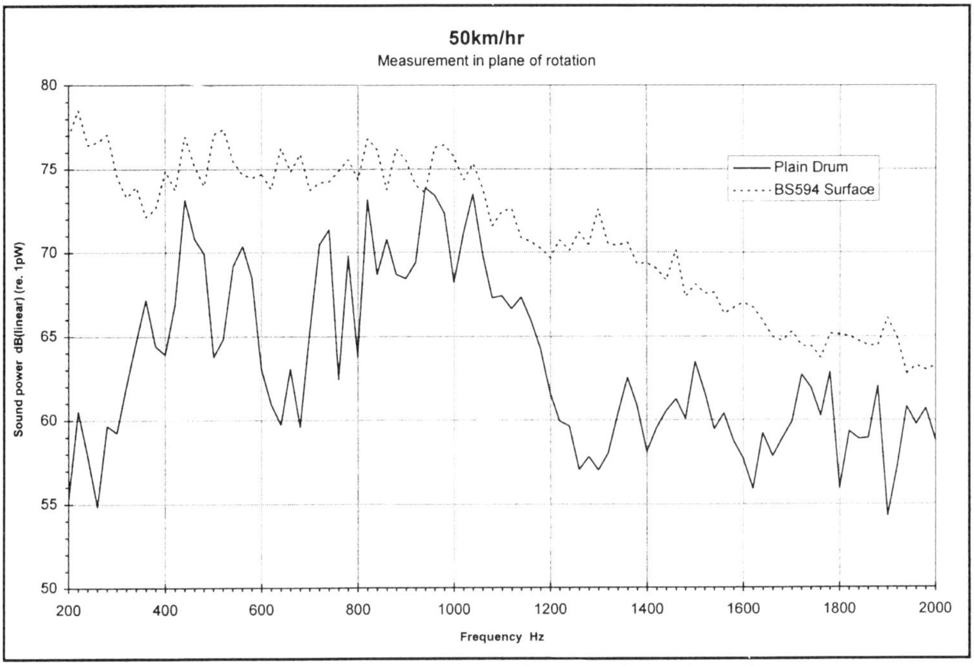

Figure 1. Sound power spectra from in-plane of rotation measurement for the plain steel drum and BS594 surface.

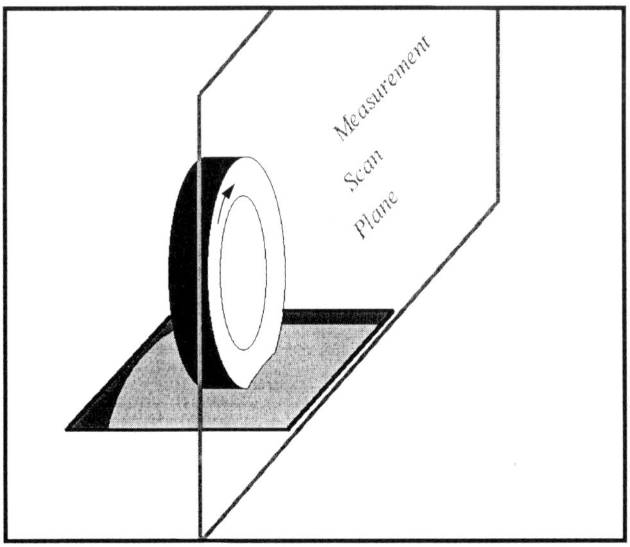

Figure 2. Schematic of measurement plane.

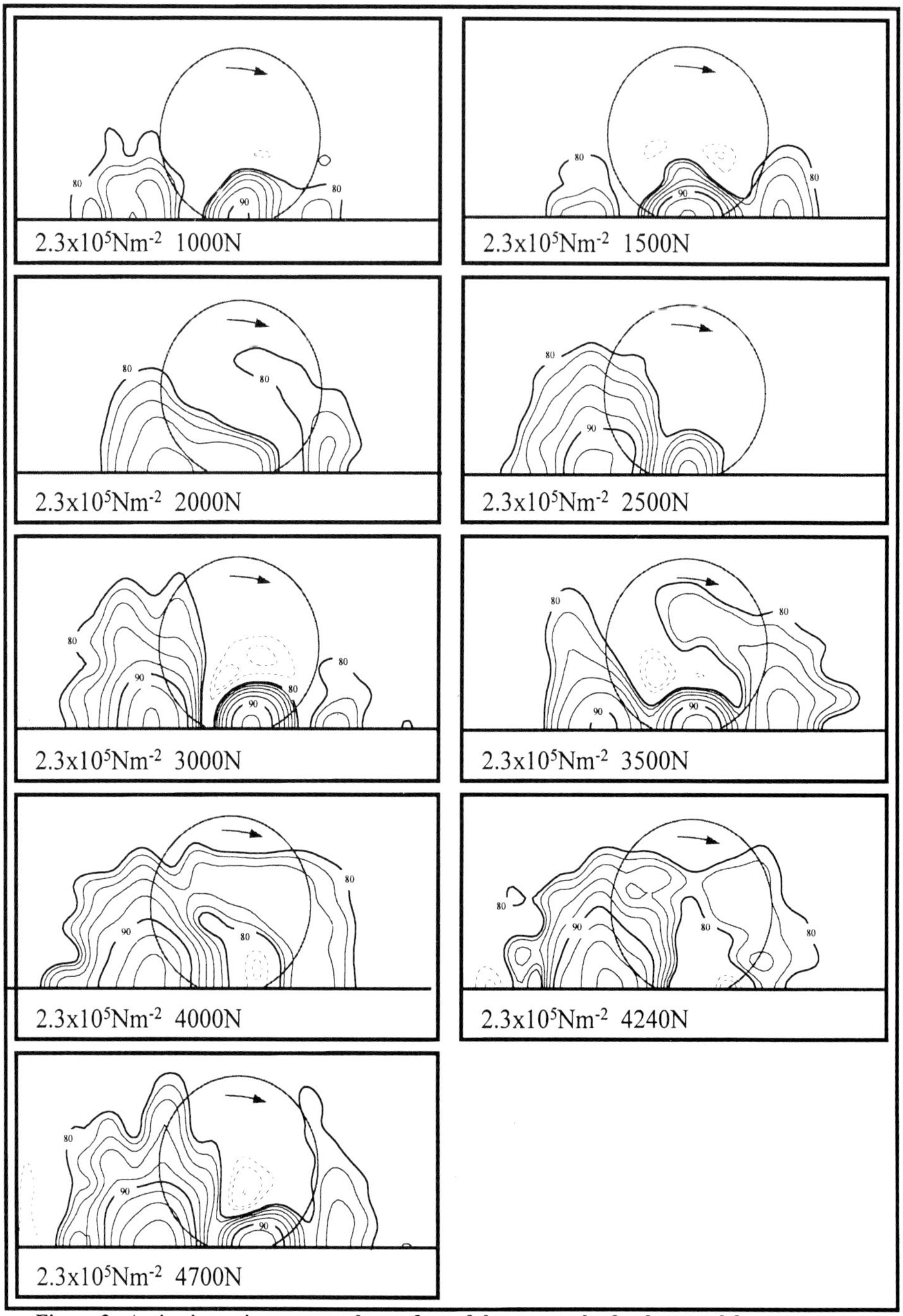

Figure 3. Active intensity maps at the surface of the tyre at the fundamental frequency.

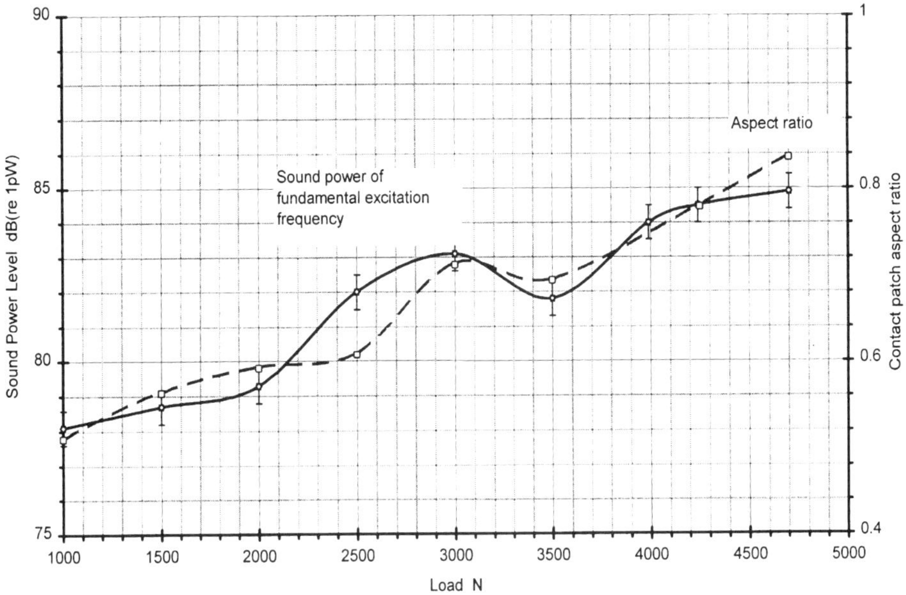

Figure 4. Sound power radiated in the plane of rotation for the fundamental frequency versus Load and Contact patch aspect ratio verus Load.

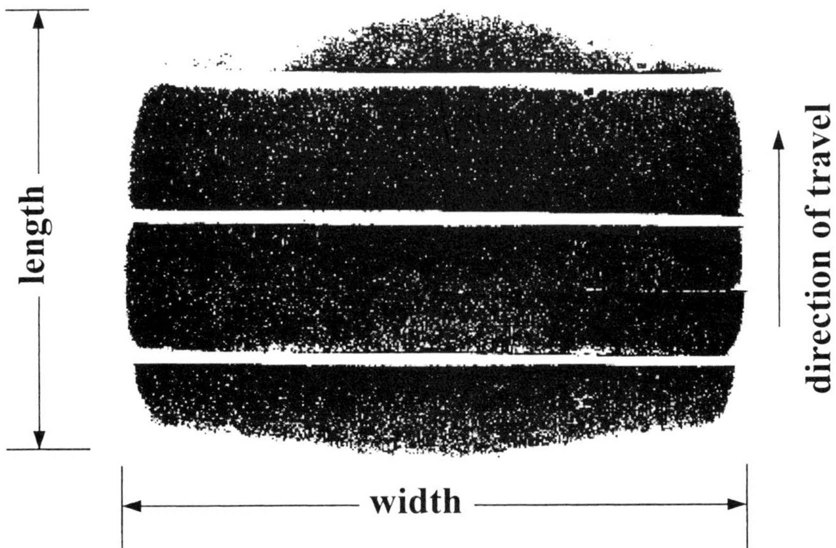

Figure 5. Static tyre contact patch shape for 2.3×10^5 Nm^{-2} **4240N load case.**

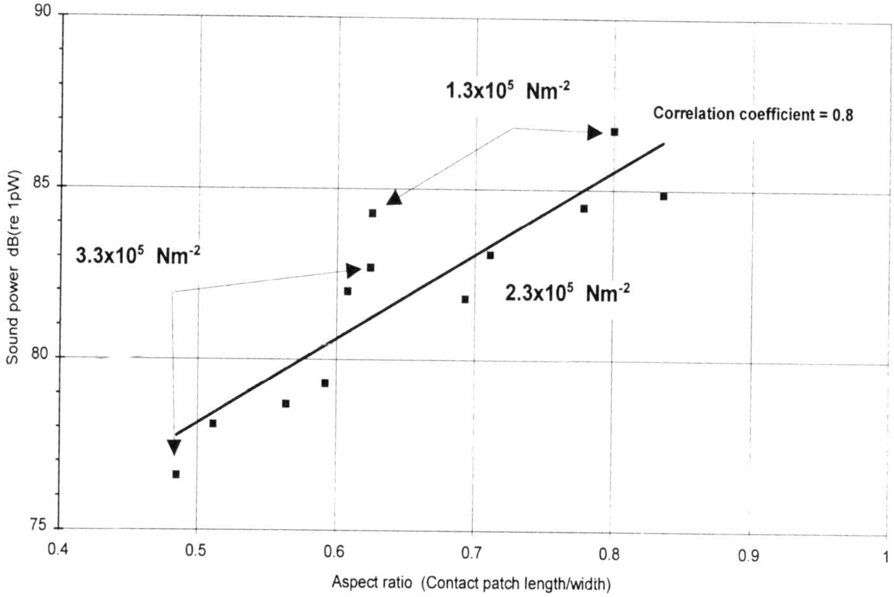

Figure 6. Sound power for the fundamental frequency versus tyre aspect ratio for all of the tyre load and pressure conditions tested.

C521/015/98

Characterizing the transfer of road-surface excited vibrations through vehicle suspension systems

D STORER BSc, PhD, MIMechE, MASME, **S GATTI** MSc, and **E PISINO**
Centro Ricerche FIAT, Turin, Italy

Synopsis

This article addresses the vibroacoustical characterisation of cars when passing over rough road surfaces and describes a technique under development for studying the transfer of vibrations through the suspension systems. Required are test-based procedures for classifying the 'transfer paths' between the suspensions and the vehicle body and for quantifying the forces transferred in each direction at each connection point across a relatively broad frequency range. Referring to an application on a current-production vehicle, it is demonstrated that the response spectra on the driver's seat guide of the vehicle determined using the forces calculated are in close agreement with direct measurements.

1. INTRODUCTION

With the aim of improving the ride-comfort behaviour of vehicles for road-surface excitation, it is necessary to understand in detail the dynamic characteristics of the suspension systems. In this context, two test-based strategies can be adopted to characterise the suspension system and the forces transferred to the vehicle body for excitations at the wheel. Firstly forces can be measured directly by performing tests on the isolated suspension and its constituent components. Secondly it is possible to apply a technique called Transfer Path Analysis, originally developed to study the aspect of powertrain-induced vibrations of the vehicle body, to the suspensions of the complete vehicle being tested on the road or in the laboratory on a road simulator (1-3). In fact the experimental research activity described in this article utilises both approaches in the study of the dynamic behaviour and vibration transmission characteristics of the suspension systems in a current-production economy-class vehicle. Whereas this study focusses on the transmission of road-surface excitation, a companion project addresses the aspect of powertrain-generated transmission through the suspensions (4).

2. TRANSFER PATH ANALYSIS

2.1 Overview of the technique

The aim of Transfer Path Analysis (TPA) is to decompose the total vibrational (or acoustical) response p_i at a point i due to one specific source into contributions from each of the n different structural transfer paths assumed to transfer energy from the source to the structure.

$$p_i = \sum_j p_{ij} \qquad \begin{cases} i = 1..t \\ j = 1..n \end{cases} \tag{1}$$

where :

p_{ij} : contribution at the ith point due to the jth transfer path
t : total number of response measurement points

The contribution of each transfer path in a structure can be determined by calculating the forces transferred by the source to the structure via the specific path and multiplying these by the corresponding transfer functions :

$$p_{ij} = h_{ij} . f_j \tag{2}$$

where :

h_{ij} : transfer function between response point i and input at point j
f_j : force transferred at point j

The points where the forces are transferred effectively define the transfer paths. Usually in practice points are used whenever possible where elastic elements, eg. rubber mounts, couple the source to the structure since this can facilitate the determination of the forces transferred.

For the complete structure, for all measurement points, equations (1) and (2) can be written as:

$$\{P\} = [H]\{F\} \tag{3}$$

The determination of the transfer functions is often performed off-line using a controlled, measurable input applied at a single point in a single direction using, for example, an impulse hammer.

Depending on the system of source and structure being investigated, different ways of estimating the forces can be attempted:

- direct measurement of the forces transferred
- indirect determination using the complex stiffness of the connecting elements [1]
- indirect calculation based on the inversion of a structural FRF matrix [2]
- indirect estimation using a representative mathematical model

Each of these methods presents a series of practical problems and in fact the determination of the forces generally represents the weak link in the method. In the direct measurement of forces transferred from the suspension to the car body, for example, the problem of inserting load cells at the different connection points is often considerable if not unsurmountable.

2.2 Indirect determination of forces

2.2.1 The complex stiffness method

In this technique, the forces transferred to the structure through the connecting elements are determined by measuring the relative displacement x_j across each element j which is then used as a kind of force transducer, the factor of calibration being the predetermined stiffness k_j :

$$f_j = k_j x_j \tag{4}$$

To determine the operational forces across each element in this way, it is usual to measure the acceleration at the two ends, compute the difference, and then integrate to determine the relative displacement.

Since TPA is performed using data in the frequency domain, the stiffness k becomes a complex function which appropriately represents the element in the direction of interest. In practice, however, it is often difficult to characterise each connecting element in each direction of interest in conditions which correspond to those encountered during operation.

2.2.1 The matrix inversion method

This method uses a matrix $[H^{(S)}]$ of structural (inertance) FRFs, with elements $h_{kj}^{(s)}$ relative to a single (force) input at point j and output (acceleration) at point k . To apply the technique, this matrix must characterise the dynamic behaviour of the structure and incorporate measurements for boths input and outputs at all points corresponding to the different transfer paths of interest. Once each element has been measured 'off-line', and the matrix $[H^{(S)}]$ constructed, it is possible to determine the vector of transferred forces $\{F\}$ which corresponds to the vector of measured operational accelerations $\{A\}$ for all of the relevant transfer paths:

$$\{F\} = \left[H^{(s)}\right]^{-1}\{A\} \tag{5}$$

In this formulation it is essential that all significant transfer paths are incorporated since otherwise errors will result in those included ie. the contribution of all forces which can influence the response acceleration measured at any point must be taken into account. In certain applications, this may present a serious problem since often it is difficult to determine all the sources of vibration and ascertain a-priori which transfer paths are important.

To avoid numerical problems associated with matrix inversion, it is conventional to overdetermine the system of equations and use Singular Value Decomposition (SVD) solution procedure.

Finally, by combining equations (3) and (5), response vector $\{P\}$ can be determined :

$$\{P\} = [H]\left[H^{(s)}\right]^{*}\{A\} \tag{6}$$

2.3 Single-source TPA

When the source of excitation can be assumed to be 'single', as is assumed for example in the case of a vehicle engine, the TPA procedure can be simplified since it is possible to assume that all the transfer paths from that source are directly correlated with a common phase reference. As a consequence, the operational spectra can be averaged in order to reduce the effects of random noise and, in order to make the final comparison with the measured response, each of the phase-referenced contributions from each path can be summed directly.

2.4 TPA for multiple uncorrelated inputs

In the example of TPA applied to a vehicle tested on a four-poster system, each wheel being excited in the vertical direction using four independently-generated, uncorrelated random signals, it is possible to perform single-source TPA independently for each applied excitation. In this case the total response can be calculated in terms of the sum of RMS levels of the total contributions due to each independent reference. For each response point, since the reference signals are uncorrelated, the averaging involved in computing the cross-spectra with respect to one reference automatically eliminates the influence of other inputs, which effectively only add random noise to the response, assuming a sufficient number of averages are performed.

2.5 TPA for multiple partially-correlated inputs

In situations for which the reference signals cannot be assumed to be either correlated or completely uncorrelated, it is necessary to preprocess the measured spectra before performing the TPA. Use is made of so-called *Principal Component Analysis* (PCA) which permits the transformation of the set of selected partially-correlated reference signals into a set of uncorrelated 'virtual sources' [3]. The TPA is then performed using the referenced cross-spectra with respect to the principal virtual sources. This type of analysis is necessary when analysing data measured on vehicles during road tests since it is not feasible to measure directly the operational inputs at each of the wheels.

2.6 General limitations of the TPA method

Apart from the difficulties associated with the measurement and practical implementation of TPA already indicated, other fundamental limitations also exist :

- Firstly the structure being characterised is assumed to be linear or linearisable. In particular when tested 'off-line' the dynamics of the system should correspond to the 'on-line' behaviour during operation ;
- Secondly there is a need to take into consideration all significant transfer paths, particularly when applying the matrix inversion method to determine the forces. This presents a serious difficulty in many applications, especially when structural transfer paths are known to exist but are difficult to monitor, or when airborne paths might be important.

In the tests on the vehicle described subsequently certain difficulties were avoided by using a four-poster system to provide independent vertical excitation at the four wheels, and by applying the TPA method to the vibrational responses at the seat guides which are less likely to be influenced by airborne transfer paths than the acoustic responses inside the vehicle.

3. ANALYSIS OF A SINGLE SUSPENSION

3.1 The McPherson suspension system

Prior to the application of the TPA methodology on a complete vehicle, a series of tests were performed on a single, front-left McPherson 'monosuspension' system, comprising the tyre, wheel-hub assembly, lower arms, strut, spring and top mount and the other bushings (Fig. 1), in order to clarify certain important aspects related to the influence of the different constituent components across the frequency range of interest, and to investigate the dependence on the level of dynamic excitation.

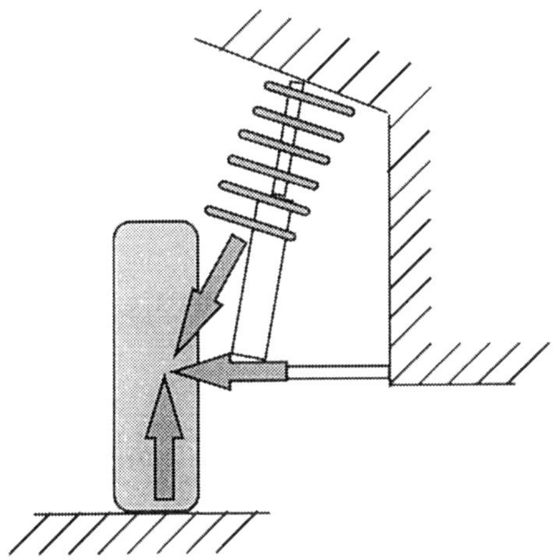

Figure 1 **Forces acting in the McPherson suspension system**

In this type of suspension, the shock absorber or 'strut' is both the principal damping element and the kinematic link between the car body and the wheel. However the geometry of the McPherson suspension causes the shock absorber to experience lateral loading which can be offset by the force of the inclined spring only in one position. This lateral loading can in turn influence the behaviour of the shock absorber, particularly as regards the friction generated by the different seals and rod guides on the rod.

The behaviour of the shock absorber is also strongly influenced by the top mount at the upper point of connection to the car body. On the vehicle tested a single mount connects both the spring and shock absorber to the car body. In practice, many factors influence the selection of the most appropriate stiffness(es) of this top mount. It is generally noted that if the mount is too flexible, the spring and/or shock absorber cannot function adequately particularly at lower frequencies, whereas if the mount is too rigid, it is less effective in attenuating the vibrations transferred by the suspension to the car body over the higher frequency range.

In the monosuspension test rig, a hydraulic actuator is used to apply vertical excitation to the wheel simulating profiles of typical test tracks. By isolating the suspension from vehicle in this way, it is possible to instrument the suspension in order to study the behaviour of each component and its influence on the complete system. In particular, the forces at the suspension attachment points can be measured directly, permitting the suspension to be characterised in terms of dynamic stiffness functions, often unfeasible in tests on the vehicle.

The obvious limitation of the mono-suspension system is that, since the attachment points are fixed, the filtering efficiency of the total system cannot be evaluated. Also, since these points on the vehicle are mobile or elastically yieldable the monosuspension, inevitably behaves somewhat differently from the vehicle, particularly at low frequencies. Nevertheless tests on the monosuspension are useful for understanding the behaviour of the suspension system particularly as regards classifying the importance of the different attachment points in terms of the forces transferred for a vertical input at the wheel.

3.2 Results of dynamic characterisation

In the dynamic characterisation tests performed, both the dynamic stiffness functions and acceleration transmissibility functions were measured in order to characterise the dynamics of the system and evaluate the filtering efficiency of the suspension.

Figure 2 illustrates dynamic stiffness functions respect to the vertical output force at the top attachment point relative to the three levels of random excitation.
Fig. 2(a) illustrates the dynamic stiffness of the top mount incorporated in the sub-assembly from which it is possible to observe the characteristic tendency of visco-elastic materials to exhibit an effective increase in stiffness at higher frequencies.
Fig. 2(b) illustrates the dynamic stiffness of the sub-assembly of mount plus spring/damper incorporated in the mono-suspension system. At higher frequencies, the effective stiffness of the sub-assembly tends towards the stiffness of the top mount indicating the increasing tendency for the system to block at higher frequencies caused by the poor functioning of the shock absorber at frequencies above approx. 20Hz.
Fig. 2(c) illustrates the dynamic stiffness function with the vertical displacement at the tyre as the input signal. At low frequencies the magnitude of the function is directly dependent on the equivalent stiffness of the tyre, spring/damper and top mount in series. At higher frequencies, the magnitude tends to increase, reaching a peak in the range 20-40Hz (depending on the level of excitation) which corresponds to the so-called 'non-suspended' resonance of the wheel hub mass between the tyre and suspension stiffnesses.

Figure 3 illustrates both the acceleration transmissibility functions and dynamic stiffness functions in the three translational directions relative to the vertical input applied at the tyre. (Using conventional notation, X is the longitudinal, Y lateral and Z vertical.)
Fig. 3(a) illustrates the acceleration transmissibility for response at the hub. Non-suspended resonances of the hub are clearly visible in the range 20-30Hz in each direction, particularly in the direction of the applied excitation. Above approx. 40Hz the vertical transmissibility falls to below 0.1 indicating reasonable attenuation of the system, whereas at frequencies above 80Hz the dynamic behaviour is largely influenced by the tyre.

Fig. 3(b) illustrates the transfer functions between the input displacement at the tyre and the output force in X, Y and Z at the dome. Of particular interest is the fact that whereas at low frequencies the vertical direction is the most important in terms of force transferred, as might be expected, above 60Hz the different directions become equally significant.
Fig. 3(c) illustrates the transfer functions relative to the output force measured at the forward lower arm attachment in X, Y and Z. At this location, the forces transferred in each of the directions are nominally equal and attain a relatively high level, particularly in the frequency range of the non-suspended mass resonances.
Fig. 3(d) illustrates the transfer functions relative to the output force at the rearward lower-arm attachment in X, Y and Z. Evidently the vertical direction is most important at this point.

4. ANALYSIS OF COMPLETE VEHICLE

4.1 Objectives

In order to understand the transmission of forces through the suspension system of a vehicle, a series of tests was performed using a conventional 4-poster test facility. TPA was applied to the complete vehicle over the frequency range 0-200Hz in order to determine the principal transfer paths with respect to the response at the driver's seat guide for vertical excitation at each wheel. A second objective was to establish if, for an input applied at the front left wheel for example, all suspension attachment points need to be included in the analysis or whether it is sufficient to consider only the transfer paths associated with the suspension of that wheel.

The principal advantage of performing tests in laboratory in this way is that the TPA analysis can be simplified by applying uncorrelated, measureable inputs in the vertical direction at each wheel. Furthermore, the same inputs can be used repeatedly, enabling more response points to be measured with relative ease as opposed to testing the vehicle on the road.

4.2 Test procedure

Triaxial accelerometers were positioned at the wheel hub, on the shock absorber rod, and in correspondence with each of the different mounts and bushings where the front and rear suspension systems are fixed to the car body. In order to analyse the effect of each excitation independently, uncorrelated random signals were used to excite the four wheels in the vertical direction while simulating the vertical response at each hub measured during road tests.

Subsequently, in order to permit application of the matrix inversion technique, a series of tests was performed to characterise the frequency response matrix of the structure with respect to each of the suspension attachment points and the seat guide. For this task an impulse hammer was used to provide excitation independently at each point and in each translational direction, the responses being measured using the accelerometers already positioned at the attachment points. Assuming that the different mounts and bushings effectively decouple these two subsystems, the tests were performed without detaching the suspension from the body, thus simplifying the test procedure significantly.

4.3 Application of TPA

Since the four inputs applied at the wheels were completely uncorrelated, the matrix inversion method for determining the forces was applied independently for each input. The measured and calculated responses at the driver's seat guide are shown in figure 4, relative to the four wheels. In each diagram, the measured response spectrum is compared with that calculated using TPA, considering firstly the transfer paths relative to all suspension attachment points and secondly only those relative to the suspension of the reference wheel.

Good agreement with the measured spectrum has been achieved in each case when using the full matrix of transfer paths in the calculation of the response. Moreover, for the front suspension on both left and right sides, across the frequency range all thc principal transfer paths are associated with the suspension of that wheel. Instead, for the rear suspensions, not only the local attachment points are important particularly at higher frequencies.

Analysing in greater depth the individual transfer paths relative to the response at the seat guide for vertical excitation at the wheels, the following observations are possible :
- For the front suspensions, the top attachment points in the vertical direction are the dominant transfer paths up to 50Hz, whereas the forward attachments of the lower arm become the most important contributors in the range 50-200Hz ;
- For the rear suspensions the rearward attachments points in the vertical nearest to the wheel excited correspond to important transfer paths up to 80Hz. Around 40Hz, the forward attachments in the vertical direction also contribute significantly whereas in the range 80-140Hz, the forward attachment point of the forward suspension lower arms become important. At higher frequencies, the vertical attachments of the rear suspension but on the side opposite to the wheel excited contribute significantly.

5. CONCLUSIONS

From the results of the tests on the monosuspension test rig, it is possible to distinguish the significance of the 'non-suspended' mass resonance which falls in the range 20-40Hz (depending on the behaviour of the spring/absorber unit and therefore on the stiffness of the top mount, the friction in the shock absorber and the level of excitation). In this range, the forces transferred at the top attachment point in the vertical direction, and the forward lower-arm attachment in all three directions are significant. At higher frequencies the force transferred in the lateral and longitudinal directions also become important.

When applied to the complete vehicle, Transfer Path Analysis technique yielded good results, due partly to the advantages of performing tests in the laboratory with controlled, measureable excitation waveforms as opposed to analysing directly road-test data. The results indicate that for excitation applied at the front wheels, the 'local' transfer paths, ie. those associated with the suspension of the wheel in question, are most significant, particularly the top attachment point in the vertical direction at lower frequencies. As observed also in the tests on the monosuspension, the lower arm attachment points become important at higher frequencies. On the other hand, as regards the rear suspensions, it is necessary to consider the forces transferred by all attachment points, not only the local ones, particularly at higher frequencies.

REFERENCES

1. **Accurate vehicle FRF measurement for indirect force determination based on matrix inversion;** Proc. of ISMA 19, Leuven, Belgium, September 1994

2. **Road/tyre noise development using noise path analysis techniques;** Proc. of ISMA 19, Leuven, Belgium, September 1994

3. **An overview of the European research project DIANA;** SAE Paper No. 971897, Proceedings of the 1997 Noise and Vibration Conference, Traverse City [MI], May 1997

4. **Structure-borne transmission of powertrain noise through the drive axles in cars;** To be presented at ISMA 23, Leuven, Belgium, September 1998

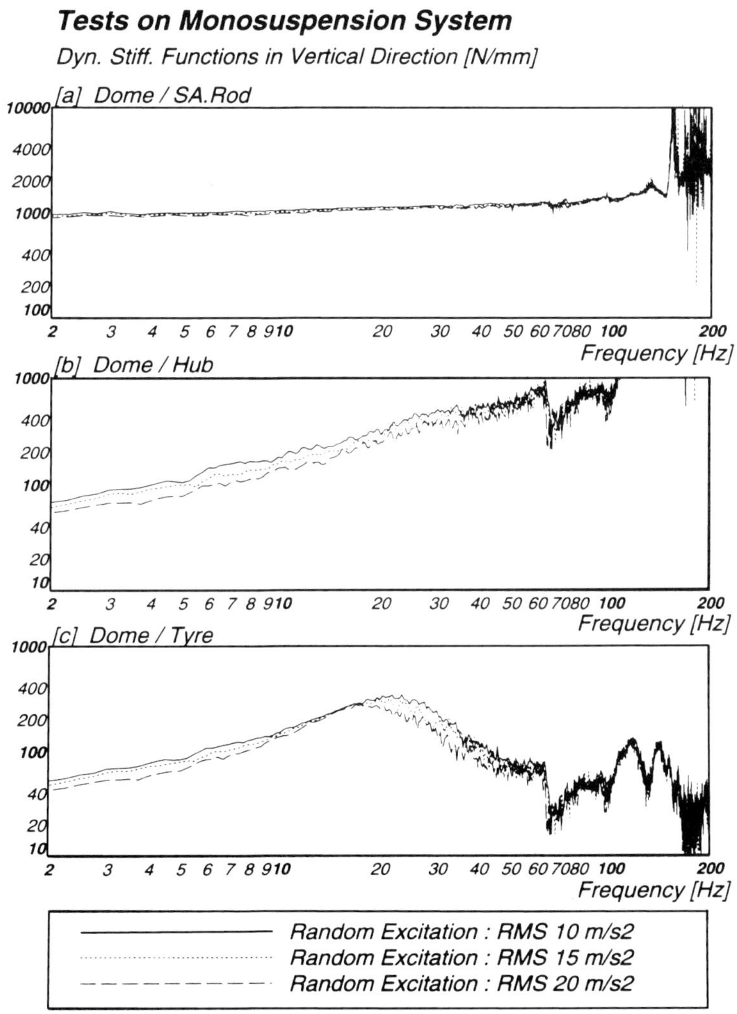

Figure 2 **Dynamic characterisation of the monosuspension system**

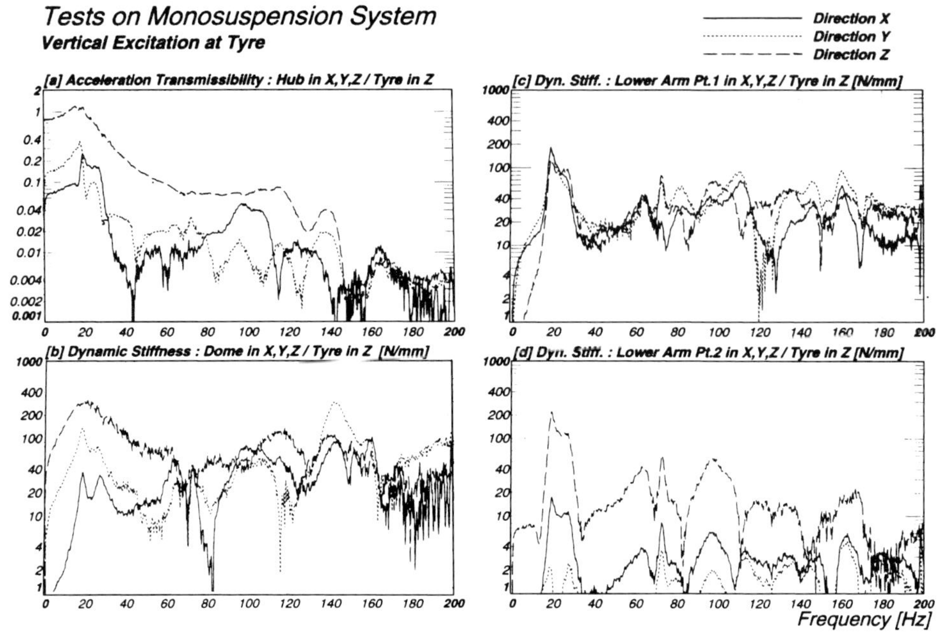

Figure 3 Acceleration transmissibility and dynamic stiffness functions measured from the monosuspension system

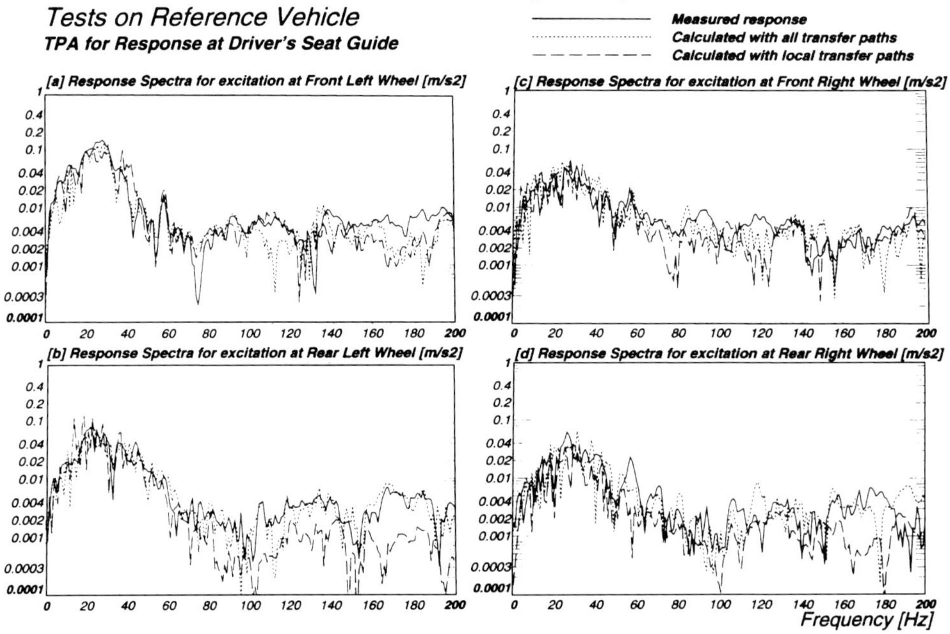

Figure 4 Application of Transfer Path Analysis to the reference vehicle tested on the four poster test facility

C521/025/98

Tyre vibrations and some issues in suspension design

R S SHARP BSc, MSc, FIMechE
School of Mechanical Engineering, Cranfield University, Bedford, UK
D J ALLISON BEng
Vehicle Dynamics, Rover Group Limited, Birmingham, UK

Synopsis
The low frequency vibrations of tyres in running contact with textured road surfaces and the participation in the vibrations of suspension components are considered in this paper. It is known that a planar rigid ring tyre model is capable in respect of tyre vibration description up to 100Hz and reasonable parametric descriptions of tyres and suspension components are available. Tyre and suspension system models are combined to give a description of one vehicle corner in which the body structure is considered seismic and forced vibration responses are calculated through frequency and speed ranges for variations in system design. The forcing mechanism used consists of sinusoidal speed variations superimposed on the steady speed of the roadway. The influences of the main design variables on the vibration peaks are observed and conclusions are drawn on how the system behaviour can be controlled by fundamental design strategies.

1. INTRODUCTION

It is not uncommon for vehicle users to find that certain road surface textures, combined with certain vehicle speeds, will cause low frequency vibrations in the cockpit, which can be both felt and heard by the passengers. Evidently, the tyre/road interaction causes the tyre structure to vibrate and these vibrations involve the suspension components to a significant degree.

Alleviation of the problem must be sought primarily via a "tuning" strategy, involving system designs in which components work together to avoid "worst case" behaviour. Such worst case conditions may involve sub-system resonances coinciding, for example. Some advantage may also be gained from ensuring that all the system modes have reasonable damping factors but the opportunities for adding energy dissipation are limited, since the tyre itself is essentially a minimally damped structure, in the interests of minimising the rolling resistance.

The problem referred to is basically a two dimensional one, involving in-plane vibrations of the tyre and longitudinal, vertical and rotational motions of the hub carrier and wheel assembly. Recent research has shown that the in-plane vibrations of a wheel and tyre interacting with the ground in typical fashion can be represented accurately in the frequency range up to 100Hz by a rigid ring tyre model (1-14). In such a model, the wheel and tyre consists of; a rigid hub; a rigid tread ring, elastically connected to the hub with two translational freedoms and one rotational one; and tread rubber blocks with vertical compliance and slip interaction with the ground, giving rise to shear force generation in the plane of the tyre. For small slip values, very little sliding of rubber with respect to road occurs and a linear force/slip law is implied. The rolling radius of the tyre may be regarded as a variable, dependent on the tyre vertical compression due to loading. The tread rubber compliance must take an effective value, normally called a "residual", which allows the model to account to some extent for the real tread ring bending in the neighbourhood of the contact region. Between the hub and the tread ring, the tyre sidewalls provide axisymmetric translational stiffness and a torsional stiffness.

In particular, the authors' previous work has involved forced vibration of a suspension/tyre/drum system, the construction of a mathematical model of that system and the use of that model in a parameter identification scheme to obtain the parameter values of the tyre, wheel and rig components (8). Parameters identified for a particular tyre are similar to those obtained independently in (6, 7) and they provide a good base for the theoretical prediction of the behaviour of appropriate systems.

The objective of the present work is to extend the prior studies to predict the vibration transmission properties of a tyre/wheel/suspension system representing one corner of a typical car. The in-plane vibrations of a hub carrier, wheel hub and tyre are modelled and excitation is provided by road speed variations. This is a simplified description of the mechanisms by which road profiling and texturing disturb the rolling tyre. Anti-dive kinematic properties are included in the suspension description and longitudinal compliance in the hub carrier location is an important feature. A key component in the system is the bush determining the longitudinal compliance and the damping associated with longitudinal hub motion, subsequently called the long bush, and parametric influences of bush properties and anti-dive geometry are conducted.

2. MODELLING AND PARAMETER VALUES

The rigid ring tyre is mounted via a free bearing to a hub carrier, which can move only in a longitudinal/vertical plane. The hub carrier is acted on by a suspension spring/damper parallel unit, through which it is preloaded in the downward direction. In the equilibrium state, the tyre sidewalls are deflected and provide an upward force which counters the spring preload and the weight of the hub carrier and hub. In turn, the tread rubber is strained and provides an upward force on the tread ring to counter the sidewall force and the weight of the tread ring itself. The hub carrier can translate horizontally and vertically and can rotate about a lateral axis but the top point on it is constrained to move only vertically. Thus, the hub carrier rotation is around this top point and it only has two degrees of freedom. A bottom point on the hub carrier is connected to a longitudinal radius arm, the geometry of which controls the anti-dive and the stiffness and damping of which represent the long bush properties. The hub carrier is illustrated in Fig. 1, while the rigid ring tyre is shown in Fig. 2.

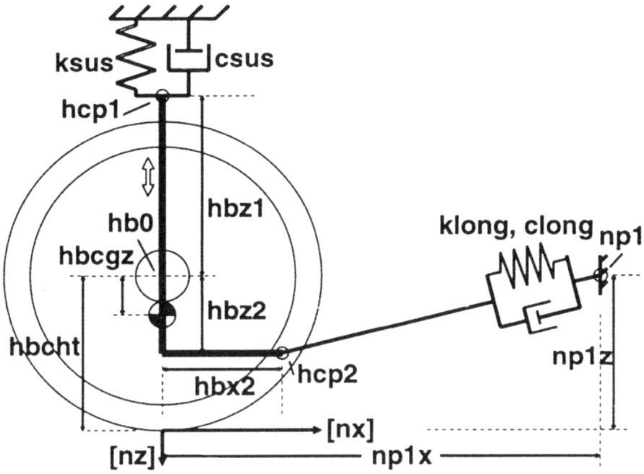

Fig. 1 Diagrammatic hub carrier with dimensions.

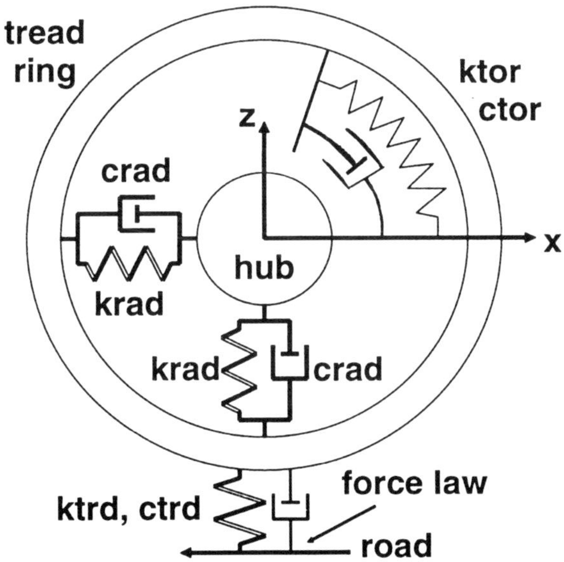

Fig. 2 Diagrammatic representation of rigid ring tyre model.

The tyre develops a longitudinal force proportional to the slip between the tread band and the ground, in opposition to the slip, with an invariant slip stiffness. The slip is defined as the difference between the road velocity and the tyre tread band velocity in the tread rubber support region divided by the modulus of the road velocity. It is a measure of the tread rubber deformation, when no sliding of the rubber across the road occurs. The tread band velocity has components from the wheel hub longitudinal translation and from the wheel spin, the latter part involving the rolling radius. The rolling radius is described as having a small linear dependence on the tyre radial deflection (15). Mathematical details of the model are constructed by the model building program, AutoSim (16-21), which, in this case, writes the linear equations of motion for the system in state space form into a MATLAB "m" file. Input to the system is provided by variations in the road velocity, such that the total road velocity is made up of a constant part and a sinusoidally varying part, having an amplitude which is 0.1% of the road velocity. This representation of the excitation of the tyre vibrations by interaction with the road is, of course, not supposed to mimic real events, which are too complex and unknown to model accurately. Its purpose is to provide excitation to the system at the tyre/road interface, with the notion built in that the forcing vigour is proportional to the speed of travel. The "m" file which AutoSim writes contains the statement "vr = -5;" , which needs to be modified to read "vr = -5*(1+0.001*vrfreq)" by hand in order to connect the input to the system as intended. "vrfreq" is the input to the frequency response calculations and it is presumed to have unit magnitude.

A set of parameter values defines a datum system design. Values in the datum set comprise some which are estimates for a realistic car suspension and some which are the tyre parameters identified in (8). In the datum system, the longitudinal hub location bush is very compliant and there is virtually no anti-dive property, i.e. the longitudinal rod is substantially horizontal. Variations in longitudinal bush stiffness and damping coefficient and in longitudinal location rod angle are employed to reveal behavioural influences. A listing of the AutoSim program for building the system model is included in the Appendix. This program contains the datum system parameter values and the definitions of the symbolic variables used, so that the program serves to specify the datum and also to fix the notation

3. CALCULATION SCHEME

The MATLAB "m" file which AutoSim writes contains the datum parameters and the linearised equations of motion of the system in the standard state space form:
$\dot{x} = [A]x + [B]u$; $y = [C]x + [D]u$.

In fact, the "m" file sets the parameters and forms the elements of the A, B, C and D matrices numerically. A check on the system stability was first made via the instruction "eig(A)". As intended, all eigenvalue real parts were negative. The "m" file was then made into a function file for calculating the gain and phase of the various output quantities specified in the AutoSim program; long bush force, vertical force - etc.; as functions of frequency, using the "bode" function.

A modest MATLAB program was then written to define a vector of road speeds and to step through speed, storing the frequency responses for plotting, in "waterfall" style. All parametric studies can be done in MATLAB, without re-running the model building program. A new set of results can be generated in about 15 minutes, using a modern personal computer.

4. RESULTS

Each system design studied gives rise to extensive results showing the various outputs as functions of frequency and road speed. A minimal set of design changes of interest is considered to include stiffness and damping variations of the long bush and anti-dive variations in the form of the inclination angle of the longitudinal hub location rod. Space limits imply the need to be very selective in the results shown. Consequently, detailed results mainly relate to the long bush force. This is the primarily longitudinal force, which would excite the body structure into vibration, depending on the driving point mobility of the body. An obvious aim is to make this force as small as possible, consistent with other needs. It is also desirable that the frequencies at which the suspension transmits forces with relative vigour should be those for which the mounting point mobility is minimal but this is beyond the detailed scope of the present inquiry.

Outputs which have been plotted are the long bush force, the main suspension vertical force variation, the wheel hub angular velocity variation, the wheel hub longitudinal acceleration and the wheel hub vertical acceleration. The vertical force, transmitted through the main suspension spring/damper unit, which is "soft" compared with the long bush in all cases, is very small in comparison with the longitudinal force. With no anti-dive, this is particularly true, since the cross-coupling mechanism, by which the longitudinal road excitation causes vertical responses, is very weak. Even with a long rod inclination angle of about 22 degrees, it remains the case that the vertical forces are small compared with the horizontal ones. Likewise, the vertical hub acceleration is very small compared with the longitudinal one. Wheel spin velocity variations are of some intrinsic interest, demonstrating circumstances in which the wheel spin response is high but is not accompanied by large forces. The long bush force responses of the datum system are shown in Fig. 3 and the ways in which these change with regular increases in the long bush stiffness and damping coefficient are demonstrated in Figs 4 to 6, while the wheel spin velocity variations for the parameter set of Fig. 4 are shown in Fig. 7.

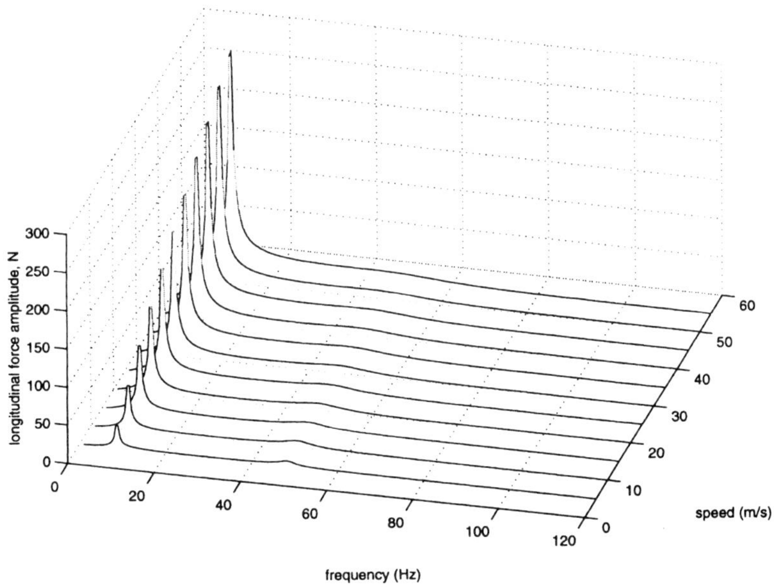

Fig. 3 Longitudinal force amplitude responses to road velocity forcing of amplitude 0.1% of mean velocity; datum system.

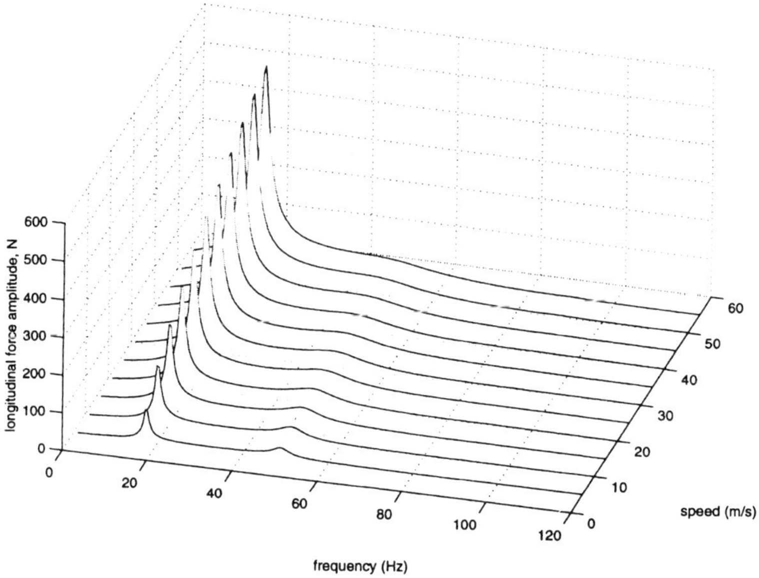

Fig. 4 Longitudinal force amplitude responses to road velocity forcing of amplitude 0.1% of mean velocity; CLONG*2, KLONG*4.

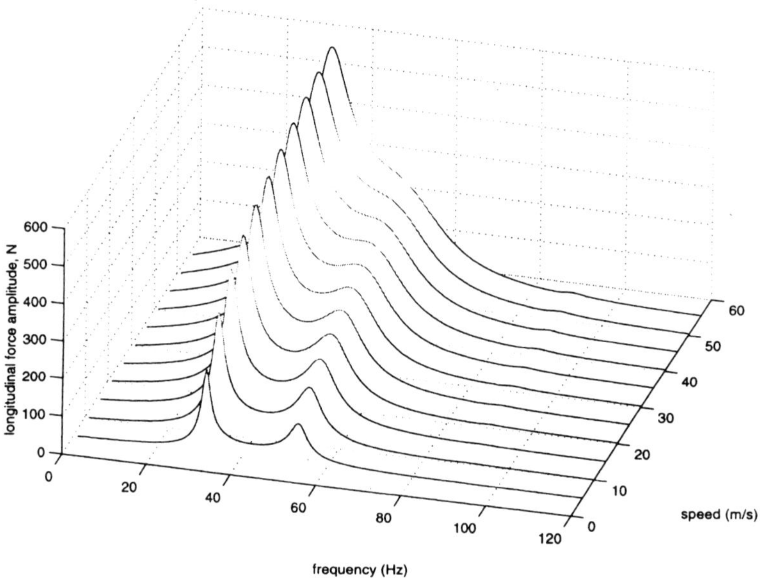

Fig. 5 Longitudinal force amplitude responses to road velocity forcing of amplitude 0.1% of mean velocity; CLONG*4, KLONG*16.

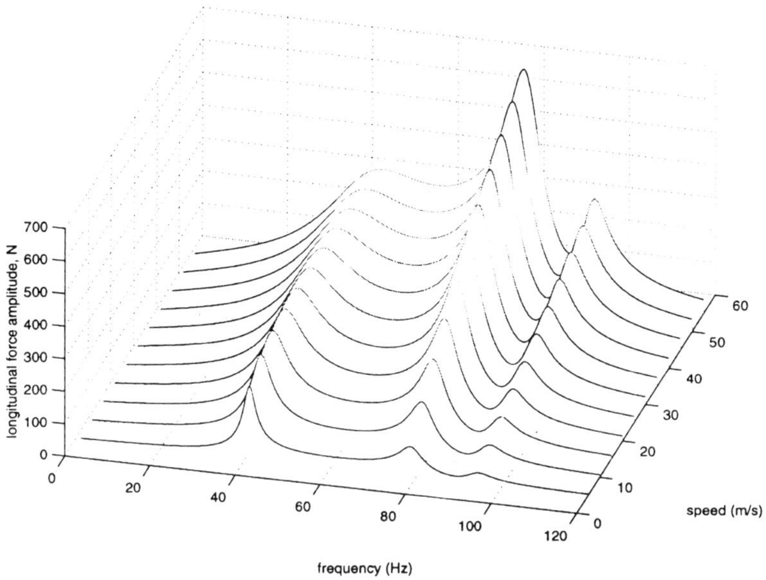

Fig. 6 Longitudinal force amplitude responses to road velocity forcing of amplitude 0.1% of mean velocity; CLONG*8, KLONG*64.

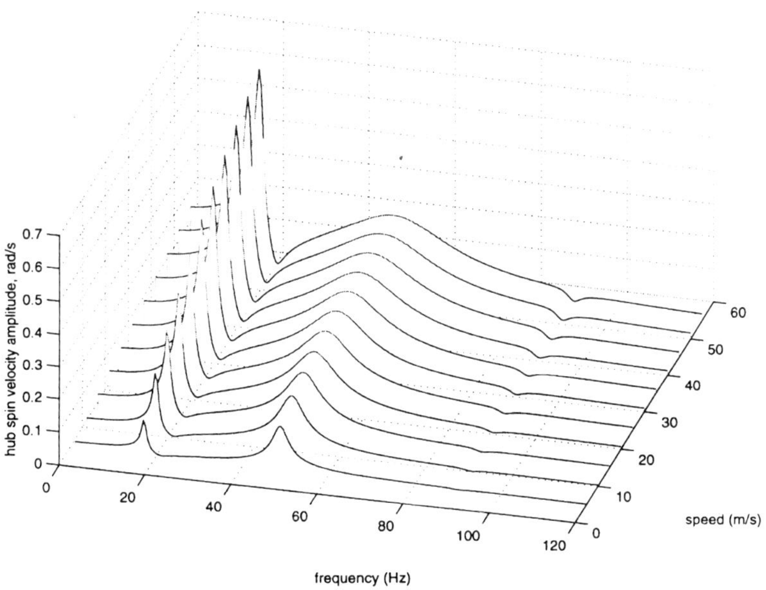

Fig. 7 Wheel spin velocity amplitude responses to road velocity forcing of amplitude 0.1% of mean velocity; system of Fig. 4.

Resonant responses are apparent in all these figures, with the sharpness of the peaks varying widely. The most compliant systems generally have the simplest response properties, comparatively little force transmission occurring away from the main suspension longitudinal mode natural frequency. The compliant long bush is an effective filter for the higher frequency vibrations, particularly involving wheel spin velocity variations and longitudinal hub accelerations.

Fig. 8 shows the long bush force for the bush parameters of Fig. 5 but with anti-dive geometry, involving a long rod declination angle of about 22 degrees. By comparison of Figs 5 and 8, it can be seen that the influence of the long rod declination on the bush force is small. The comparative smallness of the vertical force variation is illustrated in Fig. 9, for the same case as in Fig. 8. Unless the main suspension is locked by friction, these vertical forces are too small to matter much.

If the long bush damping coefficient is increased independently of the stiffness, the bush force peaks are, in general, reduced in size. This confirms the idea that the peaks correspond to resonant conditions. The amplification at resonance is reduced by increased energy dissipation in the long bush. Hydraulic bushes are energy dissipative and frequency tunable and it follows from the results that a hydraulic long bush would need to be tuned to the main resonant frequency of the suspension and tyre longitudinally, in order to be most effective. The influence of damping change alone is illustrated in Fig. 10, which should be compared directly with Fig. 4, noting the difference between the vertical scales.

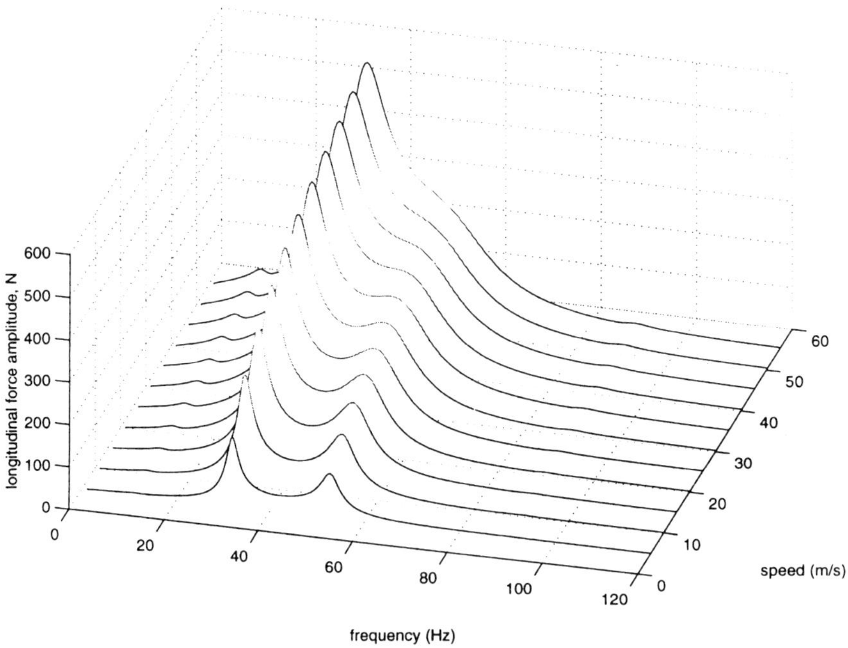

Fig. 8 Longitudinal force amplitude responses to road velocity forcing of amplitude 0.1% of mean velocity; Fig. 5 parameters with anti-dive; NP1Z = 0.012m.

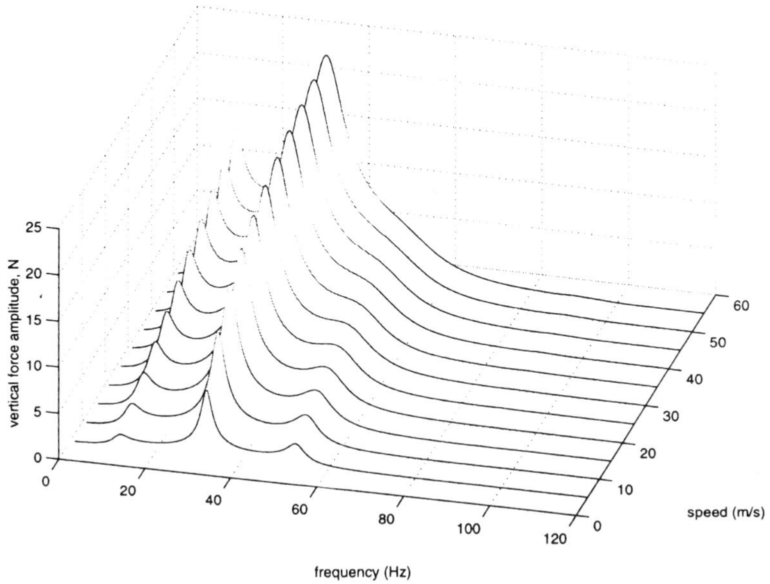

Fig. 9 Vertical force amplitude responses to road velocity forcing of amplitude 0.1% of mean velocity; Fig. 8 parameters.

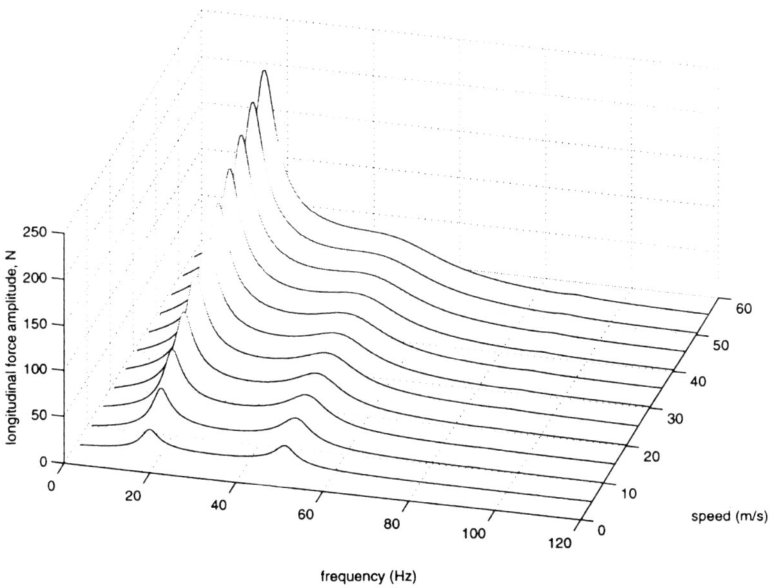

Fig. 10 Longitudinal force amplitude responses to road velocity forcing of amplitude 0.1% of mean velocity; Fig. 4 system with CLONG*4.

5. CONCLUSIONS

A mathematical model of a suspension/wheel/tyre corner of a car has been constructed in order to predict the responses of the suspension system to excitation of the tyre from the road surface in the frequency range to 100Hz. The model is closely related to that used successfully to identify tyre rigid ring model parameter values as reported in (8). Its structure and parameter set can therefore be regarded as a reliable basis for the prediction of behaviour within its design limits. Parameter trials have covered variations in longitudinal bush stiffness and damping together, in longitudinal bush damping coefficient separately and in anti-dive angle.

The responses of the system to road velocity variation excitation are dominated by resonances, which change their frequencies very little with changes in vehicle speed. The important resonant frequencies can be controlled by the long bush stiffness, in a somewhat complex manner, so that the system defies simpler analysis. High damping in the long bush helps to reduce the longitudinal force amplitudes at the resonance points. It can be expected that bad vehicle behaviour will arise when road texture and vehicle speed combine to provide narrow band forcing to the tyre, through the tyre/road contact, which is centred on a system resonant frequency as illustrated in the figures. Also, the structural response will be greatest if the long bush driving point mobility is low in the condition envisaged. The desirability of predictive capacity in relation to suspension/wheel/tyre vibrations, in addition to structural capabilities, is demonstrated.

This study is by no means exhaustive, within its envelope, and it raises questions relating to other parametric effects, more general hub carrier kinematic designs, the excitation mechanisms of tyre vibrations, the differences between individual tyres and higher frequency vibrations, for which the rigid ring ideas are unsuitable.

ACKNOWLEDGEMENT

The authors are pleased to acknowledge the support of the UK Engineering and Physical Sciences Research Council and the technical and resource contributions of their colleagues in Dunlop Tyres Limited and Rover Group, especially A. R. Williams and R. A. M. Smith.

REFERENCES

(1) Allison, D. J. and Sharp, R. S., On the low frequency in-plane forced vibrations of pneumatic tyre / wheel / suspension assemblies, Tyre Models for Vehicle Dynamics Analysis (F. Bohm and H.-P. Willumeit eds), Vehicle System Dynamics Supplement, **27**, 1997, 151-162.

(2) Gong, S., A study of in-plane dynamics of tyres, Doctoral Dissertation, Delft University of Technology, 1993.

(3) Zegelaar, P. W. A., Gong, S. and Pacejka, H. B., Tyre models for the study of in-plane dynamics, The Dynamics of Vehicles on Roads and Tracks (Z. Y. Shen ed.), Vehicle System Dynamics Supplement, **23**, 1994, 578 - 590.

(4) Zegelaar, P. W. A. and Pacejka, H. B., The in-plane dynamics of tyres on uneven roads, The Dynamics of Vehicles on Roads and Tracks (L. Segel ed.), Vehicle System Dynamics Supplement, **25**, 1996, 714 - 730.

(5) Zegelaar, P. W. A. and Pacejka, H. B., Dynamic tyre responses to brake torque variations, Tyre Models for Vehicle Dynamics Analysis (F. Bohm and H.-P. Willumeit eds), Vehicle System Dynamics Supplement, **27**, 1997, 65 – 79.
(6) Mancosu, F. and Matrascia, G., Techniques for determining the parameters of a two dimensional tyre model for the study of comfort, Proc. 15th annual meeting of Tire Society at University of Akron, 1996, Tire Science and Technology.
(7) Bruni, S., Cheli, F. and Resta, F., On the identification in time domain of the parameters of a tyre model for the study of in-plane dynamics, Tyre Models for Vehicle Dynamics Analysis (F. Bohm and H.-P. Willumeit eds), Vehicle System Dynamics Supplement, **27**, 1997, 136 – 150.
(8) Sharp, R. S. and Allison, D. J., In-plane vibrations of tyres and their dependence on wheel mounting conditions, Dynamics of Vehicles on Roads and Tracks (L. Palkovics ed.), Swets and Zeitlinger, Lisse, 1998, in printing.
(9) Zegelaar, P. W. A. and Pacejka, H. B., Dynamics of Vehicles on Roads and Tracks (L. Palkovics ed.), Swets and Zeitlinger, Lisse, 1998, in printing.
(10) Gong, S., Savkoor, A. R. and Pacejka, H. B., The influence of boundary conditions on the vibration transmission properties of tires, SAE paper 931280, 1993.
(11) Huang, S. C. and Su, C. K., In-plane dynamics of tires on the road based on an experimentally verified rolling ring model, Vehicle System Dynamics, **21**, 1992, 247-267.
(12) Huang, S. C., The vibration of rolling tires in ground contact, Int. J. of Vehicle Design, **13**, 1992, 78-95.
(13) Kim, S. J. and Savkoor, A. R., The contact problem of in-plane rolling of tires on a flat road, Tyre Models for Vehicle Dynamics Analysis (F. Bohm and H.-P. Willumeit eds), Vehicle System Dynamics Supplement, **27**, 1997, 189 – 206.
(14) Oertel, Ch., On modelling contact and friction - calculating of tire response on uneven roads, Tyre Models for Vehicle Dynamics Analysis (F. Bohm and H.-P. Willumeit eds), Vehicle System Dynamics Supplement, **27**, 1997, 289 – 302.
(15) Pacejka, H. B., Tyre factors and front wheel vibrations, Int. Jour. Veh. Des., **1**(2), 1980, 97 - 120.
(16) Sayers, M.W., Automated formulation of efficient vehicle simulation codes by symbolic computation (AUTOSIM), The Dynamics of Vehicles on Roads and Tracks (R. Anderson ed.), Vehicle System Dynamics Supplement, **18**, 1990, 474 - 487.
(17) Sayers, M. W., Computer modeling and simulation of vehicle dynamics, UMTRI Research Review, **23**(1), 1992, 1-15.
(18) Mousseau, C. W., Sayers, M. W. and Fagan, D. J., Symbolic quasi-static and dynamic analyses of complex automobile models, The Dynamics of Vehicles on Roads and Tracks (ed. G. Sauvage), Proc. 12th IAVSD Symposium, 1992, 446-459 (Swets and Zeitlinger, Lisse).
(19) Striberski, A. and Sayers, M. W., Applications of symbolic manipulation for investigating the nonlinear dynamics of ground vehicles, The Dynamics of Vehicles on Roads and Tracks (ed. G. Sauvage), Proc. 12th IAVSD Symposium, 1992, 608-624 (Swets and Zeitlinger, Lisse).
(20) Sharp, R. S., Vehicle dynamics modelling with the aid of a symbolic multibody code, Proc. Avec '96, RWTH Aachen, Vol. 2, 1996, 971-984.
(21) Sharp, R. S., Use of the symbolic modelling code AUTOSIM for vehicle dynamics, AUTOMOTIVE VEHICLE TECHNOLOGIES, Mech. Eng. Publ., Bury St Edmunds, 1997-7, 209-217.

APPENDIX: AutoSim program for model building

```
;; RIGMOD1.LSP
;;==================================================================
(reset)
(si)
(add-gravity :direction "[nz]")
(setsym *compiler* "msf\\")
(setsym *stiff-integrator* t)
(setsym *safe-sqrt* t)
(setsym *safe-divide* t)

(add-body hbcarr :parent n  :name "hub carrier"
      :joint-coordinates(0 0 -hbcht)
      :cm-coordinates(0 0 hbcgz)
      :mass mhbcarr
      :inertia-matrix (0 ihbcarry 0)
      :translate ([nx] [nz])
      :body-rotation-axes y  :parent-rotation-axis y
      :reference-axis z)
(add-point hcp1  :body hbcarr
      :coordinates(0 0 -hbz1))
(add-point hcp2  :body hbcarr
      :coordinates(hbx2 0 hbz2))
(add-speed-constraint "dot(vel(hcp1),[nx])"  :u "ru(hbcarr)")

(add-body hb :parent hbcarr  :name "wheel hub"
      :mass mhb  :inertia-matrix (0 ihby 0)
      :body-rotation-axes y
      :parent-rotation-axis y
      :reference-axis z)
(add-body trdtran  :parent hbcarr  :name "tread translation"
      :mass mtrd  :inertia-matrix 0
      :translate ([nx] [nz]))
(add-body trdrot  :parent trdtran  :name "tyre tread rotor"
      :mass 0  :inertia-matrix (0 itrdy 0)
      :body-rotation-axes y
      :parent-rotation-axis y
      :reference-axis z)
(add-line-force susp  :name "suspension force"
      :direction [nz]
      :point1 hbcarr0  :point2 n0
      :magnitude "-ksus*(xsusp-xsusp0)-csus*vsusp+sf"
      :x xsusp :x0 xsusp0 :v vsusp)
(add-point np1  :body n  :name "long suspension reaction point"
      :coordinates(np1x 0 -np1z))
(add-strut flong  :name "longitudinal reaction force"
      :point1 hcp2  :point2 np1
      :magnitude "-klong*(xflong-xflong0)-clong*vflong"
      :x xflong :x0 xflong0 :v vflong)
(add-line-force swx  :name "x-direction sidewall force"
      :direction [hbx]  :point1 hb0  :point2 trdrot0
      :magnitude "-krad*Xsw-crad*Vsw"
      :x Xsw :v Vsw)
(add-line-force swz :name "z-direction sidewall force"
      :direction [hbz]  :point1 hb0  :point2 trdrot0
      :magnitude "-krad*Xsw-crad*Vsw"
      :x Xsw :v Vsw)
(add-line-force sfsw  :name "side wall static force"
      :direction [nz]  :point1 hb0  :point2 trdrot0
      :magnitude "-sf-(mhbcarr+mhb)*g")
(setsym sftrd "sf+(mhbcarr+mhb+mtrd)*g")

(add-line-force trdfz  :name "tread rubber reaction force"
```

```
      :direction [nz]    :point1 trdtran0  :point2 n0
      :magnitude "min(0.0, -ktrd*(xtrd-xtrd0)-ctrd*(vtrd))-@sftrd"
      :x xtrd :x0 xtrd0 :v vtrd)
(add-moment swtorq  :name "sidewall torque"
      :body1 trdrot  :body2 hb  :direction [ny]
      :magnitude "-ktor*(rq(trdrot)-rq(hb))-ctor*(ru(trdrot)-ru(hb))")
(add-variables dyvars real vrfreq)

(setsym RRad "Rad-(tq(trdtran,2)+tq(hbcarr,2))*(1-eta)")
(setsym roadvel "vr*(1+0.001*vrfreq)")
(setsym linvel "@RRad*(ru(trdrot)+ru(hbcarr))+dot(vel(trdtran0),[nx])")
(setsym slipvel "(@linvel-@roadvel)/abs(@roadvel)")

(add-line-force sliplon  :name "longitudinal slip force on hub"
      :direction [nx]  :point1 trdtran0  :point2 n0
      :magnitude "-@slipvel*slipstif")
(add-moment sliptorq  :name "torque corresponding to force"
      :direction [ny]  :body1 trdrot  :body2 n
      :magnitude "fm(sliplon)*@RRad")
(set-defaults
      hbcht 0.312 hbcgz 0.0 ihbcarry 1.0 mhbcarr 25 hbz1 0.3 hbx2 0.0
      hbz2 0.1 hbx3 0.2 mhb 20 ihby 0.62 mtrd 5.5 itrdy 0.585 ksus 18000
      csus 1000 sf 4000 np1x 0.5 np1z 0.212 klong 125000 clong 125
      krad 1584000 crad 78 ktrd 328760 ctrd 262 ktor 94620 ctor 12
      slipstif 233280 rad 0.325 vr -5 Eta 0.94 "ru(trdrot)" "vr/@Rrad"
      "ru(hb)" "vr/@Rrad")
(set-names clong "damping coefficient of long bush"
      csus "suspension damping coefficient"
      crad "tyre sidewall translational damping coefficient"
      ctor "tyre sidewall torsional damping coefficient"
      ctrd "tyre tread residual damping coefficient"
      eta "slope of effective tyre rolling radius/deflection curve"
      hbcgz "distance of hub cg below wheel centre (w.c.)"
      hbcht "height of wheel hub from global origin at contact centre"
      hbx2 "distance of hcp2 in front of w.c."
      hbx3 "distance of hcp3 behind w.c."
      hbz1 "height of hcp1 above w.c."
      hbz2 "distance of hcp2/hcp3 below w.c."
      ihby "spin inertia of wheel hub"
      ihbcarry "pitch inertia of hub carrier"
      itrdy "spin inertia of tyre tread band"
      klong "stiffness of longitudinal restraint bush"
      ksus "main suspension wheel rate"
      krad "tyre sidewall translational stiffness"
      ktor "tyre sidewall torsional stiffness"
      ktrd "tyre tread rubber residual stiffness"
      mhb "mass of hub"
      mhbcarr "mass of hub carrier"
      mtrd "mass of tyre tread band"
      np1x "distance forward of np1 from global origin"
      np1z "distance up of np1 from global origin"
      rad "wheel free radius"
      sf "main spring preload force"
      slipstif "tyre/ground slip stiffness"
      vr "road surface velocity - positive along [nx]")
(setsym longsusp "fm(flong)")
(setsym suspforz "fm(susp)")

(add-out "@longsusp" "longforz")
(add-out "@suspforz" "vertforz")
(add-out "ru(hb)" "hbspin")
(add-out "tu(hbcarr,1)" "hlonvel")
(add-out "tu(hbcarr,2)" "hvervel")

(linear :u (vrfreq))
```

Test Techniques

C521/002/98

An experimental attempt to identify vibration energy flow on car exterior panels through intensity method

N KOJIMA, T HIRAYU, and **M MIKAMI**
Department of Mechanical Engineering, Yamaguchi University, Yamaguchi, Japan
H ZHOU
Department of Electronic and Control Systems Engineering, Shimane University, Shimane, Japan

The identification of excitation sources and vibration energy transmission paths can give decisive information on the noise and vibration control in a mechanical structure. In simple structures, vibration intensity analysis has been a subject of interest for this purpose. In the present study, the envelope technique, by which transient characteristics of energy flow can be analyzed, is employed to the data processing of vibration intensity measurement. The attempt is made to measure the envelope vibration intensity in car exterior panels. The transient vibration energy flow pattern is shown through this method.

INTRODUCTION

For noise and vibration controls in mechanical structures, it is necessary to identify excitation sources in the structure and transmission paths of the vibration energy. The vibration intensity technique[1] is thought to be useful for this purpose and has been studied by many researchers. Most studies on vibration intensity analysis have focused on the theoretical treatment or calculation based on the FEM and some on the measurement itself. Few works have reported on the application of the intensity technique to practical mechanical structures[2]-[4].

Vibration intensity is measured as vibrational power per unit width in a given direction. In most studies on vibration intensity, the time-averaged intensity of flexural vibration was used. The expression of the time-averaged intensity in x-direction becomes

$$I_{vx} = \left\langle Q_x \cdot \dot{\xi} \right\rangle_t + \left\langle M_x \cdot \dot{\theta}_x \right\rangle_t + \left\langle M_{xy} \cdot \dot{\theta}_y \right\rangle_t \tag{1}$$

where Q_x, M_x, M_{xy}, $\dot{\xi}$ and $\dot{\theta}$ are shear force, bending moment, twisting moment, out-of-plane velocity and angular velocity, respectively[1]. There are some difficulties in applying this intensity method to diagnostics for practical mechanical structures. The calculation of third-order derivatives in Eq. (1) from a set of vibration data measured at different points may cause error. Furthermore, transient features of vibration energy flow, which are generally observed in most practical structures, are lost due to the time-averaging process.

The envelope vibration intensity technique is one useful way to analyze transient energy flows and has been developed by the research group of the authors[(4)-(6)]. In the present study, a passenger car is selected as an example of a practical mechanical structure, and transient vibration energy flows in the exterior panels of the car are analyzed by the envelope intensity technique using the three-channel pick-up method. Time-dependent multipath-transmissions of vibration energy in the car exterior panels are discussed.

MEASUREMENT OF TRANSIENT VIBRATION ENERGY FLOW

Envelope vibration intensity

Vibration intensity is the vibrational power per unit width of cross section flowing in a structure and is a vector quantity. The time-averaged intensity is usually used to determine the characteristics of vibration energy flow through a structure. Using this method, however, the transient aspects of the energy flow are lost. The envelope vibration intensity is a practical way to identify the transient energy flow pattern.

The envelope vibration intensity in a plate transported by flexural vibrations in x-direction is

$$I_{vex}(f,t) = \mathrm{Re}(Q_{xa}(f,t)\cdot\dot{\xi}^*_a(f,t)) + \mathrm{Re}(M_{xa}(f,t)\cdot\dot{\theta}^*_{xa}(f,t)) + \mathrm{Re}(M_{xya}(f,t)\cdot\dot{\theta}^*_{ya}(f,t)) \tag{2}$$

where Q_x is shear force, M_x is bending moment, M_{xy} is twisting moment, $\dot{\xi}$ is out-of-plane velocity, and $\dot{\theta}_x$ and $\dot{\theta}_y$ are angular velocities in x and y direction, respectively. The subscript a refers to an analytical signal and the superscript * refers to the complex conjugate. Assuming that the measurements are made in the far and free field region, Eq.(2) can be reduced to

$$I_{ve}(f,t) = \frac{\sqrt{Dm}}{\pi f}\,\mathrm{Re}(\ddot{\xi}_a(f,t)\cdot\dot{\theta}^*_a(f,t)) \tag{3}$$

where m is the mass per unit area and D is the bending stiffness of the plate. The general expression of the analytical signal of an arbitrary variable χ in the present paper is expressed as

$$\chi_a = \frac{1}{\sqrt{2}}\left\{\chi + jH[\chi]\right\} \tag{4}$$

where the operator H[ ] shows the Hilbert transformation. Using the Fourier and invert Fourier transformation, Eq.(4) becomes

$$\chi_a(f,t) = F^{-1}[\sqrt{2}\,\mathrm{F}[\chi(t)]]_{f_1}^{f_2} \tag{5}$$

where f is the center frequency of the band frequency from f_1 to f_2.

Three channel method

As shown in Eq.(3), two variables, $\ddot{\xi}$ and $\dot{\theta}$ should be measured at the same point and time so as to determine the envelope intensity in the far and free field region. In the present study, a probe with three accelerometers shown in Fig. 1 was used[7]. Three accelerometers were set on a circular table distributed at 120° angle. For distance r between the center of the circular table and that of the each accelerometer, the out-of-plane acceleration $\ddot{\xi}$ at the center of the probe is calculated as the average of the three accelerations, $\ddot{\xi}_1$, $\ddot{\xi}_2$ and $\ddot{\xi}_3$.

The angular velocity of the plate in α direction can be approximated as,

$$\dot{\theta}_\alpha \simeq \int \frac{1}{3r}\{(2\ddot{\xi}_1 - \ddot{\xi}_2 - \ddot{\xi}_3)\cos\alpha - \sqrt{3}\,(\ddot{\xi}_2 - \ddot{\xi}_3)\sin\alpha\}\,dt. \quad (6)$$

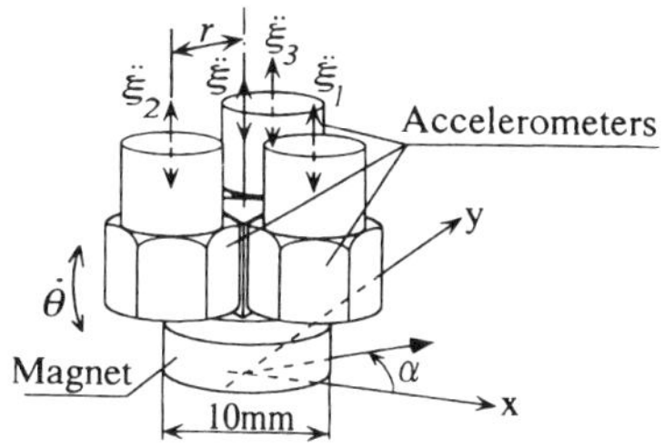

Fig. 1 Three-channel pick-up.

Thus the α direction component of the intensity is obtained through this method. The envelope intensity at the measuring point is the maximum of the α direction component with changing α.

The probe used in the present experiments is shown in Fig. 1. The distance between the center of the table and that of each accelerometer is 5 mm. The probe with a table of 10 mm in diameter has a strong magnetic bottom (10.5 kG) to fix it to the measuring points of the plate to reduce measurement time. The total mass of the probe is 7.6 g.

The application of the three channel method to measure the near field region will cause error in determination of the intensity as Eq.(3). The preliminary FEM analysis was performed to check this region. Figure 2(a) shows the mesh model of the exterior panel of the car door used in the FEM analysis. Figures 2(b)-(1) and 2(b)-(2) show the spacial distributions of the envelope intensity in a car door calculated from FEM results using Eq.(2) and Eq.(3), respectively. In the intensity calculation, the spacial derivatives were approximated through the finite difference method[8]. In Eq.(2), the third-order derivatives of the displacement ξ are required to calculate the intensity, so that 12 finite difference points are employed (termed 12 point method)[9]. In Eq.(3), only the first-order derivative of ξ is required to calculate the intensity, so that 4 finite difference points are employed (termed 4 point method).

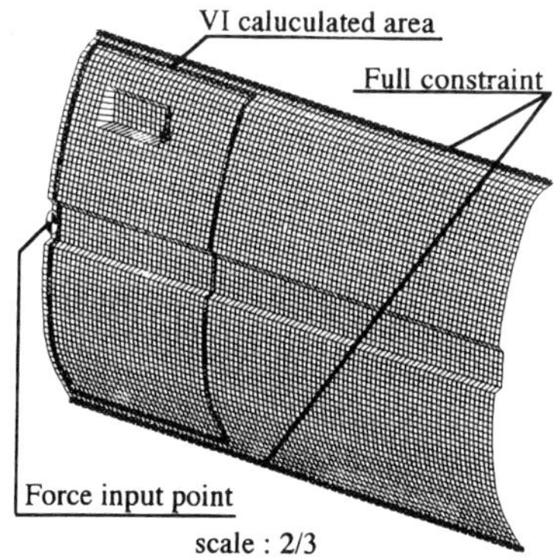

(a) FEM analysis model.

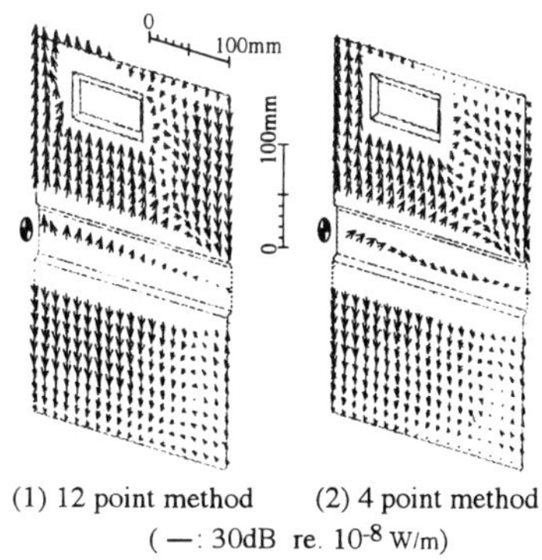

(b) Envelope vibration intensity vectors of 250 Hz 1/3 octave band at 2 msec after excitation.

Fig. 2 Energy flow patterns in a car door exterior panel by FEM analysis.

It is found that both energy flow patterns are almost identical in most regions except for the vicinities of the excitation point and the edges. Through more detail investigation, it was clarified that the Eq.(3) gives a good approximation in the analysis region at a distance greater than a quarter-wavelength apart from the excitation points or boundaries of the plate.

EXPERIMENTAL SETUP

Figure 3 shows the schematic view of the tested small passenger car. The excitation was made at an upper mount of the suspension by a shaker with a rod penetrating through a small hole of the bonnet, as shown in Fig. 3. This may simulate a vibrational power input from the suspension, such as one caused by road noise. Measurements were made by the three-channel pick-up for exterior panels at 408 points on the bonnet, at 183 points on the right front fender and at 457 points on the right front door. The distance between the measuring points was set to 40 mm. The thickness of the exterior panels is 0.7 mm.

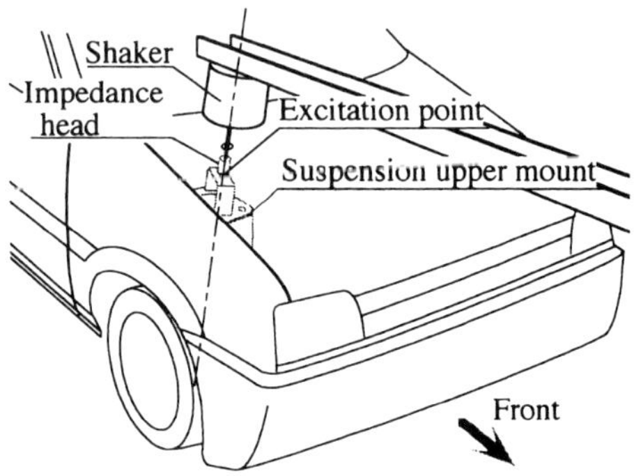

Fig. 3 Excitation method for the car.

There are press lines in the lower area of the fender and in the center and lower areas of the door. The upper area of the fender and the lower area of the bonnet have small curvature radii. The bonnet consists of the exterior panel and the interior frame combined at the edge and is attached to the body frame by a hinge at each upper corner and by a hook at the lower part. The interior frame is also connected to the exterior panel at 18 spots by rubber cement. The front door which has the exterior panel, the interior panel and a window, etc. is attached to the body by 3 hinges on the right side and by a hook on the left side. The front fender consisting of a panel is fixed to the body by 5 bolts on the upper side and by a bolts at each lower corner. These exterior panels are indirectly connected to each other through the body frame.

The excitation force and acceleration are measured by an impedance head set between the shaker and the excitation point. The vibration signals from the probe are amplified, digitized by an A/D converter and then processed by a computer. The data sampling frequency is 5.12 kHz. The data length is 1024 points. The upper limit of analysis frequency is restricted to about 1 kHz at which the vibration amplitude decreases 3 dB by attaching the additional mass of the three-channel pick-up to the car exterior panel with thickness of 0.7 mm.

RESULTS AND DISCUSSION

The right upper mount of the suspension is excited by periodic impulses with intervals of 200 msec. Figure 4 shows the excitation force, acceleration and the spectrum of the input power of an impulse. The peak force and width of the impulse are about 70 N and 1 msec, respectively. The input power in the frequency domain shows large value in the frequency range from 100 to 600 Hz.

The spectrum of the spacial average value of the time-averaged intensities for each exterior panel is shown in Fig. 5. In the front fender, the power level is higher in most frequency ranges than those in the bonnet and front door, especially in the frequency range from 250 to

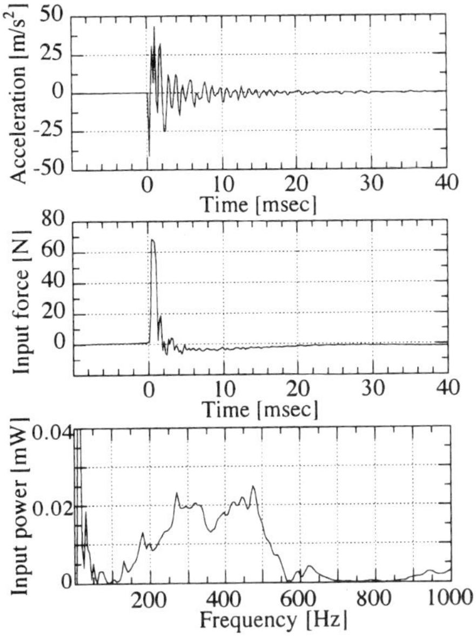

Fig. 4 Excitation acceleration and force at the excitation point and spectrum of the input power.

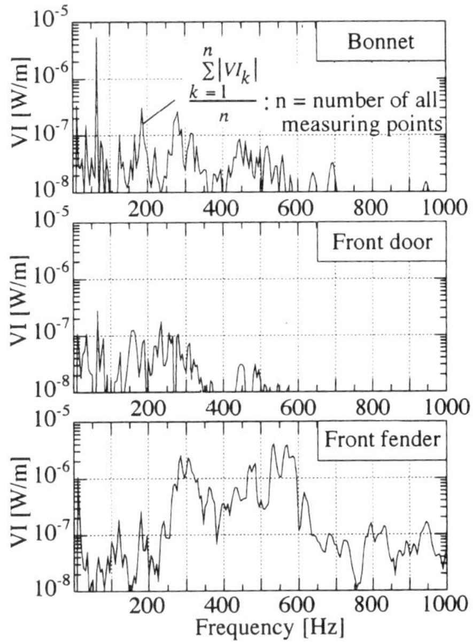

Fig. 5 Spectra of spacial average of time-averaged vibration intensity for different car exterior panels.

600 Hz. In the front door, most vibrational power is distributed below 350 Hz. It is of interest to clarify the energy flow characteristics in the front door in view of the relation of noise characteristics for the driver.

Depending on the magnitude of the averaged intensity in the front door, attention was given to two different band frequency such as 1/3 octave bands of 250 Hz and 500 Hz. The envelope vibration intensities over the exterior panels were obtained for these two band frequency components. Generally, frequencies lower than 300 Hz are important for road noises.

Figure 6 shows sequential flow patterns of transient vibration energy of 250 Hz 1/3 octave band component in the exterior panels of the car. The energy in the fender is found to flow in various patterns, such as reflection, branching, confluence and vortex. As for the fender panel, the energy enters from the upper edge. An especially large amount of energy enters from a left part of the upper edge which is close to the excitation point. This energy flows downward, reflects at the left edge and the lower left corner, and goes upward. Then it joins the energy flow from the upper area and flows to a right part of the lower edge. The energy flow entering from a center part of the upper edge goes up and down by reflection, and some of it branches also to a right part of the lower edge. The interaction of these flows generates vortex flow patterns (Figs. 6(d) and 6(g)). Figure 6 suggests that the main sinks of the energy in the fender panel exist at the lower left corner and at a right part of the lower edge.

The energy flow in the exterior panel of the bonnet appears later than that in the fender. Though one of the hinges of the bonnet is located near the excitation point, the energy does not enter near such an area. Rather, it enters from the lower area of the bonnet where the interior frame contacts the exterior panel at several points. It is interesting that the energy source for the exterior panel of the bonnet is located far from the excitation point. Though such contact points

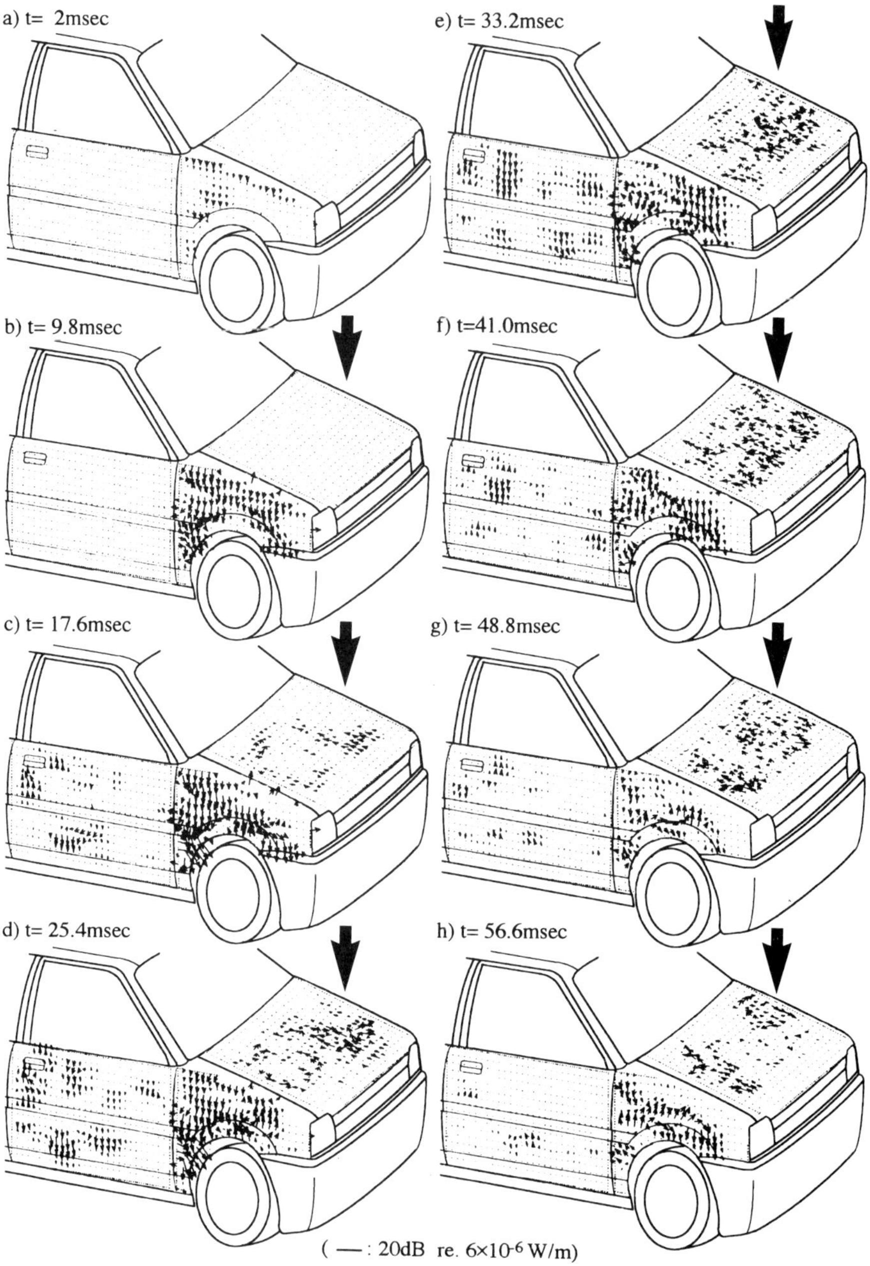

Fig. 6 Time history of envelope vibration intensities of 250 Hz 1/3 octave band in the car exterior panels excited by an impulse.

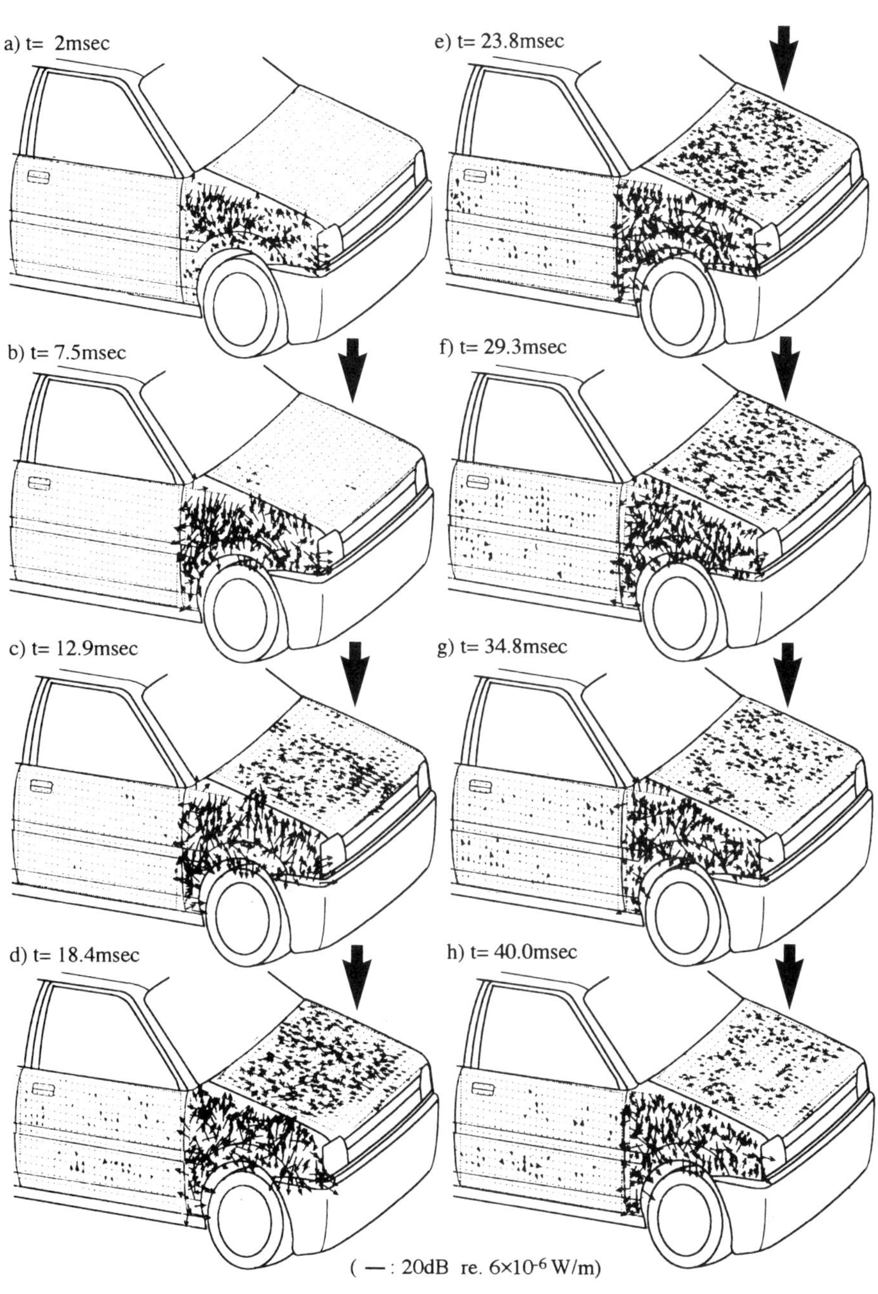

Fig. 7 Time history of envelope vibration intensities of 500 Hz 1/3 octave band in the car exterior panels excited by an impulse.

exist in the center area, too, energy sources cannot be observed clearly there. The contact points in the center area may play the role of energy sinks.

As for the exterior panel of the door, the energy flow appears as late as that in the bonnet. As in the fender, not so much energy flows in the exterior panel of the door. The energy flow is considered to reach the exterior panel through the body frame, hinges and the interior panel. The energy enters into the exterior panel from every side though it is not clearly shown in this figure. As can be seen in Fig. 6, the energy flow reflects on the center press lines and on the right of both the upper and lower edges. Finally the energy flow goes out from the left of both upper and lower edges.

Figure 7 shows sequential energy flow patterns of 500 Hz 1/3 octave band component in the exterior panels. In comparison with Fig. 6, more vibration energies are found to enter into the fender and the bonnet but less energy into the door at 500 Hz than at 250 Hz. Flow patterns in the fender at 500 Hz are basically the same as at 250 Hz but are not so orderly as at 250 Hz. As for the bonnet, the energy source exists in the center area of the lower part in the same way as at 250 Hz. The energy may reach the bonnet through the hook because it is underneath the energy source area. The lower part of the bonnet, however, seems to become an energy sink afterward. In Fig. 7, flow patterns are not clear due to the small magnitude of the intensity vectors. A detailed flow pattern on the door panel is shown in a different scale in Fig. 8. This figure shows the energy entering into the door panel from the right side.

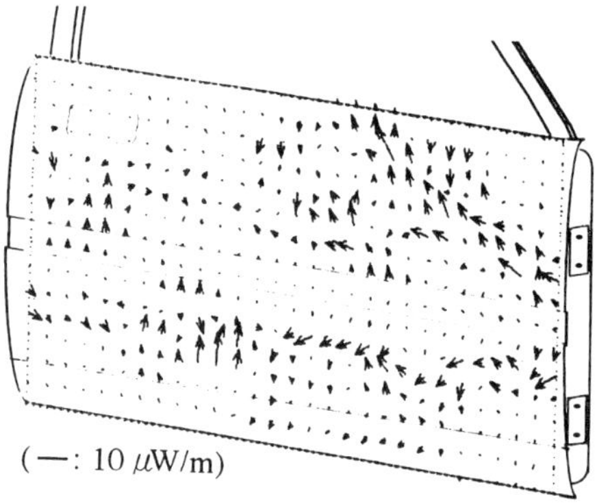

Fig. 8 Vibration energy flow pattern at 13.7 msec shortly after the energy enters the exterior panel of the door (500 Hz 1/3 Oct. band).

CONCLUDING REMARKS

We attempted to identify transient flow patterns of the vibration energy on a practical mechanical shell structure through an intensity technique. An envelope vibration intensity technique using the three-channel pick-up method was employed, and the transient energy flow patterns on the exterior panels of a passenger car were visualized successfully.

The results clearly show time-dependent multipath-transmissions of vibration energy in car exterior panels. Multiple reflection makes the energy flow pattern complicated. The results also suggest that the magnitude of energy flowing into a panel, energy transmission paths, the locations of the energy sources and sinks depend on frequency.

The application of the present technique should be informative in the diagnostics and control of noise and vibration in practical mechanical structures.

ACKNOWLEDGEMENT

This research was partly subsidized by Grant in Aid for Scientific Research Funds from the Ministry of Education, Japan.

REFERENCES

(1) Noiseux, D. U., "Measurement of Power Flow in Uniform Beams and Plates", *J. Acoust. Soc. America*, 47-1 (1970), 238-247.

(2) Kojima. N., et al., "The Use of Vibration Intensity Measurement in Identifying the Source of Impact in an Engine Structure", *SAE Paper* 891167, (1989).

(3) Zhou. H., et.al., "The Measurement of Vibration Intensity on the Cylindrical Casing of Compressor", *Inter Noise '95*, (1996), 677-680.

(4) Kojima, N., et al., "Measurement of Transient Vibrational Power Flow in a Car Door Panel Using Intensity Technique", *SAE Paper* 972035, (1997).

(5) Kojima, N., et al., "Measurement of Envelope Vibration Intensity in Plate", *Trans. JSME*. 60-578, C, (1994), 3281-3287 (*in Japanese*).

(6) Kojima, N., et al., "The Measurement of Transient Vibration Intensity Using Three Channel Method", *SAE Paper* 931273, (1993), 83-91.

(7) Zhou, H., et al., "Vibration Intensity Measurement Using Three Accelerometers", *Trans. JSME*. 59-558, C, (1993), 445-451 (*in Japanese*).

(8) Pavic, G., "Vibration Energy Flow in Thin-Walled Shells", *Proc. 3rd Int. Cong. on Intensity Techniques*, (1990), 91-96.

(9) Honda, S., et al., "Fundamental Study on Vibration Intensity", *Trans. JSME*. 60-574, C, (1994), 1941-1948 (*in Japanese*).

C521/012/98

Combining design of experiments and modelling techniques to resolve complex clutch pedal noise and vibration problems

P KELLY MSc, CEng, MIMechE
Powertrain Systems Engineering, Ford Werke AG, Germany
H RAHNEJAT MIMechE
Department of Mechanical Engineering, University of Bradford, UK
J W BIERMANN and **B HAGERODT**
Institute for Automotive Engineering (IKA), University of Aachen, Germany

SYNOPSIS

This paper contains a case study of the application of experimentation and modelling techniques to investigate and resolve clutch pedal noise and vibration issues.

To improve the system understanding, the Design of Experiment tool is used to discover the effects of the chosen factors.

This research led to the creation of a numerical model for multi-body non-linear dynamics to simulate the systems performance. The model further simplifies the design process, and enables the optimisation of future clutch systems to balance cost and function targets with the minimum of resource implications. The use of computer models that can identify the vibration levels at the concept stage, provides the designer with a tool that avoids the problem before parts are produced.

With these tools it was possible to efficiently establish a robust solution to the NVH problem.

CLUTCH SYSTEM INTRODUCTION

In all cars fitted with manual transmissions a clutch system provides the ability to break the torque transfer between the engine and driveline to allow gear changes. Defined as the component parts from the pedal to the flywheel, the clutch system is normally actuated by the use of a cable mechanism or a hydraulic system.

Figure 1 shows the layout of the clutch cable system under investigation.

During the process of disengaging and engaging the clutch on some vehicles an unwanted tactile foot vibration accompanied by a disagreeable noise (together known as whoop) can occur.

This noise, vibration, harshness (NVH) problem is generally hidden from customers by different palliatives, such as shown in Figure 2, added to the clutch system during the vehicle development phase. These masking solutions, used industry wide, are expensive in terms of time and piece cost. One thing this approach highlights is the universal lack of root cause analysis into the problem to provide an efficient cost effective, robust solution.

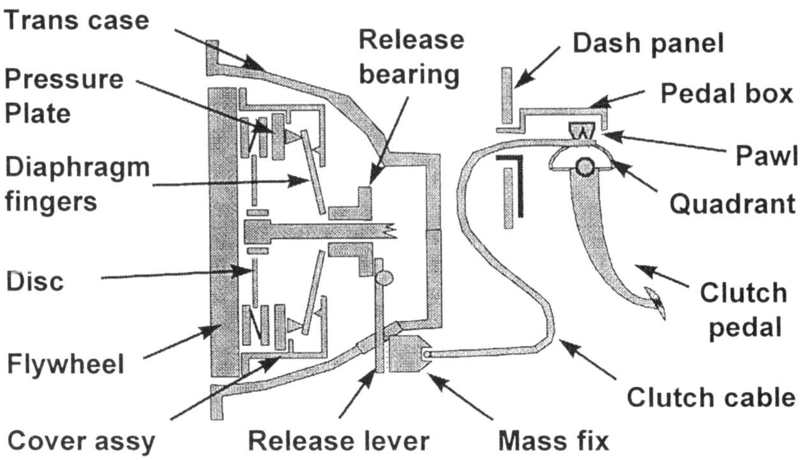

Figure 1: Parts in clutch system

Ford Motor Company led an investigation team to increase the understanding of this phenomena. Team members from IKA at Aachen, clutch supplier LuK, Germany, together with researchers from the Motor Industry Research Association at Nuneaton, and the University of Bradford, UK, were brought together to improve the fundamental lack of knowledge in this area.

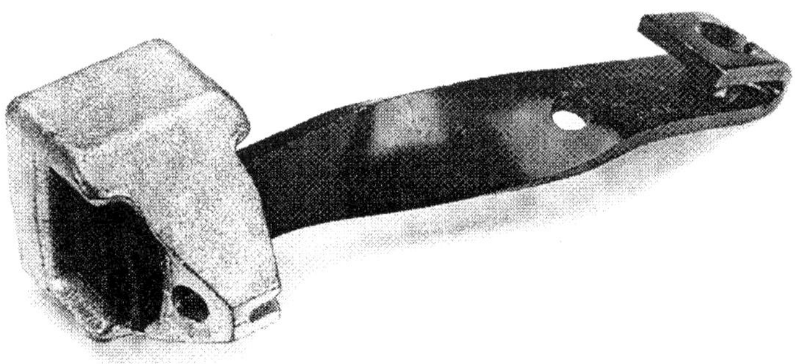

Figure 2: A typical whoop fix lever

The experimentation investigation part of this work was undertaken by IKA(1), and this paper gives details of the approach taken, the results and the subsequent analysis. The ADAMS modelling expertise was provided by the University of Bradford (2).

There are many published research papers on general torsional clutch vibration, however, only a paper for Toyota by T.Hasebe (3) deals with the specific dynamic clutch pedal noise and vibration issue tackled in this paper.

The investigation for Toyota attempted to show that the pedal vibration and interior noise are mostly affected by the dynamic characteristics of the clutch cover assembly. The research included a vehicle vibration observation and a non running engine rig experiment concerning the excitation. They stated that the clutch pedal noise and vibration path emanated from the flywheel vibration and was transmitted through the clutch and actuation system. Measurements were made at various engine speeds and different clutch positions, on a stationary petrol car with the transmission in neutral.

They found that the pedal vibration and noise peaks occurred at the just disengaged position of the pedal and with the engine running at 4400rpm. The pedal vibration consisted of first, third and sixth harmonics of engine rotation, and the interior passenger noise consisted of the third harmonic. Based on these results the research focused on the 200 to 300Hz range where the third engine harmonic had its peak magnitude.

The Toyota paper investigated only the clutch cover contribution using a static engine rig whereas this team wanted to broaden the investigation to include more of the components in the clutch system and to develop a computer model to simulate the vibration behaviour.

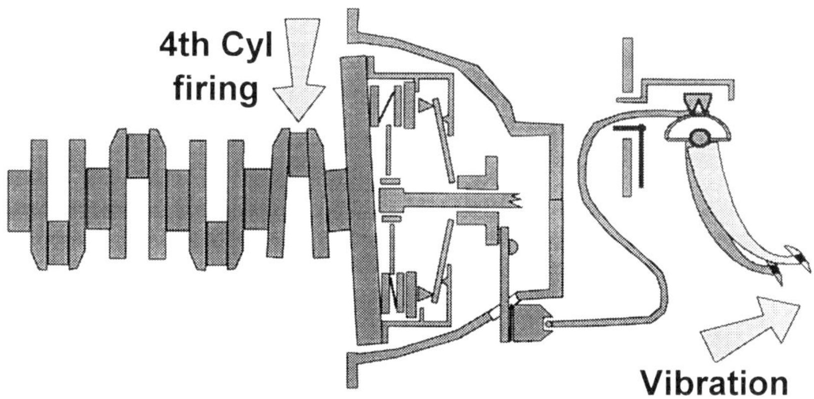

Figure 3: Crankshaft bending impulse to clutch system resulting in pedal vibration

Earlier research by Dixon and Rhodes (4) reported on ½ order excitation problems of diesel flywheels caused by crankshaft bending. The effect of this bending is an impulse excitation to the clutch system, as shown in Figure 3.

The experimentation work reported here, attempts to discover the influence of the individual component parts on the transmission of the subsequent vibration and noise.

INTRODUCTION TO EXPERIMENTATION

Traditionally, experimentation was based on making 'one-step-at-a-time'. This led to making changes without including the effect on other neighbouring systems. The problem with this approach was that design changes may have a particular effect on the system that depended on how the system was set up when the test was actually made. Thus the information gained from this approach was time consuming, usually incomplete and had a limited design shelf life.

Taguchi proposed an alternative, using statistically designed experiments, to plan the programme of testing. These Design of Experiments (DoE) meant making many design changes at once, conducting several tests and subsequent evaluations, before any actual decisions were taken. These changes were so designed to discover the effect of each design change.

During the last 10 years there has been increasing use of this type of experimentation for quality improvements. This (re-discovered) technique by Taguchi now gives the opportunity for major breakthroughs in many aspects of Quality.

OBJECTIVE OF THE DOE CASE STUDY

To investigate whoop, a complete vehicle was used to make the investigation as realistic as possible. The objective was to investigate the whole clutch system within its working environment to increase the NVH behavioural understanding. This would add to the problem solving work already performed (5). Figure 4 shows the test vehicle used. Once the factors are established the intent was to add these to the multi-body dynamic clutch model to validate the understanding of the function of the system. The model would then be used to predict further modifications to optimise the system.

Figure 4: Vehicle used for Design of Experiment investigation

A screening type design of experiments was selected, because of the many possible factors that could be involved in the transfer of whoop and limited the resources available. More specifically the major objectives were:

- To identify the important control factors (those factors that the designer can easily control) in the presence of specific noise factors (nuisance variables).
- To estimate the contribution of the control factors.
- To be able to predict the whoop performance of a system.

Noise Factors

The noise factors were determined from known influences available from previous problem solving studies. The final whoop fix solution would have to work in the presence of different clutch pedal application speeds and a range of engine speeds to be a "robust" fix.

The Noise Factors chosen were:

- Clutch application speeds - Fast (0,5 s) and Slow (1,5s)
- Engine speeds - 2000 rpm, 3000 rpm and 4000 rpm

Control Factors

To run a statistically designed experiment, the control factors have to be identified prior to the experiment. A two level experiment was chosen, meaning that two values for each control factor were selected. Normally for a first experiment, two levels for each factor are adequate. In a brainstorming session a list of factors were identified, and the feasibility and ease of producing these parts was assessed.

See Table 1 for the seven control factors and the two levels of each part that were tested.

To protect the Ford Motor Company confidentially requirements, these factors will only be referred to as letters A to G with no cross reference to which is which.

Control Factors	**Level 1**	**Level 2**
Flywheel	Standard mass	Reduced mass
Clutch Lever	Standard mass	With extra mass damper
Cable	Standard length	Longer length
Pressure Plate	Standard mass	Reduced mass
Clutch Disc	Standard stiffness	Higher stiffness
Clutch Cover	Standard stiffness	Reduced stiffness
Release Bearing	Standard 15 balls	16 balls

Table 1: The control factors used in the DoE

The concept of control factors could be easily distinguished here. The clutch supplier can easily control the clutch disc cushion spring stiffness to the required quality or production requirements but they cannot control the engine speed or speed of clutch pedal application. Thus the cushion spring stiffness is a control factor and engine speed is a noise factor. The aim therefore is to have a clutch system that is insensitive to noise factors.

METHODOLOGY FOR THE DESIGN OF EXPERIMENTS (DOE)

Traditionally experiments were conducted through full factorial designs. A full factorial design which identified all the possible combinations for a given set of factors is the most complete way of showing the important factors and their interactions, in an experiment. However, this results in a large number of tests. As most industrial experiments involve a large number of factors a full factorial is often difficult or even impossible to perform for resource reasons. In the whoop experiment there are seven factors chosen, each with two levels and two noise factors, the total number of combination were 256. For each of these runs 3 engine speeds and two pedal application speeds are recorded, a total of 1536 specific tests.

To reduce the number of experiments to a practical level, only a small set from all the possibilities, was selected. A partial factorial experiment was selected which would then act as a screen to limit the number of experiments, but would still provide the information to identify the effect of the factors.

Orthogonal arrays

The Taguchi approach to DoE provided a systematic way of conducting the experiment. Orthogonal arrays have been established by Taguchi to describe a large number of experimental situations. The arrays are matrices in which each level of factor is performed in a balanced way. That is, each level of every factor is tested an equal number of times.
The L8 orthogonal array selected for the whoop experiment is shown in Table 2.

	Control Factors							Noise Factors fast application & 2000 rpm	
Run	A	B	C	D	E	F	G	Vibration m/s^2	Noise dB(A)
1	1	1	1	1	1	1	1	7.9	76.0
2	1	1	1	2	2	2	2	7.1	73.5
3	1	2	2	1	1	2	2	3.8	68.8
4	1	2	2	2	2	1	1	2.3	72.3
5	2	1	2	1	2	1	2	5.8	75.8
6	2	1	2	2	1	2	1	3.2	73.8
7	2	2	1	1	2	2	1	6.4	74.7
8	2	2	1	2	1	1	2	4.2	72.2

Table 2: Orthogonal array used in whoop experiment

Each process configuration is called a run and the L8 array consists of 8 runs. During each run the peak value of footwell noise and the peak clutch pedal acceleration value was recorded for each of the engine and pedal application speeds. To improve repeatability the footwell microphone position was controlled by a jig and the ambient temperature was maintained by having the static vehicle tests located in an air conditioned room.

Travel measurements were based on release lever movement to minimise any errors due to clutch cable travel losses seen at the pedal.

Shown in Table 2, is the run for the fast pedal application speed at the 2000 rpm engine speed. The other engine speeds and pedal applications are also made during each run, but not shown. This procedure was repeated until all the runs were tested. It can be seen that at each run more than one factor (or design change) was made.

Looking at the vibration and noise results in Table 2, it can be seen that there is significant variation. Variation, however, would be expected even if the design had been unchanged and the purpose of the statistical analysis is to identify the variation in the response related directly to the changes in the factor levels.

At first sight this type of experimentation seems confusing, and this is often a criticism of this method, however, more can be gained from this approach than is first appreciated. Some runs produced a decrease in the response (as required) but other runs increased them again.

One complication is that the lowest response for pedal acceleration was not always the lowest for audible noise. This could be explained by lack of sufficient positioning and temperature control, in the measurements of the noise, or an indication that the paths of the noise and vibration are not the same.

When the variation in response is related to the variations in factors, an optimised combination can be found. This combination may not have been tested, but it can be calculated from the experimental data.

Factors that had an important effect on the response and those that had no effect, are evaluated and this knowledge can be put to use. Firstly the factors that have no effect can be selected for commercial reasons only . Secondly, the important factors can be focused upon and no time wasted on the less important ones.

ANALYSIS OF DOE RESULTS

Effects plots

To evaluate the impact of each factor the average value of the response or quality characteristic (acceleration and audible noise) was compared when that factor was at level 1, to the average value of the quality characteristic when the factor was at level 2. In other words the factors contrast the results according to the levels of each factor. The difference between these two factors is the effect it has on the response so that the relationship to the response can be defined. A negative answer means that changing from level 1 to level 2 had a beneficial effect.

The simple calculation to determine the effect of a factor is possible because of the balance in the experiment. Looking at the experimental array in Table 2 it can be seen that the changes are structured in a systematic way. Each factor is tested exactly four times at level 1 and four times at level 2. This property of balance is crucially important. The important advantage of this balance is that if one factor has a large effect on the response characteristic then this will cancel out when the effects of other factors are determined.

The effects of each factor in turn can be established by simple calculations. For example, by referring to Table 2, you can see that the effect of factor A can be determined by comparing level 1 from runs 1 to 4, to level 2 shown by runs 5 to 8. This is established by adding the sum of the responses for each factor and level. The difference between the sums of the responses for each level gives the effect. For convenience their effects are displayed in a response table - see Table 3 for the vibration data.

Factors	**A**	**B**	**C**	**D**	**E**	**F**	**G**
Average of level 1	5,3	6,0	6,4	6,0	4,8	5,2	5,0
Average of level 2	4,9	4,2	3,8	4,2	5,4	5,1	5,2
Absolute Effect	**0,4**	**1,8**	**2,6**	**1,8**	**-0,6**	**-0,1**	**-0,3**

Table 3: Vibration response for the 2000 rpm Fast Application in m/s^2

The same procedure was followed for Audible noise, but is not shown.

The response table can be used to pick out which factor levels are apparently best for the response characteristic, simply by choosing the levels that give the lower average response.

Vibration plots

It is useful to present numbers graphically, and the graphical version of a response table is called an effects plot (Figure 5 shows a sample for the fast application at 3000 rpm engine speed). This picture indicates the average response for each factor at each level, so that the vertical distance between points is equal to the size of the effect.

It is then easy to see which Factors contribute most and which have little effect. Factor C followed by B and D clearly have the most influence.

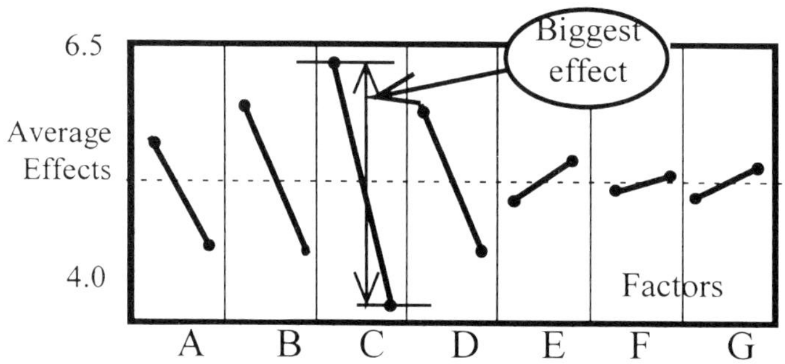

Figure 5: Effects Plot - Pedal vibration, Fast application at 2000 rpm

Combining prior engineering knowledge and the numerical information it is possible to select those factor levels which the data suggest improve the situation but which also make sense.

In the case of the cable and mass lever, prior knowledge in the one step at a time approach showed that these changes were expected to show significant improvements. This input also provided a measuring indication on how effective the other changes could be.

Audible Noise plots

A summary of the audible noise plots are not shown. The noise effects are much less distinguishable with only factor B showing any significant audible noise effect. The lack of other strong effects could suggest the noise path was not the same as the vibration path.

Daniel plots

Statistical methods should be used to determine whether the results of experimental data can be taken at face value.

Conventional statistical practice takes a sceptical view of apparent effects: only those which clearly stand out from the noise are treated as real.

In engineering, confirmation can usually be conducted very quickly, and so some of the statistical rules about the significance of factors are often relaxed. The most important 'significance test is whether the result can be repeated or not. However, one of the functions of statistical methods, is to sort out which effects are too large to be accounted for by the inherent variability of the process.

A common method to help decide which effects might be treated as real are the normal plots (both full and half normal plots). The half normal plot is often referred to as a Daniel plot, after Cuthbert Daniel, who first suggested using them.

The Daniel plots are shown in this analysis were made by :

1. Removing all the signs off the absolute effects (from Table 3) and arranging them in order.
2. Producing a scatter plot of all the values in which the "y" values are the sorted absolute effects and the "x" values are the expected ordered values for the half of the standard normal distribution (known in this case as half normal scores) and are available from published statistical tables.

Daniel plot - Vibration

To interpret these pictures, all that needs to be known is that effects which can be attributed to experimental error (and therefore not real effects) plot as a straight line pointing at the origin. Reproducible effects plot above the line, the largest being on the right-hand side of the chart. Most of the points for vibration in Figure 6 cluster around a straight line through the origin of the plot, but three points lie well clear of the line. These three effects C, B and D would be then regarded as real vibration effects and, in the absence of other evidence, the rest of the points only as the results of random variation.

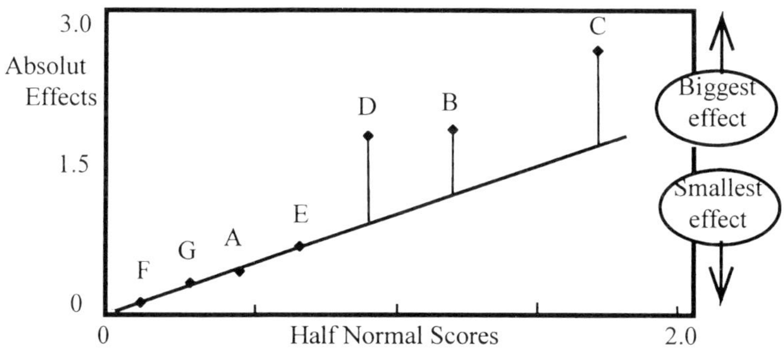

Figure 6: Daniel plot for Vibration at 2000 rpm and Fast application

Daniel plot - Audible noise

The Daniel plot confirmed the effects plot analysis that there was little difference in noise behaviour at 2000 rpm. For higher speed conditions C, F and G have an effect. (These plots are not shown due to space limitations).

The team initially thought it would be difficult to recognise apparent audible noise differences subjectively heard by the operator by using only one figure from a noise trace. The quality or signature of the noise would be lost. The experimental results obtained confirmed that one value did give sufficient information to identify some differences.

However, the audible noise results obtained did not reflect the drivers perception as well as the vibration results.

It is important to know if a factor has no effect or only a small effect and this can be just as important as finding a large effect. When managing a design there are usually good reasons for leaving factors alone unless there is a clear advantage to be gained from changing them. Here, because they are not important contributors, factors E, D and A can be disregarded for future noise analysis. The selection of these factors can then be chosen based on commercial or other considerations.

The optimum outcome could even be a selection that was not tested in the experiment.

With the knowledge and understanding of the system gained in the experiment it is also much easier for any future changes to be made without always re-testing.

THE MULTI BODY DYNAMIC MODEL

The advantage of developing a detailed model of the clutch system is evident. When such a model is validated against experimental evidence it can be subsequently employed to carry out many DoE studies through virtual prototyping simulations. This approach is becoming more commonplace in industry as the cost of development of experimental rigs, condition monitoring and analyses can be prohibitive when a large number of factors are to be investigated, many of which interact with one another. However, traditional mass-spring models are rather simplified and do not represent the physical reality of the system geometrical assembly and its constraints. Multi-body dynamics methodology can be employed to create "life-like" models of mechanisms which include all system inertial properties, sources of compliance, as well as the system constraints that ensure the ideal function of the mechanism.

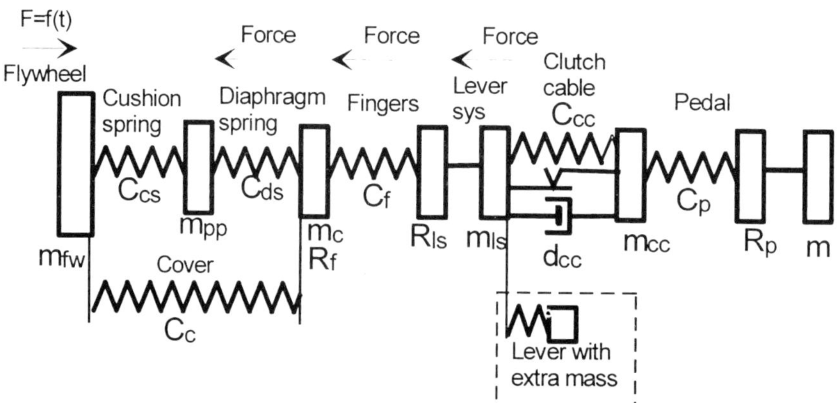

Figure 7: Clutch System Model

A commercially available code; ADAMS is employed for this purpose and a generic model for cable-operated clutch system for diesel passenger cars is developed. A simplified schematic representation of the model is shown in Figure 7.

There are 17 rigid parts, as shown in Table 4, in the multi-body model. Each part accounts for 6 unconstrained degrees of freedom. The constraints imposed between the various clutch system components are modelled using appropriate scalar constraint functions that represent joints.

Table 4 also lists the constraint functions employed in the multi-body model. Each of the 98 constraints is represented by an algebraic function. Each degree of freedom is represented by a differential equation of motion (102 in this multi-body model). Therefore, there are 4 unconstrained independent degrees of freedom in the model. These are the translational motion/floating of the flywheel/cover assembly, the axial motion of the release bearing, the articulation of the release lever and the movement of the clutch pedal.

The input is given by a Dirac force function of short duration with an amplitude of 100 N applied directly to the flywheel. This simulates the impact during flywheel bending.

Rigid parts	**Part Name**	**Typical mass (kg)**	**Constraint**		
			Function	**Type**	**No. of**
	ground	-	-		
1	housing	20	ground	fix	6
2	flywheel	4.5	ground	translational	5
3	friction disc	0.5	flywheel	fix	6
			hub	fix	6
4	hub	0.5			
5	pressure plate	2.0	housing	translational	5
			bearing	coupler	1
6	cover	1.5	flywheel	fix	6
7	bearing	0.2	housing	translational	5
			release shaft	curve-curve	2
8	input shaft	2.0	housing	fix	6
9	release shaft	0.5	housing	cylindrical	4
10	lever	0.5	release shaft	fix	6
11	vib. absorber	0.5	lever	translational	5
12	cable guide	0.05	housing	cylindrical	4
13	cable lever	0.05	vib absorber	spherical	3
			cable guide	translational	5
	cable pedal		pedal box	translational	5
			quadrant	rack-pin	1
14	cable quadrant	0.05			
15	pedal box	2.5	ground	fix	6
16	quadrant	0.1	pedal	fix	6
17	pedal	0.5	pedal box	resolute	5
	Σ constraints				98

Table 4: The parts in the multi-body model

A simulation run of the vibration model is undertaken initially without the 1 kg "Diehl fix" mass-damper. A sample of the spectrum of the pedal angular velocity is shown Figure 8. There are 4 significant frequencies with their side-bands. The contribution in the region of 130 Hz coincides with the 4th engine order at the engine speed of 2000 rpm (the condition existing for this simulation). The 4th order contribution has also been monitored in the form of noise in the footwell area by a microphone pick-up at IKA(1). The contribution at 16.7 Hz is due to the forcing frequency at the troublesome half-engine order (4). The peaks occurring between the half engine order and the 4th order contributions are the full and half order harmonics of engine speed. The whoop frequency (i.e. the natural frequency of the system) is observed at 250 Hz and agrees well with the experimental rig findings and the vehicle tests. This coincides with the 8th engine order. It is now well understood that the 2nd, 4th, and 8th engine orders occur as the main contribution of crankshaft combined torsion-deflection modes in 4 cylinder engines, particularly in diesel engines with a higher torque output. The existence of the fundamental natural axial rigid mode of the clutch system within the 8th engine order, therefore, gives rise to resonant conditions.

RESULTS AND DISCUSSIONS

The important factors for the vibration transfer have been identified and validated in the model using the case with and without the Diehl fix lever fitted, as shown in Figure 8. Without the extra mass and stiffness fitted to the vehicle, resulted in a actuation natural frequency of 250Hz and 525 Hz when fitted. The modelling results of 250 and 450 Hz gave good correlation to these measurements.

This means that the resonant conditions may not appear up to an engine speed of 3500-4000rpm which is acceptable for the diesel engine assessed.

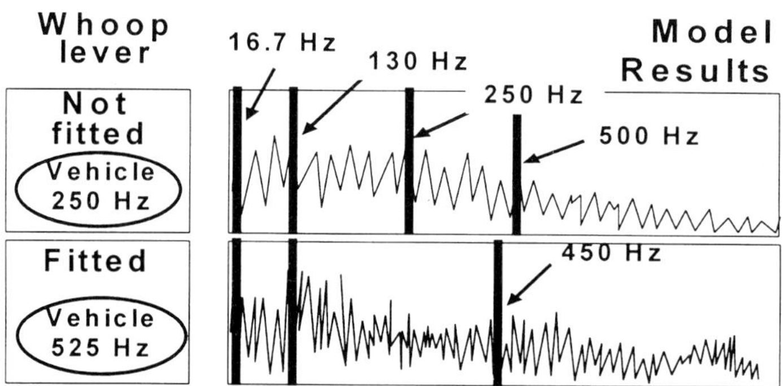

Figure 8: Comparing vehicle to modelling data

Therefore, a realistic and detailed model for the clutch whoop problem has been devised and validated against experimental evidence. This model can now be employed in future studies for investigation of whoop and design of new clutch systems.

The audible noise phenomenon of whoop was not satisfactorily determined during this study and further noise path analysis is required.

REFERENCES

(1) Kelly, P., Biermann, J.-W., Hagerodt, B., Clutch Pedal Dynamic Noise and Vibration Investigation using Taguchi Methods, 6th Aachen Colloquium, Automobile and Vehicle Technology, Germany, 22nd October 1997

(2) Centea, D., Rahnejat, H., Kelly, P., Non-linear Multi-Body dynamic analysis for the study of In-Cycle vibrations (Whoop) of cable operated clutch systems, Simulation, Diagnosis and Virtual Reality, ISATA, Florence, Italy, 18 June 1997

(3) Hasebe, T., u.a., Aisin Seiki and Toyota, "Experimental study of reduction methods for clutch pedal vibration", SAE Transactions, Paper 932007, 1993

(4) Dixon, J., Rhodes, D. M., The generation of half orders by structural deformation, International Conference Vehicle NVH & Refinement, IMechE Conference Paper, C487/032/94, MEP,1994, pp.9-17

(5) Kelly, P. and Rahnejat, H., "Clutch pedal dynamic noise and vibration investigation", in Rahnejat, H. and Whalley, R.(eds.) Multi-Body Dynamics: Monitoring and Simulation techniques, MEP Publications Ltd., March 1997

C521/013/98

Quantifying the noise emission of engine oilsumps, valve covers, etc. using artificial excitation

P J G van der LINDEN
LMS International, Leuven, Belgium
J SCHNUR and **T SCHOMBURG**
Mercedes-Benz, Stuttgart, Germany

P_r	=	pressure at point r (Pa)
Q_r	=	volume velocity in point (m^3/s)
V_i	=	velocity in point i (m/s)
x^2	=	autopower spectrum or $\lvert x \rvert^2$
F_i	=	force at interface point i (N)
$\underline{A}$	=	vector of baseplate accelerations (m/s^2)
$\underline{F}$	=	vector of interface forces (N)
H^{PV}	=	frequency response function between point velocity excitation and point pressure response ($Pa.s.m^{-1}$)
H^{AQ}	=	vibro-acoustic transfer function : acceleration response due to volume velocity excitation ($m^{-2}s^{-1}$)

SYNOPSIS

Oilsumps, valve covers and distribution covers belong to the main noise radiating parts on modern engines. These parts have in common that they undergo velocity excitation at their fixation to the main engine structure.

A method was developed to quantify the noise emission of the individual part using artificial excitation. This measurement technique allows the assessment of the parts and their mounting to the engine for efficient variant comparisons.

The difficulty lies in the simulation of near to rigid boundary conditions of those parts and in the effective velocity excitation. A combination of reciprocal acoustic excitation and inverse force identification is used. The objects are excited by volume velocity sound sources. On the basis of a reciprocity relationship and inverse force identification the effective noise emission is quantified for spatially uncorrelated velocity excitation.

A comparison of aluminium, plastic and rubber decoupled oilsumps is shown as an example.

1. OVERALL PRINCIPLE

The primary operational excitation, inside engines, works on the main structure. This structure will start to vibrate, also at the interface of the test component. The test component is assured to the essentially more flexible and lighter than the main structure, where it is mounted on. The vibration field at the interface is therefore almost entirely determined by the structural dynamic of the main structure. This leads to the following conclusions :

1) The best pure boundary condition simulation of the test object is rigid or clamped;
2) The noise radiation of the test object should be related to the velocity at the interface rather than the interface forces.

A velocity excitation is therefore to be simulated (fig. 1). To do this, we simplify the continuous interface vibration field to a set of discrete points at which the velocity is controlled. For a simulation of the complete true velocity field a half bending wave length discretisation is needed. But for oilsump of engines this becomes impractical; above 1000 Hz the number of points to consider would become too high.

Both to reduce the number of interface points and to come to a practical excitation simulation, we assume that the velocity excitation can be simplified to uncorrelated motion predescribed at a limited set of points equally spread around the interface (1), (3). This approximation is reasonable for the broadband engine excitation, analysing in third octave bands, as long as these bands contain several modes.

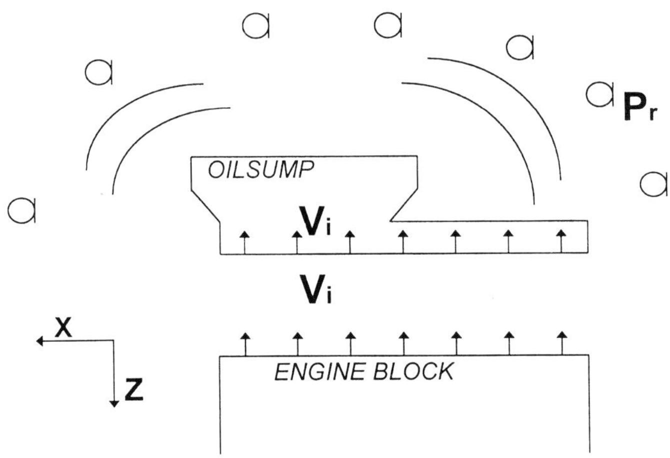

Fig. 1 : Sketch of the object and discretisation principle

The noise due to velocity at one interface point (in one direction) is quantified in a velocity-acoustic transfer function (fig. 2) :

(1) $$H_{ri}^{PV} = \left.\frac{P_r}{V_i}\right| Q_{r=0} \qquad (\mathrm{Pa.s.m^{-1}})$$

The sound pressure at any point in the sound field can then be quantified by :

(2) $$P_r^2 = \sum_i H_{ri}^{PV\,2} \,.\, V_i^2 \qquad (\mathrm{Pa^2})$$

To assess test objects, independent of the operational excitation, an equal velocity amplitude $/V_i^2/$ is assumed. And to quantify the noise emission proportional to sound power we consider an even spread of pressure points in semi-free field at one meter distance from the test object.

The interface and hemisphere averaged pressure due to the interface's velocity excitation becomes :

(3) $$/\,\widetilde{H}^{PV}\,/ = \left(\frac{1}{r}\sum_r \sum_i H_{ri}^{PV\,2}\right)^{1/2} \qquad (\mathrm{Pa.s.m^{-1}})$$

2. RECIPROCITY AND INVERSE FORCE IDENTIFICATION

The point velocity excitation is difficult to achieve in practice. But luckily, the objects and sound fields are generally linear, which means one can apply reciprocity :

(4) $$H_{Vi}^{PV} = \frac{P_r}{V_i}\bigg/_{Q_{r=0}} = \frac{F_i^*}{Q_r^*}\bigg/_{V^*_{i=0}} \qquad (\mathrm{Pa.s.m^{-1}})$$

* denotes the reciprocal cas

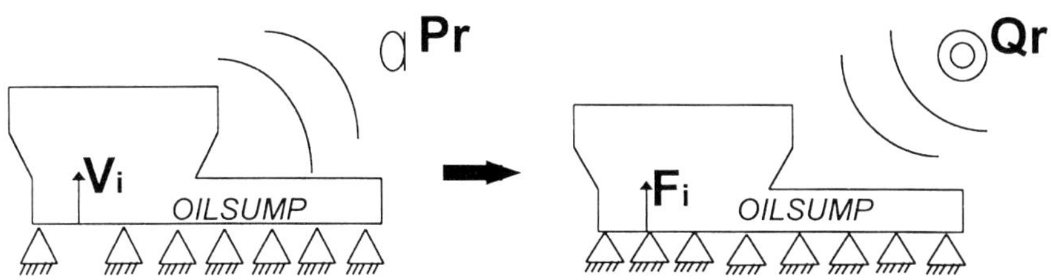

Fig. 2 : Clamped object single point velocity excitation and reciprocal acoustic excitation

This means that acoustic excitation with point sources in semi-free field of the clamped object will provide the same desired velocity acoustic transfer functions, as long as it is possible to measure the forces in the interface.

These forces are not measured directly because of instrumentation difficulties and the risk of modifying the system. Inverse force identification is used on the basis of acceleration measurements.

The test object is therefore mounted at its interface on a plate. The impedance of this baseplate at the interface is finite, but still needs to be essentially higher than the impedance of the object. A large number of vibro-acoustic FRF (H^{AQ}) are thus measured by accelerations on the plate under acoustic excitation. After this the test object is removed and using structural excitation, the transfer functions (H_{ki}^{AF}) between the force at each interface point (i) and all acceleration signals (k) are determined.

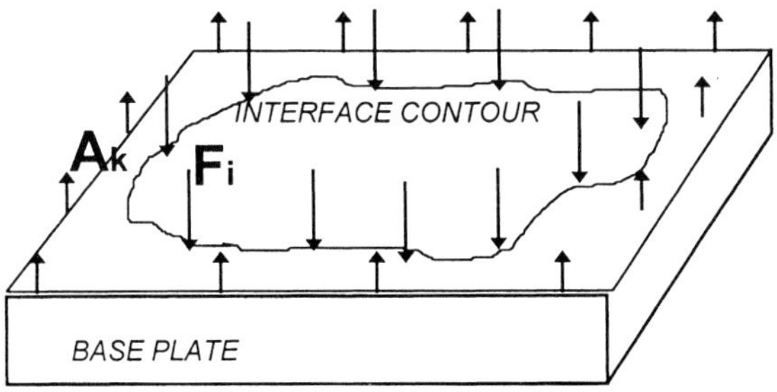

Fig. 3 : Sketch of the baseplate, force excitation at the interface and accelerations on the base plate

All mechanical transfer functions are combined in a matrix. The inverse (or pseudo-inverse) of this matrix relates the measured accelerations in the vibro-acoustic FRF (H^{AQ}) to the interface forces (in the velocity-acoustic FRF H^{FQ} or H^{PV}) (2), (3).

$$(5) \qquad \underline{H}^{FQ} = \left[H^{AF}\right]^{-1} \cdot \underline{H}^{AQ} \qquad (N)$$

The measured volume velocity to acceleration transfer functions are thus translated to velocity acoustic transfer functions.

3. TEST CONFIGURATION

A thick damped aluminium plate is suspended close to the floor of a semi-anechoic test site. Eight accelerometers are mounted on the bottom of the plate. Two mid frequent volume velocity sources are positioned, alternating, at six points on a one meter hemisphere (4). (This is a simplification of a standard sound power measurement in semi-free field, where the microphones are replaced by volume velocity sources.)

Fig. 4 : Photo of the test set-up; sound source above test object and baseplate covered with rubber

The oilsump variants are either glued, or bolted, directly onto the baseplate or the engine's bedplate is first glued onto the baseplate to provide the correct interface. The input accelerance functions of the baseplate and the oilsump can be seen in figure 5. The clamped boundary condition is sufficiently approximated, but still allowing some motion of the plate to be measured by the accelerometers.

4. RESULTS

A number of checks were performed on the intermediate results and on the final averaged velocity acoustic transfer.

The condition of the force identification matrix was checked. A condition limit was applied to improve the force estimations below 1000 Hz, where the number of excited modes in the baseplate is small. But this force estimation could be a study on its own, and is therefore not elaborated here (2), (3).

The next check is a comparison of the observed H^{PV} to two idealised situations which can be easily calculated. The first is the H^{PV} of a point source with the interface's velocity level which is relevant for the frequencies below 1000 Hz where the oilsump noise radiation is not yet determined by small plate nodes. The second idealised H^{PV} is derived from a 100% radiation efficient oilsump with the interface velocity level which is relevant for higher frequencies; above 1000 Hz . These confirmed that the correct order of magnitude is found.

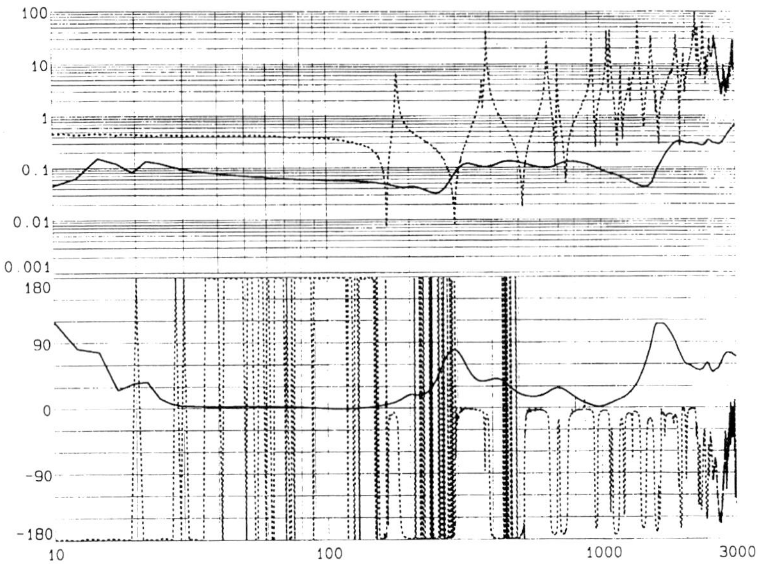

Fig. 5 : Input accelerances, oilsump (dashed line) and baseplate (solid line)

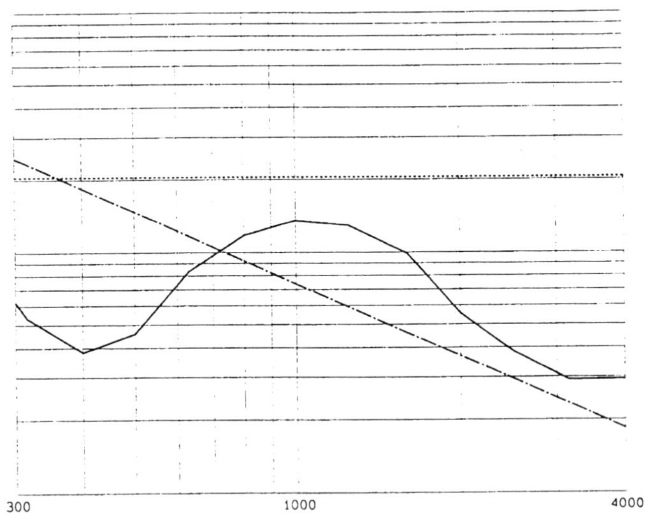

Fig. 6 : Averaged velocity acoustic transfer ($Pa.s^2.m^{-1}$) of an oilsump (solid line) and the point source reference curve (dashed) and the 100% radiation efficiency reference curve(dash-dot)

The averaged acoustic velocity transfer should not be sensitive to the exact location and number of interface velocity points. This is demonstrated in figure 7, where six or eight discretisation points were used.

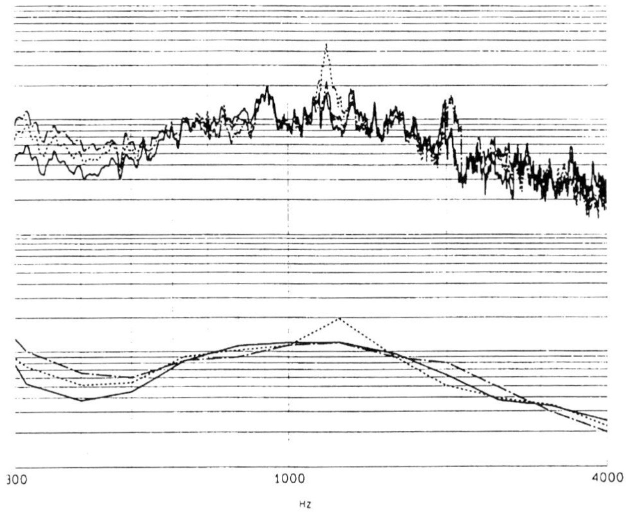

Fig. 7 : Averaged velocity acoustic transfer (Pa.s^2.m^{-1}), eight points (solid), six points (dashed), six other points (dashdot) Top window narrow band, lower window third octave bands

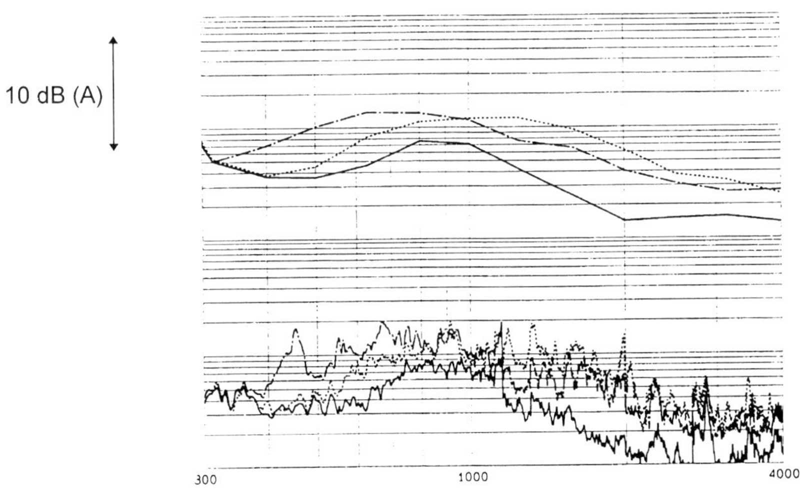

Fig. 8 : Averaged velocity-acoustic transfer (Pa.s/m) of the bare baseplate (solid line) and two oilsump variants (dashed/dash-dot)

Figure 8 shows three curves, the averaged velocity-acoustic transfer of two oilsumps and the average velocity-acoustic transfer of the baseplate without test object. The oilsump curve should be higher than the bare baseplate to be sure of the observed differences between oilsump variants.

The acceleration levels on the plate are higher with object than without for most of the targeted 300 to 3000 Hz frequency range. This is acceptable, but near to critical.

The low acceleration response under acoustic excitation (At 110 dB SPL) is part of the limitation. The other limitation is more principle; the interface of the oilsump to the engine, if it is closed by a perfectly damped material of minimal surface, will still radiate noise. (The damped aluminium baseplate is close to such a perfect termination)

Figure 8 shows the comparison of a fixed aluminium oilsump to a rubber gasket decoupled variant. Figure 9 shows the same aluminium oilsump compared to a fibre reinforced plastic variant.

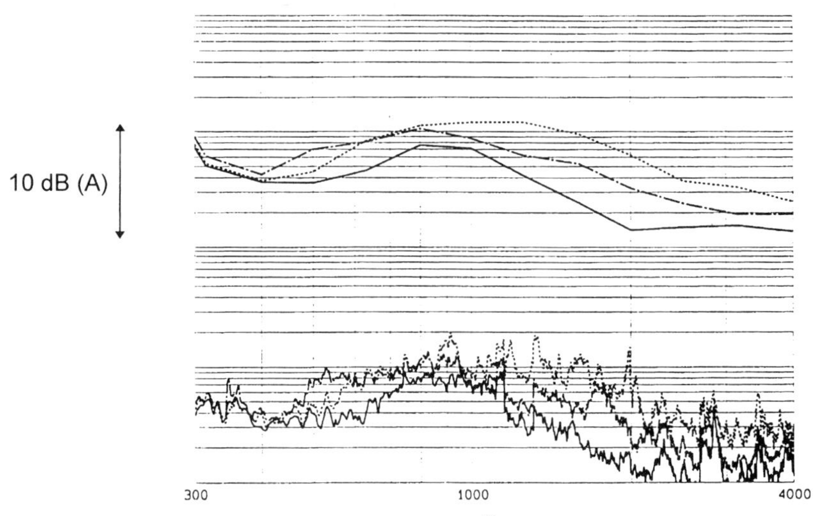

Fig. 9 : Averaged velocity acoustic transfer (Pa.s/m)of the bare baseplate (solid line) the alu. oilsump(dashed) and the plastic variant (dash-dot)

5. CONCLUSIONS

The objective was a procedure for correct assessment of oilsumps, valve covers etc. including their mounting method.

The developed procedure approximates both the clamped boundary condition and the velocity excitation.

The average velocity-acoustic transfer approximates the objects' noise emission independently from the engine main structure dynamics and independent from the engine operating condition.

An estimation of the operational sound pressure due to the oilsump (or other) on a testbed can be obtained on the basis of the measured averaged velocity of the interface.

(6) $$/\,\tilde{P}\,/^2 \;=\; /\,\tilde{V}\,/^2 \,.\, /\,\tilde{H}^{PV}\,/^2 \qquad (Pa^2)$$

The measurements proved to be practical. The analysis complexity was significant, but could be handled. Vehicle measurements of various oilsump variants have confirmed the observed differences of the variants in the averaged velocity acoustic transfer.

6. REFERENCES

(1) Airborne source strength measurements on the basis of pressure indicator inversion applied to an engine valve cover, P.J.G. van der Linden, K. De Meester, J.P. Le Martret, Fourth International Congress on Sound and Vibration, 1996, St-Petersburg (Russia).

(2) Accurate vehicle FRF measurements for indirect Force Determination based upon matrix inversion, W. Hendricx, ISMA 1994, Leuven (Belgium).

(3) Inverse and reciprocity methods for machinery noise source characterisation and quantification, J.W. Verheij, Fourth International Congress on Sound and Vibration, 1996, St-Petersburg (Russia).

(4) Omnidirectional volume velocity point sources; how can they be approximated, P.J.G. van der Linden, Internoise 1997, Budapest (Hungary).

C521/016/98

Airborne noise transfer path analysis

R STRYCZEK and **R FREYMANN** AIAA, DGLR
BMW AG, Munich, Germany

ABSTRACT

A newly developed experimental-numerical method is described allowing the identification, analysis and optimization of airborne noise transmission paths in the field of passenger transportation systems. The method allows the separation of the structure and airborne interior noise contributions. The computer aided testing procedure is based on an indirect sound pressure level measurement approach detected by means of a multi-channel microphone system. The procedure can be applied to the detection of airborne driveline noise paths as well as to wind noise transmission paths. Due to the possibility of a clear analysis of the noise transfer phenomena, a ranking and a thorough optimization of the different noise transfer paths can be achieved.

1 INTRODUCTION

Acoustics have evolved to an increasingly important topic in the field of passenger transportation systems. Accordingly it is not surprising to note that the different manufacturers are placing a lot of effort in activities for improving the noise behavior of their vehicles.

During the last decade a variety of key methodologies have been developed and brought to practical application in the scope of the acoustic development process of vehicles. In this context just a few of these new analysis and optimization tools should be mentioned here dealing as for instance with structural dynamic [1,2], noise transfer path [3,4], acoustic cavity [5,6] and coupled vibro-acoustic [7,8] aspects.

When putting together the different pieces of the overall acoustic puzzle being available right now it can be noticed that one complex of the entire acoustic task, dealing with the airborne noise propagation, remains nearly unaddressed. It is true that in the scope of thus related investigations, procedures such as the well-known „windowing method“ or the „complete driveline disconnection approach“ are commonly used. But the practical value of these methodologies is significantly reduced by the time and experimental effort required for their practical application.

Since state of the art and future vehicle development processes are requiring fast and reliable development tools in the field of acoustics, focus was pointed on the development of a new efficient procedure to allow the analysis and optimization of airborne noise transfer paths. The newly developed procedure as well as its application to a real vehicle will be described in the following chapters.

2 DESCRIPTION OF THE METHOD

As is shown in **Fig.1,** the noise produced inside a vehicle in any real operating condition is formed by structural-borne and airborne noise contributions. The fact of this superposition of different noise contributions is the main obstacle for the analysis of the pure airborne noise transmission characteristics on a movable complete vehicle. Accordingly a practically applicable and efficient approach with regard to airborne noise transmission investigations is requiring the separation of the airborne and structural-borne noise contributions.

This can be achieved when assuming that no major noise sources are located inside the passenger compartment of the vehicle. If this assumption is fulfilled the vehicle can be considered as being formed by two subsystems; a first subsystem consisting out of the passenger compartment including its air volume and the structural parts of the enclosure, the second subsystem being related to the „rest of the vehicle“ and a volume of air around the entire vehicle (**Fig. 2**). According to the assumption made above all noise sources, such as the engine, the driveline, the muffler system, the auxiliary components but also the aeroacoustic excitation are located within subsystem II and are external to subsystem I.

If the vehicle is operated, a noise field will be generated in subsystem II. The transfer of this noise field into subsystem I is considered as the airborne noise transmission.

The key idea of the new procedure now is to recreate the noise field produced in subsystem II when the vehicle is operated by a fictitious loudspeaker configuration. Thus the loudspeaker configuration is creating the same noise field as during the operation of the vehicle. If this can be achieved there is no need to further explain that on the basis of this defined acoustic excitation a thorough airborne noise transfer path analysis can be achieved.

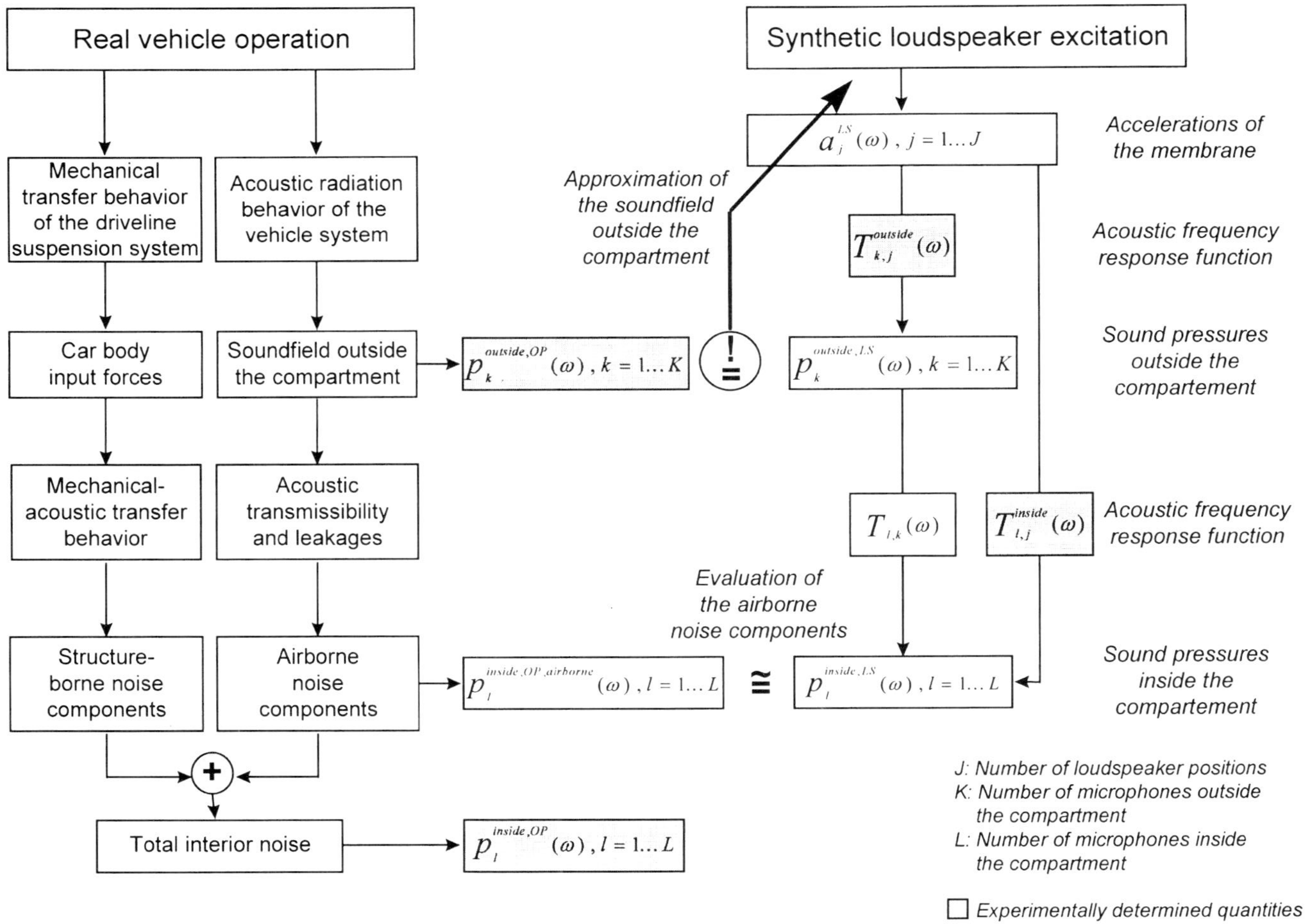

Fig. 1: Blockdiagram of the airborne noise transfer path analysis

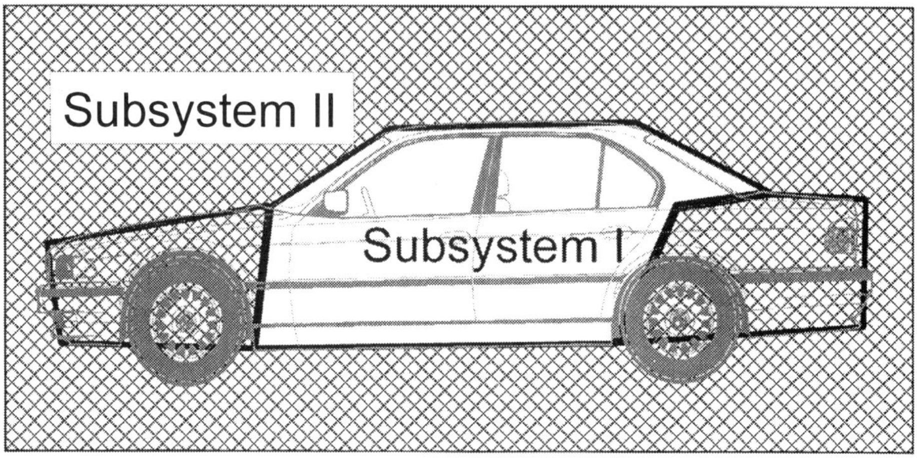

Fig. 2: Complete vehicle and subsystems considered

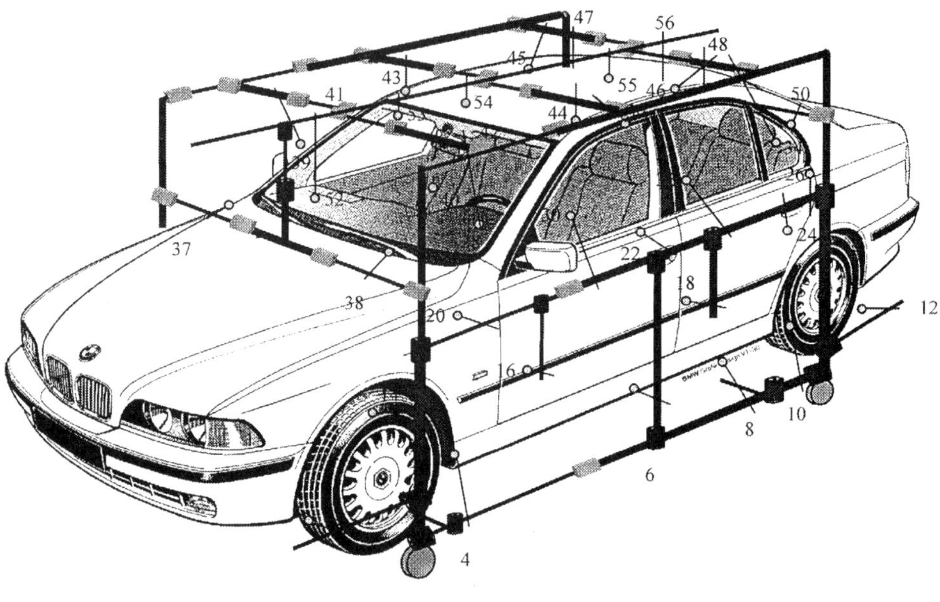

Fig. 3: Test rig for multi-channel microphone system

Accordingly, the different steps of the airborne noise transmission analysis can be summarized as follows:

a) operate the vehicle in a steady operating point in a laboratory and determine the sound pressure field in subsystem II,

b) reproduce the sound pressure field by a loudspeaker configuration, the vehicle being passive,

c) analyze the airborne noise transmission characteristics for the passive vehicle case where all structural-noise paths are inactive.

In the following sections the three different steps of the procedure will be described in detail.

2.1 Determination of the acoustic pressure field

To investigate the noise transmission from the entire driveline system to the passenger compartment, the vehicle is operated on a roller dynamometer installed in a (semi-)anechoic acoustic testing room.

To determine the acoustic pressure field in subsystem II a multi-channel microphone system is used. Microphones are installed all around subsystem I, as for example in the engine compartment, below the car floor, along the muffler system but also inside the remaining air volume around the entire car, which forms a part of subsystem II. This can easily be done when a test rig (**Fig. 3**) is used to support all the microphones which are exterior to the vehicle contour. Moreover microphones are placed into subsystem I at the standardized head positions of the passengers.

Operating the vehicle in a defined steady driving condition will allow to determine the sound pressure spectra

(1) $$p_k^{outside,OP}(\omega) \quad ,(k = 1,2,\ldots,K),$$

related to the different microphone positions k within subsystem II and

(2) $$p_l^{inside,OP}(\omega) \quad ,(l = 1,2,\ldots,L),$$

addressing the (interior noise) microphones within subsystem I.

In these notations the index OP characterizes the operating condition of the vehicle. In practical passenger car applications the total number of microphones used will be in the range of K=100 and L=4.

2.2 Reproduction of the sound pressure field

The first step of the pressure field reproduction process entails the excitation of the air volume within subsystem II by a loudspeaker at J different positions. The loudspeaker excitation is realized sequentially in the J different locations. Exciting the air volume at a defined position j allows to generate the following frequency response behavior between the acceleration $a_j^{LS}(\omega)$ of the loudspeaker membrane and the responses of the microphones $p_k^{outside,LS}(\omega)$ and $p_l^{inside,LS}(\omega)$ located within subsystems II and I, respectively:

$$(3) \qquad T_{k,j}^{outside}(\omega) = \frac{p_k^{outside,LS}(\omega)}{a_j^{LS}(\omega)} \quad ,(k = 1,2,\ldots,K\,;\, j = 1,2,\ldots,J),$$

$$(4) \qquad T_{l,j}^{inside}(\omega) = \frac{p_l^{inside,LS}(\omega)}{a_j^{LS}(\omega)} \quad ,(l = 1,2,\ldots,L\,;\, j = 1,2,\ldots,J).$$

It is well understood that because of practical reasons the number of loudspeaker positions considered will be lower than the number of microphones installed, say $J \ll K$. On the basis of a numerical approach it is now possible to identify a fictitious loudspeaker configuration which does allow to reproduce the sound pressure field within subsystem II as generated by the vehicle operation. The excitation vector of the loudspeaker configuration (Fig.1) can be approximated by the equation

$$(5) \qquad \boldsymbol{a}^{OP}(\omega) \approx \boldsymbol{PINV}\left(\boldsymbol{T}^{outside}(\omega)\right) \bullet \boldsymbol{p}^{outside,OP}(\omega)\ ,$$

with

$$(6) \qquad \boldsymbol{PINV}\left(\boldsymbol{T}^{outside}(\omega)\right) = \left[\left(\boldsymbol{T}^{outside}(\omega)\right)^T \bullet \boldsymbol{T}^{outside}(\omega)\right]^{-1} \bullet \left(\boldsymbol{T}^{outside}(\omega)\right)^T$$

being the pseudo-inverse of the matrix outlined in equation (3).

Having identified the excitation vector $\boldsymbol{a}^{OP}(\omega)$ now allows to determine the sound pressure levels $p_l^{inside,LS}$ produced by the airborne noise component during vehicle operation within subsystem I, which is formed by the passenger compartment, on the basis of Eqs.(3) and (4) as follows:

$$(7) \quad \boldsymbol{p}^{inside,OP,airborne}(\omega) \approx \boldsymbol{p}^{inside,LS}(\omega) = \boldsymbol{T}(\omega) \bullet \boldsymbol{p}^{outside,OP}(\omega) \quad ,$$

with

$$(8) \quad \boldsymbol{T}(\omega) = \boldsymbol{T}^{inside}(\omega) \bullet \boldsymbol{PINV}\left(\boldsymbol{T}^{outside}(\omega)\right).$$

$\boldsymbol{T}$ is a frequency dependent complex matrix with dimensions *L x K*, describing the transfer characteristics between the acoustic pressure fields of subsystems I and II. Formally, $\boldsymbol{T}(\omega)$ is neither depending on the properties of the loudspeaker used nor on the number or the location of the loudspeaker positions considered.

2.3 Airborne noise transmission analysis

The transfer matrix $\boldsymbol{T}$ can be considered as the acoustic transmissibility matrix related to the passenger compartment. In this context the different coefficients $T_{l,k}$ of the $\boldsymbol{T}$ -matrix provide a ranking about the acoustic sensitivity of a (microphone) position *l* inside the passenger compartment with regard to an external acoustic excitation at the position *k* of subsystem II. Accordingly high values of $T_{l,k}$ are a first indication of possible acoustic transmission deficiencies. But on the other hand the $\boldsymbol{T}$ -matrix does not provide the total information about the importance of the various noise transmission paths, since in accordance with Eq.(8), the significance of an airborne noise path is given by the product of $T_{l,k} \cdot p_k^{outside}$.

3 EXPERIMENTAL INVESTIGATIONS

The experimental-numerical procedure was applied to a real vehicle. **Fig. 4** shows the averaged sound pressure levels $\boldsymbol{p}^{outside,OP}$ up to a frequency range of 1 kHz which were experimentally determined at a defined operating condition of the vehicle. The positions of the various microphones within subsystem II are plotted as the nodes of a folded grid of the vehicle contour. As can be noticed, the highest pressure levels are detected in the engine compartment and near the muffler exhaust. Further it can be proved that the distribution of the microphones is dense enough to provide a sufficient description of the sound field. This means that the spatial variations of the sound field are smooth compared to the mean distance between the microphones.

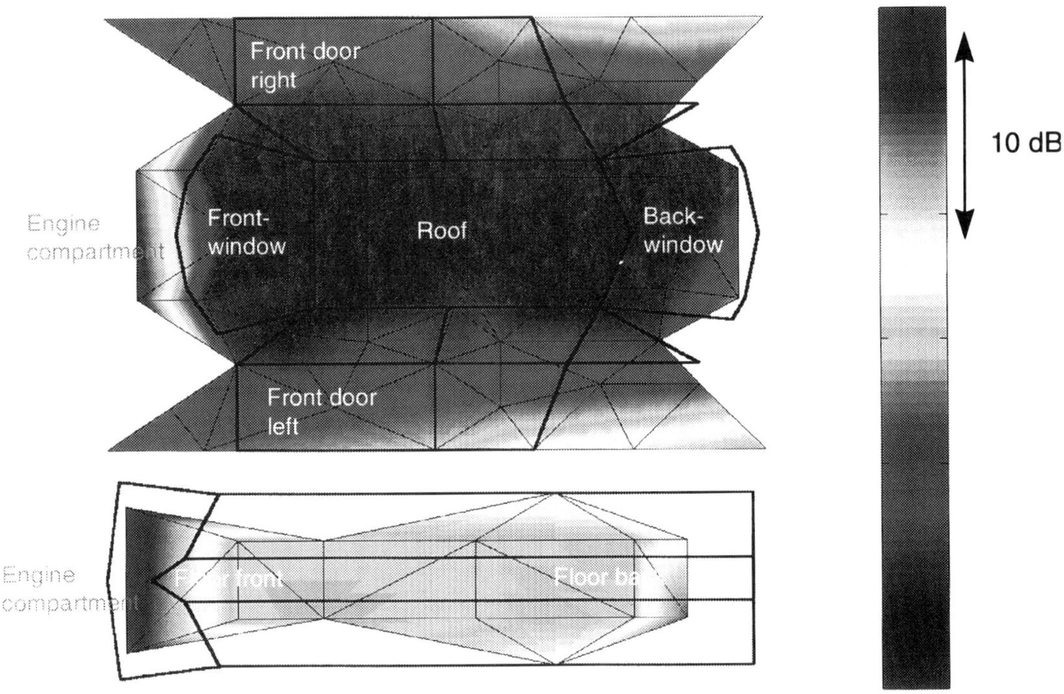

Fig. 4: Sound pressure levels at a defined operating condition

Evidently, there is an interrelation between the spatial variation of the sound field within subsystem II and the number of microphones and loudspeaker positions required to determine the transfer matrix $\boldsymbol{T}$ according to Eqs. (7) and (8). As a consequence of these interrelations, the frequency of the phenomena being investigated must not exceed an upper frequency boundary, since the variation of the sound field typically increases at higher frequencies. However, the influence of structure borne noise also decreases at higher frequencies, which makes the separation of these components less important. Preliminary investigations to this point show that with about 100 microphones distributed around the passenger compartment, a frequency range up to 5 kHz can be covered with sufficient accuracy.

To determine the transfer matrix $\boldsymbol{T}$, the matrices of transfer functions $\boldsymbol{T}^{inside}$ and $\boldsymbol{T}^{outside}$ have been measured. As an example, **Fig. 5** visualizes the RMS-value of the first six rows of the transfer matrix $\boldsymbol{T}^{outside}$ up to 1 kHz together with the corresponding position of the loudspeaker. From these measurements, the transfer matrix $\boldsymbol{T}$ can be derived. **Fig. 6** visualizes one line of this matrix again as the RMS-value up to 1 kHz. The asymmetry in the plot results from the line referring to the microphone in the front left (driver position) of the compartment. Obviously, an acoustic excitation on the left side of subsystem II results in higher sound pressure levels at the driver position than the sound pressure levels produced by an equivalent excitation on the right side.

Combining figures 4 and 6, the sound pressure at the driver position, which is related to the airborne noise transmission at vehicle operation was determined on the basis of Eq.(7). The result is shown in **Fig 7**. At a first glance, this picture is more similar to the measured sound pressure levels of Fig. 4 than to the acoustic transmissibility matrix of Fig. 6. This important result indicates that the knowledge of the sound field outside the compartment at real vehicle operation is very important for achieving a thorough optimization of the sound isolation package. The sound pressure levels of subsystem II show a spatial dynamic range of more than 40dB. Therefore, windowing measurements in acoustic reverberant rooms with a nearly homogeneous sound field can yield misleading hints if not interpreted thoroughly.

Finally, this analysis has also been performed on a defined modification of the vehicle by removing a part of the sealing material at the left front door. As **Fig. 8** indicates, the airborne noise transfer path analysis clearly detects this defect by showing an additional transfer leakage.

It should be mentioned, that such investigations on structural modifications of the vehicle can be performed very quickly, since most of the originally measured quantities remain unchanged. Only the matrix of transfer functions from the loudspeaker configuration to the microphones inside the vehicle compartment - according to Eq.(4) - is influenced by a modification of the acoustic transfer behavior of the vehicle compartment. This means that all measurements of the sound pressures related to subsystem II requiring the multi-channel microphone system need only be performed once.

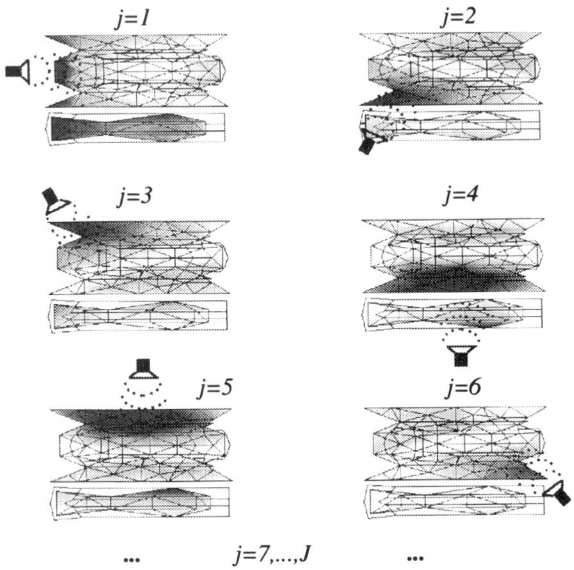

Fig. 5: Visualisation of the transfer behavior $T^{outside}_{k,j}$

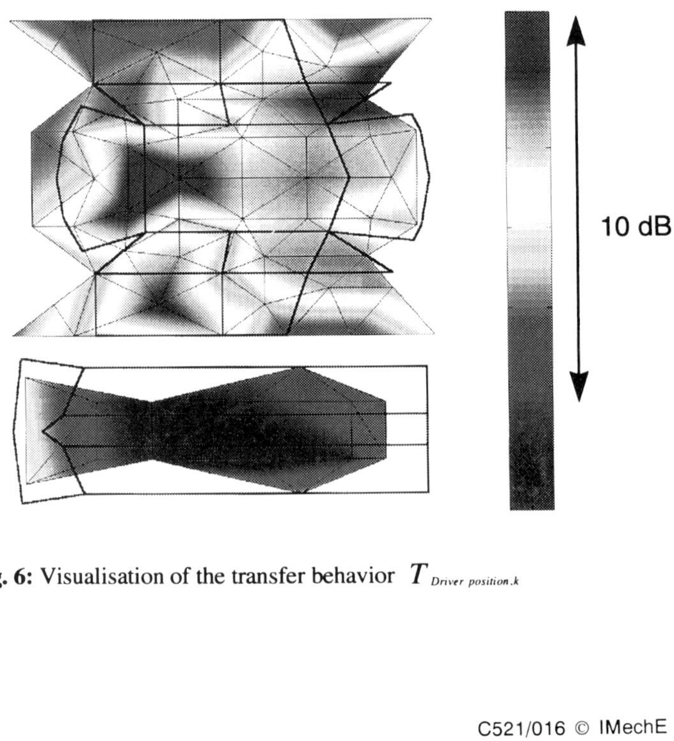

Fig. 6: Visualisation of the transfer behavior $T_{Driver\ position,k}$

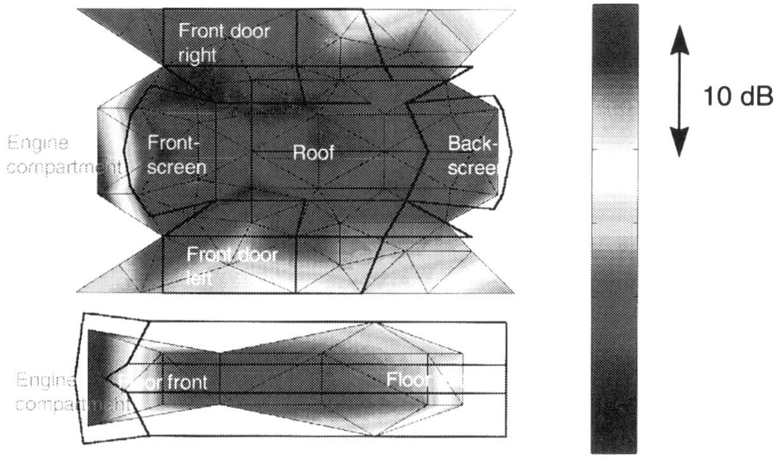

Fig. 7: Sound pressure related to airborne noise transmission

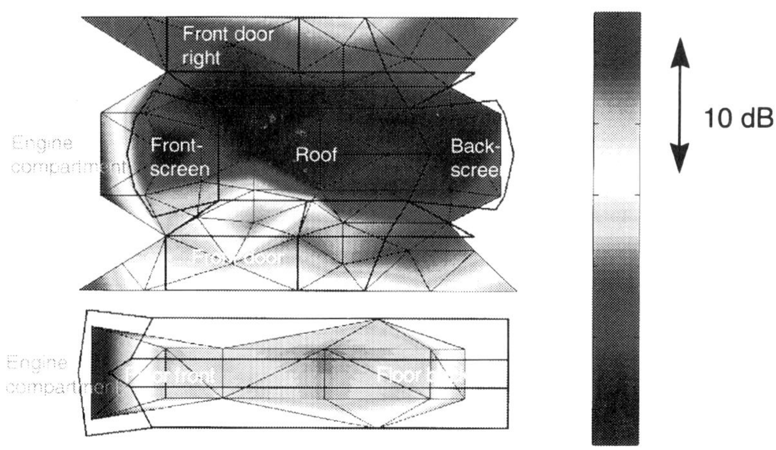

Fig. 8: Sound pressure with defect sealing at the left front door

Accordingly, if the investigations are extended to other operating conditions of the vehicle, both matrices of frequency responses (Eqs. (3) and (4)) can be re-used. In this case, only the sound field according to Eq. (1) during this new vehicle operation has to be measured.

The sound pressures $p_l^{inside,OP}$ (according to Eq. (2)) measured inside the vehicle compartment during real vehicle operation complete the picture. Specifically, they refer to the total interior noise and are as such formed by the superposition of all airborne and structural borne noise transfer paths (Fig.1). A comparison of these measurements with the evaluated airborne noise components can yield a ranking of the different noise transfer phenomena. It was detected in the special case investigated, that when operating the vehicle on the roller dynamometer, roughly 20 % of the interior noise below 1 kHz is airborne. This relatively small fraction is due to the fact that in these measurements wind noise is not included. But there is no principal obstacle against the extension of the developed method also with regard to the consideration of wind noise.

4. CONCLUSION

The described experimental-numerical method allows the identification, analysis and optimization of airborne transmission paths in the field of passenger transportation systems. Together with the well known transfer path analysis of the structure borne noise, now a systematic decomposition of the vehicle interior noise into all its main contributions is feasible. For both phenomena, the result of the transfer path analysis enables a localization of the main paths and a ranking between its excitation and the corresponding transfer behavior of the car body. This decomposition further entails the possibility to perform an optimization on the component level of a vehicle.

REFERENCES

[1] Hörnlein, H. Schittkowski, K. (Hrsg.)
Software systems for structural optimization
Birkhäuser Verlag Basel, Boston, Berlin 1993

[2] Freymann, R. Stryczek, R., Haberstok, C
Holomodal, a new computer-aided testing tool in the field of structural-acoustics
Proceedings of the Unikeller Conference
Zürich 1997, S. 8/1-8/22

[3] Stryczek, R.; Freymann, R.
An Experimental-Numerical Method for Analysis and Optimization of Structure Borne Noise Transmission
VDI Berichte Nr. 1153, 1994, pp. 207-220

[4] Freymann, R. Stryczek, R.
Noise path analysis and optimization
Proceedings of the Unikeller Conference
Zürich 1995, pp.13/1-13/27
[5] Cummings, A.
The effects of a resonator array on the sound field in cavity
Journal of Sound and Vibration 1992, pp. 25-44
[6] Freymann, R.; Beer, R.
Akustische Modalanalyse
Fortschritte der Akustik DAGA 93, 1993, pp.1092-1095
[7] Freymann, R.; Stryczek, R.; Spannheimer, H.
Dynamic Response of Coupled Structural-Acoustic Systems
Journal of Low Frequency Noise & Vibr., Vol. 14, No. 1, 1995, pp. 11-32
[8] Morand, H.J.-P., Ohayon, R.
Fluid stucture interaction
Masson, Paris, Milan, Barcelona (1995)

C521/020/98

An investigation into the effect of door seals on noise generated in the passenger compartment

C PENG BSc, Msc, PhD, CEng, MIMechE
Department of Engineering, Harper Adams Agricultural College, Newport, UK
D MORREY MA, PhD, CDipAF, CEng, MIMechE
School of Engineering, Oxford Brookes University, Oxford, UK

SYNOPSIS

A numerical model of a family saloon car was generated using the boundary element method, with both wind and road tyre noise sources. Sound pressure spectra were generated for three different sealing conditions. Experiments were carried out on the car by measuring the noise levels inside the car, at different travel speeds, for the same sealing conditions. Frequency spectra of the sound pressure were obtained, and articulation indices were calculated. The results showed good agreement between the numerical and experimental results, and the increase in sound pressure level caused by the introduction of gaps into the seal was clearly observed.

1 INTRODUCTION

Given the recent advances made in the reduction of engine noise, the major noise sources for the passenger compartment are now road noise and wind noise. A significant factor in reducing wind noise is the effectiveness of the seals used on the doors and tailgate. During normal operation, the door seal can lose contact with the vehicle body and hence introduce air gaps due to aerodynamic loads on the door, the flexure of the body shell and misalignment of the door and the vehicle body. This can result in a significant change in the noise level inside the passenger compartment.

The prediction of noise levels inside a car using different seal configurations and under different sealing conditions is highly desirable at the prototype stage. This would allow the automotive design engineer to assess and improve the performance of the sealing system in order to optimise the acoustic field in the passenger compartment.

The work described here has been undertaken as part of a larger project investigating the influence of noise transmission through automotive seals on the noise level in the passenger compartment. The research has been undertaken in two stages: (1) an experimental investigation of noise transmission through door seals in a running vehicle; and (2) the development of a numerical method to predict the influence of the gap on the noise transmission into the passenger compartment. The numerical part of the work will be described here, along with some of the experimental results.

2 PREVIOUS WORK

There have been significant improvements in reducing engine noise and road tyre noise over the last 20 years. As the noise generated by the engine and tyres is reduced, wind noise, or aerodynamic noise, is increasingly recognised as an important noise source for automotive vehicles. Wind noise usually dominates at travel speeds above 97 kph (60 mph). Measured results (**1**) show that at a travel speed of 116 kph (72 mph) the wind noise inside a car ranged from 62 to 78 dBA, whilst at 177kph (110 mph) the wind noise ranged from 72 to 87 dBA. The noise level outside a car can rise up to 124 dB at a travel speed of 150 kph (93 mph). At high travel speeds, aerodynamic noise is the dominant noise source, and this overrides the noise from the engine and exhaust at the door seals. Only at the bottom of the door does road tyre noise override the aerodynamic component of the noise (**2**).

Wind noise in automotive vehicles can be considered to be a mixture of monopole, dipole and quadropole noise sources. At low Mach numbers, as in the case of automobiles, the primary source can be considered as a monopole, or if this does not exist, then the noise source will be a dipole (**3**). Vehicle measurements have shown that there is a linear increase in interior noise level with travel speed, at a rate of 12 to 16 dB per doubling of the travel speed (**2, 4**). This indicates that wind noise in automotive vehicles can be considered as a mixture of monopole and dipole sources. Gaps caused by poor sealing behave like monopole noise sources, whereas an opened window or sunroof is more likely to be a mixture of dipole and quadropole noise sources.

Automobile aerodynamic noise is generated by the flow of air around the exterior of the car. When there is a vortex present in the air flow this can act as a dominant noise source. Examples are at locations such as the front windows, radio antennae and wing-mirrors. The vortex created at the A-post acts not only as a dominant aerodynamic source, but also generates negative aerodynamic pressure of the front window (**5**). This negative pressure pulls the front door outwards, which may cause the door seal to lose contact with the body frame and to generate gaps between the seal and the body frame.

Exterior aerodynamic noise can enter the passenger compartment in two ways: through car body panels by pressure fluctuation on the panels, and directly through gaps and holes in the body. Noise transmission through gaps can be a dominant source when these exist. Measurements have showed that up to 6 dB noise reduction can be achieved by simply applying a watershield layer in the front door assembly in order to seal off those holes and gaps which are made for the electric wiring and speakers (**6**).

Jung and Oh (**4**) carried out measurements on a vehicle with the doors in the taped and un-taped conditions; the taped condition being taken as an "almost ideal" sealing. The measurements showed that the noise level changed with sealing condition. They found that a 1 mm gap could increase the noise level at the A-post near the rear-view mirror by up to 4 dB.

3 PREDICTION OF NOISE TRANSMISSION THROUGH DOOR SEALS

Because of the complex nature of modern door seals and the reflective interior surfaces of the car passenger compartment, it was decided to model the situation under consideration, namely that of noise transmission into the interior of a vehicle, using the boundary element method (BEM). In a BEM model, the boundary surface is meshed into a number of elements upon which boundary conditions such as impedance, pressure or velocity, are applied.

For problems involving sound transmission through a cavity, multiple zones are required, with infinite zones to represent any external regions. In this project the noise transmission through the door seal of a family saloon car was to be modelled. This was represented by the acoustic mesh shown in Figure 1. This consisted of three acoustic zones: the interior zone which is the cavity inside the vehicle, the exterior zone outside the car, and the zone of the seal. The interior zone was defined as a finite zone filled with air. On the interior surface of this zone, an impedance boundary was applied, representing the acoustic properties of the trim materials in the passenger compartment. The exterior zone was defined as being infinite and filled with air, in order to represent the atmosphere outside the car. The exterior surface of the vehicle was modelled as a surface of high impedance, in order to simulate the reflective exterior body panels. The zone of the seal was defined as a finite zone with interfaces with both the interior and exterior zones. The material inside the zone of the seal was assumed to be solid rubber, ignoring the influence of the bubbles used inside most modern door seals, and the zone itself was a tunnel-like cavity with a small cross-section (15 mm x 20mm). All the elements used were linear.

Two types of noise source were incorporated into this model. These were road tyre noise, and the wind noise that would be generated by a vortex at the A-post. The road tyre noise sources were modelled as point noise sources, located just outside the vehicle, with a sound pressure level of 105 dB, as measured on a saloon vehicle travelling at 100kph in the paper by Van den Brulle and Ulrich (**2**). The vortex was modelled as a cone along the A-post with a surface pressure of 8 N/m^2 (108 dB SPL), using values suggested in the same paper.

Three different models with different sealing conditions were analysed:

(i) the front door was fully sealed;

(ii) a gap was introduced in the seal, 45 mm long, 15 mm wide, and 20 mm deep was introduced at the middle of the A-post; and

(iii) a gap was introduced in the same location, but only 1.5 mm wide.

The boundary element method gives a solution in terms of acoustic properties (sound pressure, intensity, velocity) at points on the surface mesh. In order to compare the simulation with experimental results, internal points were chosen at locations where microphones were to be positioned in the experimental tests.

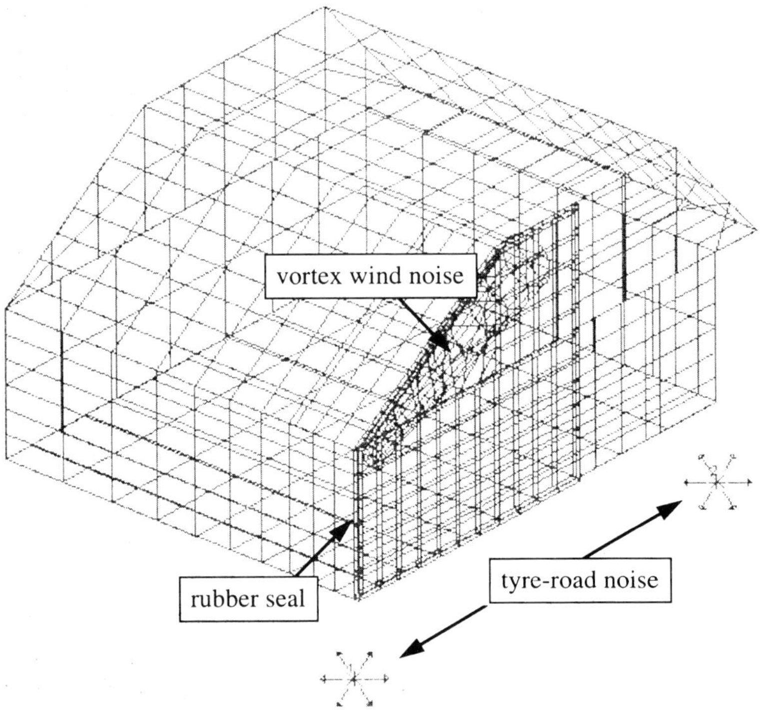

Figure 1 Boundary element mesh of car body with vortex wind noise and point road tyre noise sources highlighted

The BEM model was generated using the BEASY software package (**7**). This calculates the resulting sound field at a single frequency, and therefore it was necessary to run the model at all frequencies within the frequency range of interest. The maximum frequency was set at 1000 Hz. An A-weighting curve was applied to the predicted sound pressure levels in order to be able to calculate the total sound pressure level.

4 EXPERIMENTAL MEASUREMENT OF NOISE TRANSMISSION

In order to verify some of the predictions made using the boundary element method, experimental tests were carried out. Measurements were made with the family saloon car on a test track at constant road speeds. Microphones were positioned inside the front door frame, as detailed in Table 1.

Microphone position	Location
1,2	A-post
3,4	Top of door frame
5,6	B-post
7	Bottom of door frame

Table 1 Details of microphone locations for experimental tests

Measurements were made under various different sealing conditions:

(i) the door was taped, to simulate a perfect seal between the door and frame;
(ii) the door was un-taped; and
(iii) portions of the seal were removed, firstly along the A-post and then along the top of the door frame.

The gaps were introduced in (iii) by cutting out and removing a 45 mm length of the rubber seal at the appropriate locations. Tests were carried out at constant speeds of 80, 95 130 and 160 kph, driving along identical sections of track. Each test was carried out for one minute and repeated three times. The measured sound signals were recorded on a multi-channel tape recorder, located on the back seat of the vehicle. The signals were played back and digitised in the laboratory using a National Instruments SCXI data acquisition system. The resulting data was analysed using a PC to estimate the total weighted and un-weighted sound pressure levels, the third octave sound pressure spectrum, and the articulation index.

5 RESULTS AND DISCUSSION

5.1 Boundary element method

Figures 2, 3 and 4 show the predicted A-weighted sound pressure spectra for the seven microphone locations in the three cases:

(i) the front door fully sealed;
(ii) a 1.5 mm wide gap in the middle of the A-post; and
(iii) a 15 mm gap in the same location.

For all sets of results, both the vortex wind-noise source and road tyre noise source were present.

The resulting curves show quite clearly that the sound pressure level at location 2 (in the middle of the A-post, close to the gap) increases when a gap is introduced, and that the increase is greater for the wider gap. The increase is larger in the frequency range from 50 to 400 Hz. In addition the sound pressure level at the location of the driver's ear shows a significant increase over the same frequency range. There was virtually no change at the other microphone locations.

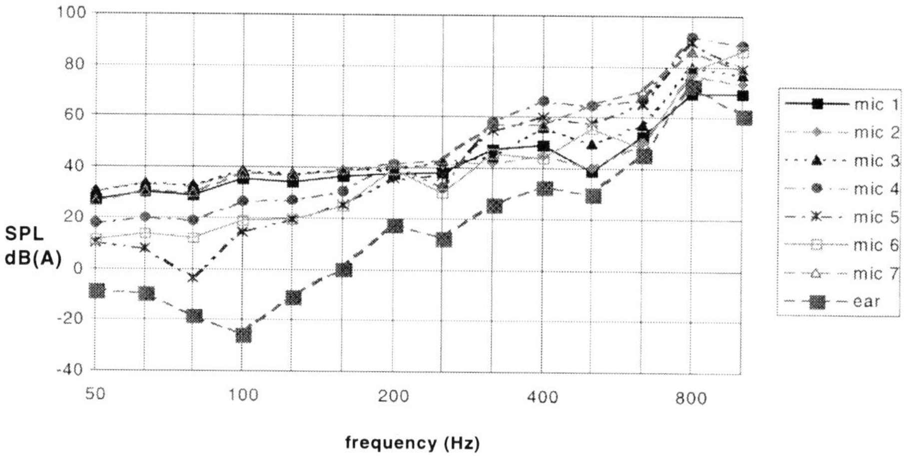

Figure 2 Predicted A-weighted sound pressure spectrum at microphone locations around the door and at the driver's ear with the vehicle fully sealed

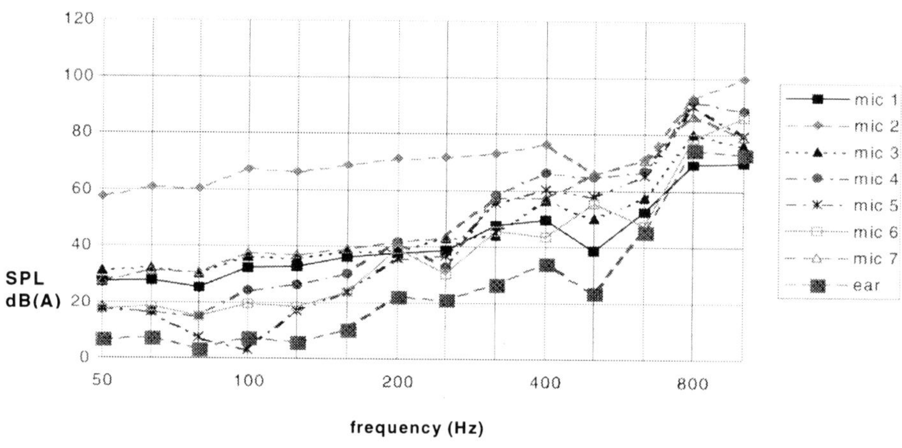

Figure 3 Predicted A-weighted sound pressure spectrum at microphone locations around the door and at the driver's ear with a 1.5 mm wide gap

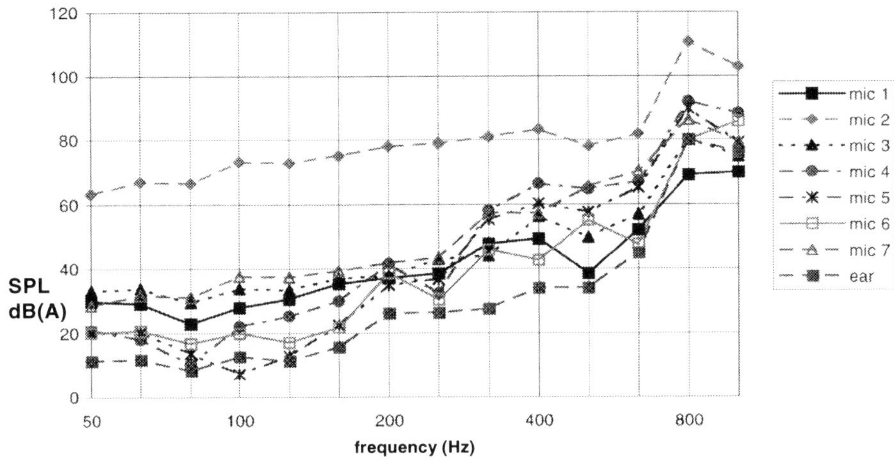

Figure 4 Predicted A-weighted sound pressure spectrum at microphone locations around the door and at the driver's ear with a 15 mm wide gap

5.2 Experimental measurements

Figures 5 and 6 show plots of the A-weighted total sound pressure level at the locations measured around the door at speeds of 80, 95, 130 and 160 kph, for the two cases of:

(i) the door fully taped;
(ii) a portion of the seal removed along the A-post.

The results at the corners of the door were obtained by interpolation.

These results give very similar total sound pressure levels to those predicted by the boundary element models. In addition it can be seen that introducing a gap in the seal along the A-post has increased that sound pressure level at microphone number 2, as predicted by the numerical model. There is also an increase at microphone number 4 which was not predicted by the numerical model. In addition the graphs show that the sound pressure levels at all locations increase with road speed.

It is interesting to note that the maximum sound pressure level in the fully taped condition was at the bottom of the door (microphone number 7). This was also the case in the un-taped condition. This was probably because of the close proximity to the road noise source, and also because a single seal, rather than a double seal, was used on this part of the door.

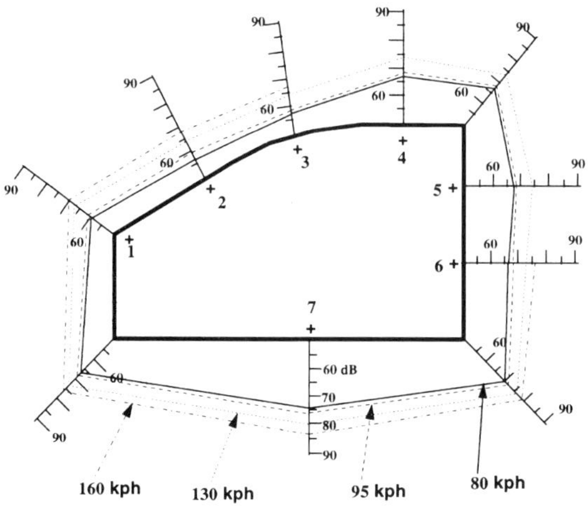

Figure 5 Measured A-weighted sound pressure levels at microphone locations around the door with the door fully taped

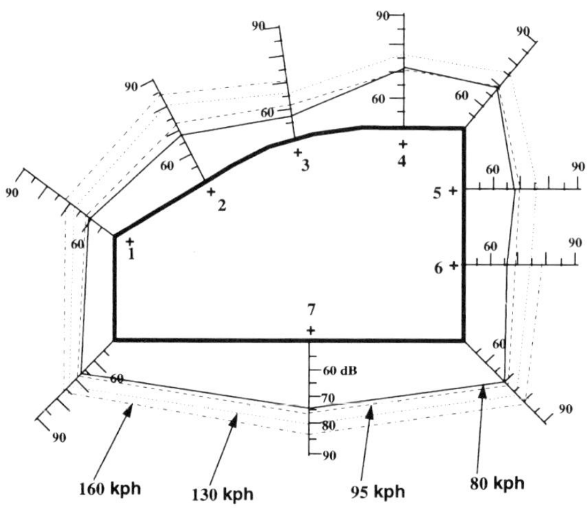

Figure 6 Measured A-weighted sound pressure levels at microphone locations around the door with a portion of the seal removed along the A-post

Figure 7 shows the third-octave band analysis of the measured sound pressure level for different travel speeds at microphone location 2, with a portion of the seal removed along the A-post. Although this plot shows a similar trend to the predicted frequency spectra shown in Figure 4, there is very poor agreement in the low frequency range up to approximately 300 Hz, and the total sound power level is lower than that predicted by the BEM model. It was felt that this might have been caused by an uncertainty in the SPL values used for the noise sources, and by using an over-simplified model of the gap introduced into the seal, which did not include sufficient detail regarding the interface between the door and the door frame.

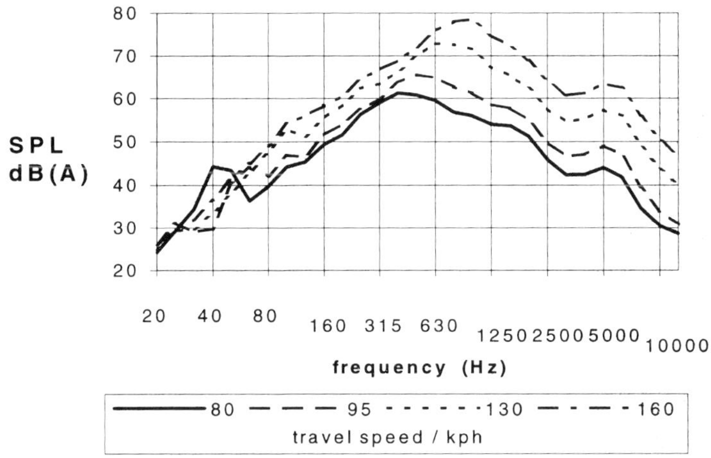

Figure 7 Third octave band analysis of the sound pressure level measured at microphone location 2 with a portion of the seal removed along the A-post

6 CONCLUSION

A numerical model of a family saloon car was generated using the boundary element method. Road tyre noise sources were included as point sources, and the wind noise which would be generated by a vortex at the A-post was included as a cone along this part of the door frame. Sound pressure spectra were generated for three different sealing conditions: with the door fully sealed, and with both a wide and narrow gap introduced into the seal. The results showed an increase in the total sound pressure level at the location near to the gap, with a greater increase for the wide gap.

Experimental tests were carried out on the car by measuring sound pressure levels at seven microphone locations inside the door frame, for different travel speeds, and for different states of sealing of the front door. The three states considered were as follows: the door was fully taped; the door was left un-taped; gaps were introduced into the seal along the A-post, and on the top of the door frame. Frequency spectra of the sound pressure levels were obtained, and the total sound pressure levels were plotted for the different microphone locations at different travel speeds. There was an increase in sound pressure level with road speed, and the total sound pressure level increased at the location near to the gap, as predicted by the boundary element model. The frequency spectra showed similar trends between the numerical and experimental results, but there was poor agreement at low frequencies.

The experimental results showed that the maximum sound pressure level occurred at the bottom of the door. This was probably because this location was closest to the road tyre noise sources, and because a single seal was used in this region of the door frame.

This work has shown that the boundary element method can be a useful tool in the analysis of the influence on the interior noise levels in a motor car of wind noise, road tyre noise, and noise transmission through door seals.

7 ACKNOWLEDGEMENTS

The authors would like to acknowledge the support of the collaborators on this project, the Rover Group at Cowley, in terms of both technical input and the use of test facilities and hardware.

REFERENCES

1. Bucheim, R, Dobrzynski, W, Mankua, H, and Schwabe, D, Vehicle Interior Noise Related to External Aerodynamics, International Journal of Vehicle Design, 1982, Vol. 3, No. 4, pp. 398-410.

2. Van den Brulle, P J, and Ulrich, A, Determination of the Sound Pressure Distribution Along the Doorslits of a Passenger Car, Internoise 1981, Vol. 2, pp. 722 - 729.

3. George, A R, Carr, J F, and Callister, J R, Recent Advances in Understanding Automobile Aerodynamic Noise, CEAS/AIAA, Paper 95-004, 1995, pp. 1-16.

4. Jung, W W, and Oh, S J, The Influence of Vehicle Elements to Aspiration Wind Noise, Society of Automotive Engineers, 1995, 950624.

5. Watanabe, M, Harita, M, and Hayashi, E, The Effect of Body Shapes on Wind Noise, Society of Automotive Engineers, 1978, 780266.

6. Zorea, S I, Aeroacoustic Source Assessment of Vehicle Door Gaps, Internoise 1995, pp. 289-292.

7. Computational Mechanics Inc., BEASY Version 5.0 User Guide, 1994.

C521/022/98

An interactive subjective assessment method for recorded sound

R M S MAUNDER BEng
Ricardo Consulting Engineers Limited, Shoreham-by-Sea, UK

1. SYNOPSIS

This paper presents a new, interactive subjective assessment method, which allows quick surveys of people's opinions of the quality of recorded sounds.

By carefully selecting the sound stimuli which are to be assessed, the subjective benefits to different people of different vehicles can be discovered. This connection between physics and opinion is the key to the design of sound quality, resulting in a product that appeals strongly to the target customer.

2. SOUND QUALITY ENGINEERING

Engineering a particular sound quality is not a simple task. At a fundamental level, people's opinions and attitudes define the quality of a sound. Therefore, quantitative (or objective) physical measurements of sound level, however sophisticated, cannot fully account for all variability in the subjective responses to sounds. On the other hand, a qualitative (or subjective) assessment, in which people are asked about their opinions, allows the engineer to develop a full understanding of the qualities which make a vehicle's sound distinctive and attractive.

Recently, effective tools have become available to design and engineer sound quality, producing a range of recorded sounds that represent different vehicles or build conditions. Assessing the relative subjective merits of such a range of sounds can reveal the benefits of the physical differences that produced the different sounds. The most cost-effective changes can then be implemented in production, optimising people's opinions of the sound quality of the vehicle.

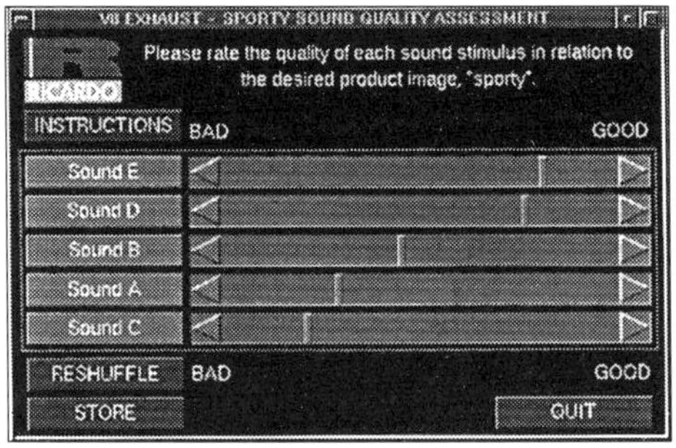

Figure 1. Interactive subjective assessment

3. SUBJECTIVE ASSESSMENT

Listeners' opinions of vehicle sound have been quantified in the past using various subjective assessment methods, such as:

- In-vehicle numerical rating (1).
- Recorded sequence numerical rating (2).
- Semantic difference (3).
- Paired comparison (4).

This paper introduces a new, interactive, method which provides a quick and cost-effective way of surveying a group of people's opinions about sound. All of these methods are compared and contrasted in the Appendix.

The interactive subjective assessment is implemented on a Silicon Graphics Indy R4600 computer as shown in Figure 1 (a version running on a PC is under development). On-screen push-buttons operated by the mouse pointer play each sound stimulus, and corresponding sliders are used to register opinion scores graphically rather than numerically.

3.1 Collection of sound stimuli

3.1.1 Recording system

The recording system used to collect sound stimuli should reproduce as accurately as possible the acoustic environment experienced by the user of the product, as poor recording fidelity may colour listeners' opinions. Transducer locations should be chosen to emphasise the acoustic feature or quality of interest.

Binaural recording systems can reproduce convincing virtual acoustic images complete with realistic spacial localisation (5). To achieve this realism, these systems use a dummy head with in-ear microphones, and a DAT recorder giving about 90 dB signal-to-noise ratio. Playback is through high quality headphones with special equalisation. These systems typically incorporate calibration facilities, allowing sound stimuli to be presented to listeners at realistic levels, both relative to one another and in absolute terms.

3.1.2 Vehicle operation

Operating conditions should be chosen to emphasise the acoustic feature or quality of interest. Irrelevant acoustic features can distract listeners and should be avoided.

The duration of sound stimuli should be chosen with care. If a sound stimulus is too short, listeners may not have enough time to form an opinion. Conversely, if a sound stimulus is too long, listeners' memories of earlier sound may fade. This can bias opinions towards the latter part of a stimulus and inhibit comparisons with previous stimuli.

Non-stationary sounds, such as engine speed transients, may exhibit many changes in sound quality over their duration. These stimuli are more difficult for listeners to remember and thus to assess. On the other hand, the use of carefully chosen non-stationary sounds can allow succinct summary assessment of different operating conditions.

3.1.3 Digital filtering

Once recordings have been made, digital filters may be applied to the recordings to enhance or counteract acoustic features of interest. In effect, such filters can simulate potential modifications to the product, allowing early assessment of the subjective benefits of these modifications. Alternatively, a "target" sound can be derived for use in vehicle development. Software and hardware to implement such filters is available from several proprietary sources.

3.2 Preparing listening assessments

Automotive manufacturers have a desired image in mind from the initial conception of a product. A statement of desired sound quality (such as 'sporty', 'luxurious', or 'unobtrusive') can be derived from this concept. During a listening assessment, listeners can use such a statement of desired sound quality to decide upon the opinion scores they will give.

Panels of listeners may be selected from the following groups of people:

- Expert noise engineers (likely to be experienced in making subjective judgements, so may give opinion scores which more accurately represent their opinions).
- Other engineers.
- Sales/marketing staff.
- Current customers.
- Product-using population in general (likely to be inexperienced in making subjective judgements, so may give opinion scores which less accurately represent their opinions).

The experience of the listeners in making subjective judgements gives an indication of the number of listeners required to achieve high confidence in the group's mean opinion scores. As a rule of thumb, at least twenty listeners should carry out the assessment. When using more experienced assessors the number may be reduced.

3.3 Listener analysis

In the software, listeners are not required to play the sound stimuli any particular number of times. A measure of each listener's thoroughness is given by the total number of plays of the stimuli during the assessment. A listener may, in certain circumstances, be rejected from further analysis on the grounds that he has not played the stimuli enough times to form a reasonable opinion.

No fixed value can be suggested for the lower limit. If the sound stimuli in an assessment are very distinct and easy to rate, the listeners may be happy with their responses after very few plays of each sound. Conversely, if the sound stimuli are difficult to distinguish, many plays may be needed for the listeners to be happy about their responses. In the software, the limit is determined interactively by the use of a graphical slider, and would be set to reject listeners with <u>unusually</u> low numbers of plays.

3.4 Analysis of mean opinion scores

The group of listeners' mean opinion scores is calculated by taking the average of the individual listeners' opinion scores. Confidence intervals are calculated for each stimulus, describing the range of opinion scores in which the true mean opinion score is likely to lie. If the confidence intervals of two stimuli overlap, no significant difference in their mean opinion scores can be inferred. However, if their intervals do not overlap, the mean opinion scores of the stimuli are significantly different. Large confidence intervals can be an indicator either of too few listeners, or of differing opinions amongst the panel of listeners (particularly in the case of a single large confidence interval).

The 95% confidence interval for a particular stimulus' mean opinion score is calculated by multiplying the standard deviation of all listeners' opinion scores for that stimulus by:

$$\frac{t_{L-1}(5)}{\sqrt{L}}$$

where L is the number of listeners, and $t_d(x)$ denotes the x percent point of the Student t distribution with d degrees of freedom. If preferred, 90% or 99% confidence intervals can be displayed.

Typically, a numerical scale (zero to ten) is used alongside the graphical sliders, where each number has a clear meaning (for example, most companies use five to mean "borderline"). These absolute scales are commonly used in the automotive industry for subjective assessments, so listeners are likely to be familiar with their use.

However, inexperienced listeners may not be familiar with the meanings of the different numbers and could find this type of scale confusing. Such listeners would prefer to be presented with a purely graphical scale with no graduations. The disadvantage of this is that different listeners could arbitrarily use different regions of the scales to represent their opinions. In other words the mean and standard deviation of each listener's opinion scores could be different, but the differences could be meaningless. In these circumstances it seems sensible to normalise each listener's opinion scores as follows:

- The positions of the graphical sliders are de-coded onto a numeric scale.
- Their mean is subtracted.
- They are divided by their standard deviation.

This calculation gives a set of opinion scores for each listener with mean of zero and standard deviation of one. These characteristics ensure that all listener's opinion scores are comparable to each other.

3.5 Further analysis of opinion scores

The analysis of mean opinion scores assumes a hypothesis that all listeners' opinion scores are estimates of a single underlying "true" opinion score of a particular sound stimulus. This hypothesis clearly would not hold if opinions were to differ amongst the panel of listeners. Indeed, were the engineer to believe mean opinion scores calculated under these circumstances, he may develop a sound quality compromise between two extreme positions, thereby actually pleasing nobody.

The opinions of the target market must be identified clearly. Therefore, a way of comparing opinion scores from different listeners is needed. Two alternatives are available: the first is simply to examine individual listeners' opinion scores, the second is to examine independent opinion classes.

Firstly, individual plots of each listener's opinion scores can be examined and compared. This allows visual or qualitative comparisons of opinions, revealing similarities or differences between the responses of different listeners. In addition, extra weight can be given to the opinions of key listeners.

Secondly, principal component analysis can be used on the matrix of all listeners' opinion scores, to determine orthogonal "virtual" opinion score sets (or principal components), each showing an independent class of opinion. Principal components with large weights account for most of the variability in the listeners' opinion scores. Those that have small weights account for small variations only, and may be ignored.

Individual listeners whose opinion scores are strongly positively correlated ($r > 0.7$) with a principal component are said to "agree" with that opinion class. Conversely, individual listeners whose opinion scores are strongly negatively correlated ($r < -0.7$) are said to "disagree".

4. CASE STUDY

An automotive manufacturer wanted to develop an appropriate sound quality for their new sports car. A set of five interior noise recordings were made with different exhaust systems fitted to the vehicle. These recordings were presented in a subjective assessment whose instructions are shown in Figure 2. The desired sound quality in this case was 'sporty'.

Twelve listeners carried out the assessment, as shown in Figure 3. Four of these (listeners 7, 10, 11 and 9) were rejected because they played the stimuli fewer than ten times in total.

Mean normalised opinion scores, together with corresponding 95% confidence intervals, are shown in Figure 4 for the eight remaining listeners. The mean opinion score of sound E is the highest, slightly above that of sound D. However, the confidence intervals of sounds E and D overlap, so no significant difference can be inferred between their mean opinion scores. Sound A has a particularly large confidence interval which could be evidence of differing opinions of this sound. Sounds B and C are clearly disliked.

Figure 5 shows individual opinion scores for the twelve listeners. By visual inspection, a group of listeners (1, 2, 3, 4, 5, 8, 9, 10, 11 and 12) can be identified as preferring sounds E and D. Listeners 6 and 7 do not conform to this opinion, both ranking sound E amongst their least favourite.

In the principal components analysis shown in Figure 6, the majority of listeners (1, 2, 3, 4, 5, 8, 9, 10 and 11) agree with principal component 1, in other words they like sounds E and D significantly more than sounds B, A and C. However, listeners 6 and 12 maintain a quite different opinion; that sounds A and D are the most preferable. Listener 7 is alone in the opinion that sounds D and B are the best. No other principal components are particularly strong, with weights below 15% of principal component 1. Therefore, there are just three important classes of opinion about these stimuli.

In conclusion, if all listeners' opinion scores are taken into account, sound E and D are the most highly preferred in the subjective assessment of sporty sound quality. The majority of individual listeners agree with this opinion. However, if listeners 6 and 12 most closely represent the target market for the particular vehicle then the manufacturer would be wise to consider implementing sound A or sound D in preference to sound E.

5. CONCLUSIONS

This paper has presented a new, interactive subjective assessment method, which allows quick surveys of people's opinions of the quality of recorded sounds.

By carefully selecting the sound stimuli which are to be assessed, the subjective benefits to different people of different vehicles can be discovered. This connection between physics and opinion is the key to the design of sound quality, resulting in a product that appeals strongly to the target customer.

6. REFERENCES

1. Elliott, D.R., Crowther, P.J., and Bennett, C.J., "Development Of A Practical Method For The Subjective Appraisal Of Total Vehicle NVH Refinement" - SAE 951285, 1995.

2. Russel, M.F., Sekowski, M., and Nikokiroulis, N., "Subjective Assessment Of Diesel Vehicle Noise" - I.Mech.E., 1992, C389/044

3. Fujita, K., Nishiyama, T., and Hayakawa, Y., "Research On Sound Quality Evaluation Methods for Exhaust Noise" - Japan SAE Vol. 9, No. 2, April 1988

4. Feng, B., Wakefield, G.H., Good, M.D., Otto, N.C., and Rebandt, R., "On The Perceived Roughness Of Automotive Powertrain Sounds" - SAE 951312, 1995.

5. Genuit, K., "The Importance Of Psychoacoustics And Binaural Signal Processing For Measuring Noise Quality" - Paper Number 10.4, ATA Conference on Vehicle Comfort, Bologna, 1992.

APPENDIX

COMPARING SUBJECTIVE ASSESSMENT METHODS

The first distinction to be made when comparing subjective assessment methods is between direct (in-vehicle) assessment and indirect (recorded) assessment.

In-vehicle assessments allow sound quality to be assessed directly. The total vehicle subjective balance may be assessed, incorporating visual and tactile qualities. The assessor may also report unexpected concerns. However, these assessments have the following disadvantages:

- Difficult or expensive for a large number of listeners to assess many vehicles.
- Considerable delay between assessments.
- Bias from non-acoustic qualities of the vehicles.
- Absolute opinion score scales not sensitive to small differences.

In general, indirect subjective assessments of recorded sounds are susceptible to inaccuracies caused by lack of recording fidelity (ameliorated by the use of high quality recording systems). However, subjective assessments of recorded sounds have the following advantages:

- Many listeners can assess the sounds cost-effectively.
- Insignificant delay between presentations of sounds.
- Archived recordings can easily be included.
- Assessments are 'blind' so responses cannot be biased by non-acoustic qualities of vehicles.

Table 1 describes the specific advantages and disadvantages of various subjective assessment methods for recorded sounds.

Table 1. Subjective Assessment Methods for Recorded Sound

Method	Description	Relative Ratings	Absolute Ratings	Advantages	Disadvantages
Recorded Sequence Numerical Rating	• Sounds are rated on an absolute numerical scale. Sounds are recorded in a fixed, unbiased sequence on tape.		✓	• Subjective assessment tape and response sheet are easily shipped.	• Absolute rating scales are not sensitive to small differences.
Semantic Difference	• Sounds are rated on a series of absolute numerical scales, each representing a pair of opposing adjectives. Sounds are recorded in a fixed sequence on tape.		✓	• Recognises the inherent multi-dimensionality of perception of sound quality. • Subjective assessment tape and response sheet are easily shipped.	• Absolute rating scales are not sensitive to small differences. • The series of absolute rating scales may be confusing for naïve listeners.
Paired Comparison	• Sounds are compared in pairs. Listeners are asked which of the two sounds they prefer. Pairs of sounds are recorded in a fixed sequence on tape.	✓		• Relative binary decisions are simple for the listeners. • Subjective assessment tape and response sheet are easily shipped.	• Merit ratings are calculated from binary preferences, so lack resolution.
Interactive	• Sounds are rated on graphical sliders. Sounds are available on computer for repeated presentation on demand.	✓	✓	• Comparisons are simple for the listeners, as they can use a combination of relative and absolute rating methods. • Ratings are highly resolved. • Flexible for the listeners, as there is no prescribed number or sequence of stimulus presentations.	• Being computer-based, software portability is an issue.

VEHICLE X SPORTY SOUND QUALITY

Vehicle X is a product intended to have a sporty image. This subjective assessment is designed to determine the best sound quality to convey this product image, by presenting a range of sound stimuli to you for rating.

You may listen to a sound stimulus by pressing the appropriate grey button. You are asked to rate the quality of each stimulus in relation to the desired product image, 'sporty', using the corresponding slider. Note that the left end means BAD quality and the right end means GOOD quality. You may listen and alter your opinion score for each stimulus as many times as you like.

Once you have listened to all the stimuli and initially rated them, press **RESHUFFLE** to rearrange them into ranked order. Then, you can check the relative differences between closely-rated stimuli.

Once you are happy with all your responses, press **STORE**, then enter your name when prompted. Press **CANCEL** if you want to stop listening without storing any opinion scores. Thank you for your time.

Figure 2. Listener instructions

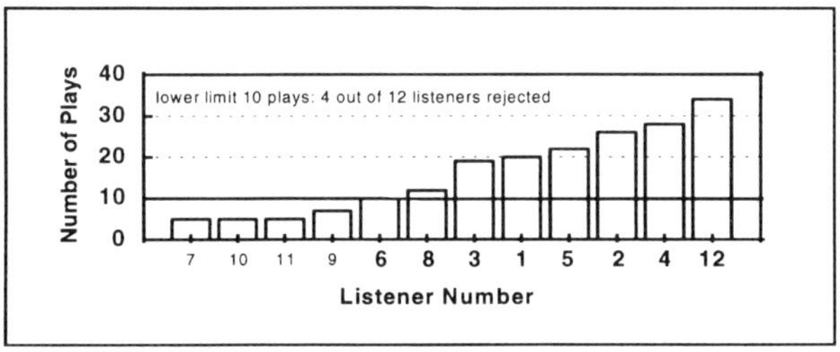

Figure 3. Listener analysis

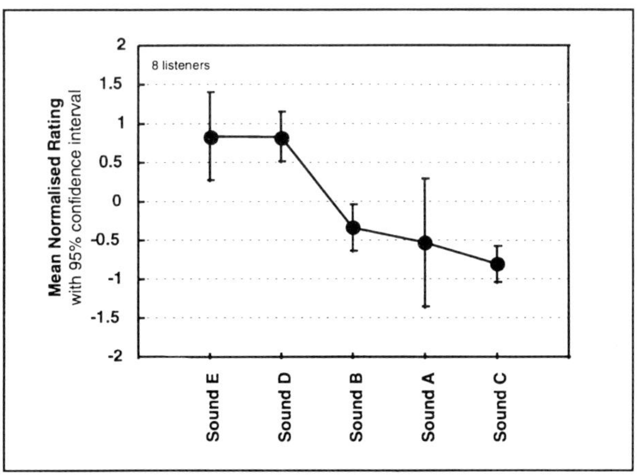

Figure 4. Analysis of mean opinion scores

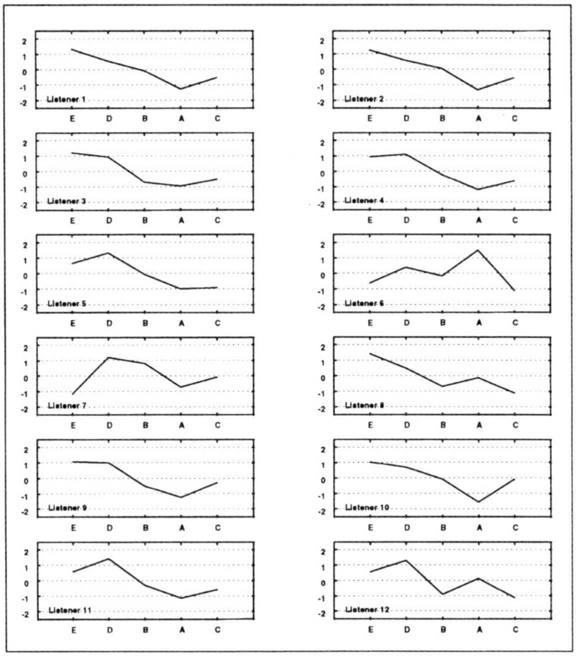

Figure 5. Individual opinion scores

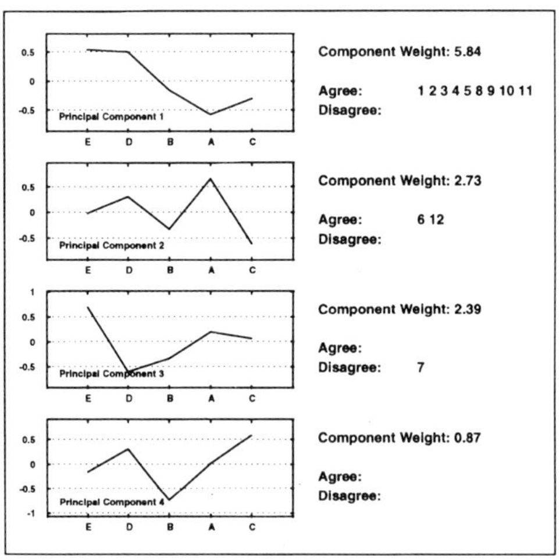

Figure 6. Principal component analysis

Authors' Index